The Castle

Broken Hays

Beaumont

HARVEY AND THE OXFORD PHYSIOL-OGISTS

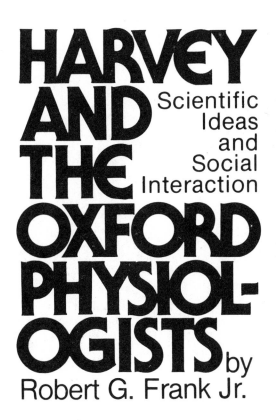

HARVEY AND THE OXFORD PHYSIOL-OGISTS

Scientific Ideas and Social Interaction

by Robert G. Frank Jr.

UNIVERSITY OF CALIFORNIA PRESS
Berkeley • Los Angeles • London

University of California Press
Berkeley and Los Angeles
California

University of California Press, Ltd.
London, England

Library of Congress Cataloging in Publication Data

Frank, Robert Gregg, 1943–
Harvey and the Oxford physiologists.

Includes index.
1. Physiology—England—History.   2. Harvey,
William, 1578–1657.   3. Physiologists—England—
Biography.   4. Oxford. University. I. Title.
[DNLM: 1. Physiology—History—England.
2. Physiology—Biography. QT11 FE5 F8h]
QP21.F76   599.01'0942   79-63553
ISBN 0-520-03906-8

Printed in the United States of America

to my parents

CONTENTS

PREFACE

ACKNOWLEDGMENTS

1 THE CHALLENGE: HARVEY'S CIRCULATION AND ITS UNSOLVED PROBLEMS

Traditional Physiology: Explanations of Heart, Lungs,
Blood, Heat, and Spirit   2
The Discovery of the Circulation   9
Harvey on Heat, Respiration, Vital Spirits, and
Blood Color   12
Harvey's ''Tradition'': Approaches, Techniques, and
Natural Philosophy   16

2 HARVEY'S LATER WORK AND ENGLISH DISCIPLES

Harvey's Early Disciples in England   22
Harvey's Circle at Oxford   25
Harvey's Later Writings—Lost and Surviving   30
Harvey's Changing Concepts of Blood, Heat, and
Respiration   38

3 THE SCIENTIFIC COMMUNITY IN COMMONWEALTH AND RESTORATION OXFORD

The University Background of Oxford Science   45
Medical Education at Commonwealth and
Restoration Oxford   48

The Oxford Experimental Philosophy Club   **51**
The Bonds of a Scientific Community   **57**
Table: Oxford Scientists and Virtuosi, 1640–1675   **63**

4 OXONIANS AND NEW APPROACHES TO PHYSIOLOGY

A Corpuscular Foundation for Biological and
   Chemical Phenomena   **93**
Nathaniel Highmore: Harveian and Atomist   **97**
The New Physiology in Oxford Medical Teaching   **101**
Ralph Bathurst and the Problem of Respiration   **106**
Building a Tradition   **113**

5 ROBERT BOYLE ON NITER AND THE PHYSICAL PROPERTIES OF THE AIR

Early Concepts of the Aerial Niter   **117**
Robert Boyle and the Properties of Niter   **121**
Boyle, Hooke, and the Vacuum Pump: the Physical
   Properties of the Air   **128**
Boyle and Hooke Defending the Physical and Chemical
   Properties of the Air, 1660–1665   **134**

6 NEW EXPERIMENTS ON RESPIRATION, 1659–1665

Boyle and Vacuum Pump Experiments on
   Respiration—Oxford, 1659   **142**
Further Experiments in Respiration—Boyle, Hooke, and
   the Royal Society   **148**
The Mechanical Theme and Hooke's Experiment
   of 1664   **154**

7 OXONIANS ON ANIMAL HEAT AND THE NATURE OF THE BLOOD, 1656–1666

Willis and Fermentation   **165**
Manipulating the Blood: Injection and Transfusion,
   1656–1666   **169**
"Collegium Anatomicum Oxoniense" and the Nature
   of the Blood   **178**
Lower's *Vindicatio* (1665): a Defense of
   Oxford Harveianism   **188**

8 A DISCUSSION AMONG FRIENDS: RESPIRATION WORK AT LONDON, 1667–1669, AND RICHARD LOWER'S *TRACTATUS DE CORDE*

Hooke, Lower, Needham and an ''Experimentum Crucis'' on Respiration	195
Fracassati's Experiment and the Floridness of Blood	205
Richard Lower and the *Tractatus de Corde* (1669)	208
Lower on Respiration and Blood Color	213
Collaboration and Research	217

9 NITER, NITER EVERYWHERE

Willis and the ''Explosion'' Theory of Muscle Contraction	221
John Mayow's Early Life and the *Tractatus duo* (1668)	224
Willis on Animal Heat and Muscular Motion	232
Clouds of Niter	237

10 FIRE AND LIFE: JOHN MAYOW'S *TRACTATUS QUINQUE* (1674) AND A GENERAL PHYSIOLOGY OF ACTIVE PARTICLES

Thomas Willis and the ''Flamey Soul''	248
Boyle and Hooke on Flame and Air	250
John Mayow and his *Tractatus Quinque* (1674): the Chemistry of Nitro-Aerial Particles	258
Nitro-Aerial Particles in Physiology	263

11 THE DECLINE OF THE OXFORD TRADITION

Fragmentation, Deaths, Distraction, and Dispersion	278
Ideas and Social Interaction	287

REFERENCES AND DATES	295
NOTES	299
INDEX	357

PREFACE

This book had its origins some years ago in what I perceived as a contrast between past and present—the past and present of science. A laboratory scientist of the twentieth century would, if pressed to give a formal definition of his discipline, respond with a recitation of the current state of ideas in cellular immunology, nucleic acid biochemistry, or low temperature physics. Yet as I listened to colleagues in the sciences, their "shop" conversation seemed dominated, not by the abstract state of principles in a field but by much more immediate concerns: the success or failure of a man's own experiments, what techniques he was using and how they differed from those of other workers, news of results from his laboratory and those of friends, or of who was working with whom, and where. It seemed, in other words, that the dominant perception of a contemporary scientist was one of science, not as product, but as process. He would, of course, have mastered the facts and theories found in "the literature," but his primary involvement would be with the acquisition of new knowledge, and with the ideas, experiments, personalities, and institutions that were bound up in this activity. James Bryant Conant, a chemist, scientific administrator, and educator, once went so far as to suggest that the significance of science lay not in its intricate and interlocking structures of explanation, but in its dynamic character, its continuous process of speculation and experiment fostering yet more speculation and experiment. When theories, principles, and laws were "embalmed" in texts, and no longer the object of investigation, he suggested, they were not science but dogma.[1]

Yet until very recently, historians of scientific disciplines, in recounting past developments, often seemed to treat only the products of science: explanations. This is all the more curious, because in the nineteenth and early twentieth century, such historians were almost invariably avocational scholars, men who were practicing scientists themselves, often of some eminence in their fields. It was almost as if such men were loath to believe that scientific ideas in the past had been proposed and modified by much the same kind of creative interaction as they observed daily in their own careers. Or rather, being accustomed in their professional lives to thinking of the past only as prologue to the present, such men often failed to read into past literature the dynamic qualities of conjecture and experiment, of question and response, of interaction with colleagues, that had dominated their own lives.

This study was conceived under the rather different assumption that the historian can, if the appropriate evidence survives, recapture even for a period as distant as the seventeenth century, that sense of process that animates the science of the recent past.

My subject—the cluster of ideas concerning air, respiration, heat, blood, and heartbeat—has been of interest to men since long before it was written in Genesis that life is in the blood, and that God "formed man of the dust of the ground, and breathed into his nostrils the breath of life."[2] Such preoccupations, whether in religion, myth, or science, long preceded "physiology" as an independent science. They reflect an irreducible perception of man's life as being intimately bound up in certain actions of his body, failing which even for a few minutes, life itself ceases. A man breathes air, his heart beats, his body gives off warmth, and his vessels are filled with life-sustaining blood. One may be deprived of food and water, of sex, of senses and motion, even of rationality itself, for periods ranging from many hours to many years. But let these paramount functions stop for even so few moments, and life passes into death.

This book tells the story of a particularly fascinating era in the development of those ideas, the half-century that followed the announcement in 1628 by William Harvey of the single most important discovery in the history of the physiological sciences—the circulation of the blood. Thereafter, from the late 1640s to the middle 1670s, several generations of English anatomists and chemists completely refashioned our knowledge of the function of the human body. These men worked out a host of new explanatory schemes: of air and its properties; of how and why respiration is carried out; of metabolic heat and its inorganic analogue, combustion; of the properties and functions of the blood; of the motion of the heart; of muscle contraction and its desiderata; and of the life of the fetus in the womb. No one man attempted to deal with all the extant physiological problems. No one man came to completely satisfactory conclusions. But the totality of this work was an accomplishment of a magnitude unsurpassed again in the medical sciences for almost a century.

These generations included many of the most famous men in English science and medicine, such as Thomas Willis, Robert Boyle, Richard Lower, Robert Hooke, and John Mayow, as well as men such as Christopher Wren and John Locke, who later gained fame for their accomplishments outside of science proper. These men, and a host of others, had differing scientific talents and interests— scientific "personalities" if you will—that are not only of interest in themselves but also directed and shaped their scientific accomplishments.

Such a confluence of important ideas and fascinating personalities has naturally been of interest, both to modern physiologists and chemists seeking the intellectual roots of their disciplines, and more recently to historians inquiring into the growth of crucial ideas in biology. Michael Foster and Francis Gotch, respectively the professors of physiology at Cambridge and Oxford at the turn of the century, wrote on what the former called "The English School in the Seventeenth Century."[3] In the years before and after World War II they were followed by such physiologists and chemists as Fulton, Franklin, the Hoffs, Patterson, McKie, and Partington.[4] In the 1960s professional historians of science, such as Wilson, Goodfield, and Mendelsohn, traced out the development of early modern ideas on respiration, heat, and blood as part of an examination of the general process of conceptual change in the medical sciences.[5] Harvey, the progenitor of this revolution and the patron saint of physiology, has been the subject of scores of books and articles by dozens of writers. Clearly the English physiologists have not lacked for scholarly attention.

Nor has their ambience. During the seventeenth century, and most especially from the 1640s on, science in England took on a societal importance that ultimately found expression in the foundation of the first formal scientific academy, the Royal Society, in 1660, and of the first scientific periodical, the *Philosophical Transactions*, in 1665. Since the mid-eighteenth century, when Thomas Birch first published the "Journal Book" of the Royal Society for 1660–1687, the institutional and cultural environment of English science has been the subject of a steady stream of books and articles.[6] These include contributions, such as those of Merton and most recently Webster, on the relations between religious motivations and the pursuit of science.[7]

Yet the sum of this scholarship is less than the whole historical picture. Excellent though such studies are, they fail to bring together, on the one hand, the careers of scientists and the character of their institutions, with, on the other, an examination of the nature and adequacy of the explanatory schemes put forward by these men. Somewhere in the midst of biography, bibliography, institutional history, and history of scientific ideas, the sense of science as process is lost. In telling the story of these English physiologists of the mid-seventeenth century, I hope to recapture that sense.

It is, in one respect, a story of the reception and elaboration of a single crucial innovation: Harvey's discovery of the circulation. In 1628 he had argued that both logic and experiment demanded that the blood must circulate perpetually throughout the body. But he could give no reason *why* it should do so. Indeed, given the prevailing Galenic system of physiological explanation, there was every reason it should not. That system had to be replaced with one compatible with the newly discovered facts of the circulation. The generations of English chemists and physiologists active in the mid-seventeenth century saw themselves as completing this Harveian intellectual revolution.

They were, and saw themselves as, Harveians in other ways. Many were physicians who saw Harvey as a laudable example of a prominent practitioner who also conducted original investigations to advance the science whose art they practiced.

Harvey's methodology of dissection and experiment became for them the model procedure for the solution of physiological problems. This experimental imperative, reinforced by the values of experimentalism derived from chemistry and from Bacon's writings, became the dominent methodological theme in the reconstruction of physiology.

But important as Harvey was in inspiring this reshaping of physiology, it is in no real sense a direct and unaltered continuation of his ideas. The new physiology was based on an atomic and chemical view of matter. Harvey, an Aristotelian to the core, was deeply suspicious of such innovations in natural philosophy, and—had he lived to see the outcome—would have been distinctly uncomfortable with the results. Mine is also therefore the story of the successful application of atomic hypotheses and new chemical modes of thinking to the remaking of physiology.

Finally, and perhaps most significantly, these generations of English physiologists shared some rather unusual common characteristics. Almost all lived and worked in Oxford during the 1650s, 1660s, and in some cases up into the 1670s. A large number of them had been educated at one school, Westminster. Many of them occupied important positions in the institutional life of the university, positions that brought them into daily contact. They attended scientific meetings, courses, and lectures together at Oxford. They met and talked in its colleges, taverns, and coffeehouses. When personal and professional commitments took them beyond the bounds of the Thames and Cherwell, they corresponded with one another, on both social and scientific topics. They constituted, not only an intellectual community concerned with particular problems in physiology and chemistry but also a remarkably diverse and cohesive social community.

In telling their story, I have attempted to recreate both communities—intellectual and social—and to show the subtle, reciprocal relations that existed between the two. To do so has required many different kinds of sources: printed scientific works, the proceedings of scientific societies, correspondence, diaries, notebooks, lecture notes, university registers, college muniments, even wills, library catalogs, and census records. Each reveals a strand—sometimes important, sometimes quite trivial—of the lattice that bound together ideas, men, and institutions. I have particularly taken pains to ascertain the friendships, activities, and movements of these men, for these are the foundations of historical fact upon which the structure of a community, both intellectual and social, is based. To do so, I have made much special use of correspondence, largely that of Robert Boyle and Henry Oldenburg.[8] More significant than its use in establishing such relations, however, is the insight correspondence provides into a scientist's subtly changing perceptions of his own work and that of his colleagues.

My narrative is perforce rich in detail—both of the profusion of scientific ideas and of the multifarious activities of scientists. Yet it is only in such detail that one can see the larger patterns of how the nature, pace, and direction of explanatory change is founded upon the qualities of interaction within a scientific community. Although there have indeed been in the history of science numerous examples of the solitary investigator who struggles successfully with refractory nature, such isolated geniuses are not as much the rule as adulatory biographies would lead us to believe. Rather, a detailed examination of the emergence and transformation of scientific ideas shows how they are conceived, shaped, tested, modified, and even discarded in ways that reflect the personalities, talents, assumptions, and patterns of communications of the scientists whose mutual interests they represent.

In sum, I have attempted to tell the story of the Oxford Harveians in a way that evokes a sense of growth, of process, similar to that perceived by a contemporary scientist in viewing his own activity.

ACKNOWLEDGMENTS

Much of the research for this book was carried out in three magnificent libraries. Without their helpful staffs and rich collections of seventeenth-century books I could not possibly have brought this work to completion: the Bodleian Library, Oxford; the Widener Library, Cambridge, Massachusetts; and the Henry E. Huntington Library, San Marino, California. My indebtedness to these great research facilities, especially to the Huntington Library, is readily and gratefully acknowledged, although not so easily discharged. Numerous other public and private libraries in the United States and Great Britain kindly provided me access to rare books and unique manuscript materials: in Oxford, the Radcliffe Science Library and the library of Christ Church; in London, the British Library and those of the Royal Society, of the Royal College of Physicians, and of the Medical Society of London; in Sheffield, the University Library; in Calne, Wiltshire, the library of the Marquis of Landowne; in Washington, D. C., the Folger Shakespeare Library; and in Los Angeles, the William Andrews Clark Memorial Library and the Biomedical Library of the University of California. My work in these many locales has been most generously supported by the National Institutes of Health, first by a Predoctoral Fellowship and later by National Library of Medicine Research Grant No. 02435.

Everett Mendelsohn of Harvard University and Marie Boas Hall of the Imperial College of Science and Technology, University of London, have read this work in more recensions than I care to recall. I wish to thank them for their advice, support, and forebearance. My frequent—and sometimes lengthy—visits to Oxford were enlivened by the good colleagueship of Alistair Crombie, Charles Webster, and the presidents, fellows, and students of Wolfson College. I shall always cherish the memory of their hospitality.

This work has passed from sometimes inchoate rough drafts to a finished form only through the intervention of expert assistance. Mary Neff, of the Department of History, University of California, Los Angeles, lent her intelligence and resourcefulness to the library research and read the manuscript more than once for structure and style. Lynn Ross typed it with a speed and accuracy little short of amazing. Maureen Cameron skillfully transformed my index slips into a final typescript.

Most of all, my wife Myra has been a source of constant aid, encouragement, and unfailing good cheer. From her help many years ago in the chill basement of the Royal Society, to her most recent firm reminders that I must, *finally*, get this book finished, she has good-naturedly accepted a continuing intrusion into our lives. Her support has meant more to me than she can know or I can adequately express.

San Marino, California   *R. G. F.*
May 1979

THE CHALLENGE: HARVEY'S CIRCULATION AND ITS UNSOLVED PROBLEMS

1

W hen William Harvey published his first book, *Exercitatio anatomica de motu cordis et sanguinis in animalibus* (1628), he was fifty years of age. He was a former Scholar of Gonville and Caius College, a Bachelor of Arts of the University of Cambridge (1597), and a Doctor in Medicine and Philosophy of the University of Padua (1602). He was a Fellow and one of six Elects of London's Royal College of Physicians. He was the Treasurer of the College, had thrice served as one of its Censors, and since 1615 had been the holder of its endowed Lumleian Lecture in Anatomy. He had been Physician Extraordinary to two British monarchs: the recently deceased James I, and his son, Charles I. He had an extensive practice among the aristocracy and at court, and numbered earls and dukes among his friends. He was the Physician to the oldest and largest of the great London hospitals, St. Bartholomew's.[1] By temperament he was deeply respectful of the ancient philosophers and physicians, most especially Aristotle. It is scarcely conceivable that such a man would challenge any established order, be it social, professional or intellectual. Yet his book began the overthrow of a medical and physiological system—concepts of man's functioning and dysfunctioning—that had endured largely unchanged for almost a millennium and a half.

Harvey did not set out to make a revolution. It emerged from the logic of his scientific work. Beginning as early as the mid-1610s, he carried out a number of anatomical investigations, the findings of which were intimated in his Lumleian lectures of 1616, the "Prelectiones Anatomie Universalis."[2] These show that he had dissected not only human cadavers, but also a large number of animals (including such exotica as an ostrich from the king's menagerie), and had conducted some simple vivisectional observations on household animals such as dogs. As a result of these inquiries, Harvey had become especially interested in one particular question: given what was known of the anatomy of the heart and arteries, how could one explain the nature and function of their perpetual motion? In the course of investigating this very specific physiological question, Harvey came, via observations and experiments, to a series of conclusions that were at variance with commonly accepted opinion.

He discovered that the heart's structure and motion were such that it continuously transferred blood along a unidirectional path out of the vena cava, through the right auricle, right ventricle, pulmonary artery, lungs, pulmonary veins and left auricle into the left ventricle, which expelled the blood forcibly by its contraction into the aorta and arteries, thereby causing the pulse. He had discovered that the heart's motion resulted in the *pulmonary transit* and the *arterial pulse*. Such findings were not in themselves new. They had been discussed and debated by the three generations of anatomists between Andreas Vesalius and Harvey, most especially in the work of the Italian anatomist Realdo Colombo and that of his followers.[3]

Harvey's innovation lay in perceiving—by an intellectual process that is still the subject of debate by historians—that his conclusions had a necessary and hitherto unseen consequence. To keep the veins from being emptied of blood, and the arteries from being choked with it, the blood in the periphery had necessarily to pass from the arteries into the veins, and thence return to the heart in a circular motion. Harvey marshalled the arguments for his new theory of the heart and blood, supported them with numerous experiments, and published them as a thin quarto in the autumn of 1628. Such, in the barest outline, are the origins of the circulation of the blood.[4]

What for Harvey had begun as a limited piece of rather orthodox anatomical investigation became, in the 1630s and 1640s, a challenge to the entire corpus of accepted scientific explanation on the function of the animal and human body. To comprehend why, it is necessary to understand Galenic physiological ideas as they were taught and accepted in the medical and scientific world of early seventeenth-century Europe.

TRADITIONAL PHYSIOLOGY: EXPLANATIONS OF HEART, LUNGS, BLOOD, HEAT, AND SPIRIT

These ideas were not, in a strict way of speaking, an independent system. Few textbooks were explicitly physiological. Rather, explanations of function were woven into anatomical compendia and were assumed in works on disease and therapy. Nor were such physiological concepts in the literal sense purely Galenic. They were founded upon his anatomical and medical works, but had undergone a number of significant changes of content and emphasis.[5] Galenic ideas first reached the Latin west in the twelfth and thirteenth centuries, simplified and systematized by Arab writers such as Avicenna. Beginning around 1520, medical humanists sought to restore and publish the original Greek texts of Galen's more important anatomical works, and to make these the basis for a purified interpretation of Galenic physiological ideas. By 1600 this ''true'' Galenism had been further modified by the results of seventy years of original anatomical research in France, Germany, and especially Italy. But purified text and new findings were both perceived and organized along the lines inherited from the medieval compendia. Thus Harvey, in contradicting certain parts of Galenic physiology, was challenging a system of ideas whose intricate structure had been elaborated over many centuries, and whose factual foundations had been newly strengthened.

The physiological concepts embedded in the Galenic anatomical tradition are best seen in two texts widely used throughout Europe in the early seventeenth century, André du Laurens's *Historia anatomica humani corporis* (1600) and Caspar Bauhin's *Theatrum anatomicum* (1605). Both were certainly well known to Harvey. He used Bauhin's text as the basis of his Lumleian lectures, and he was intimately familiar with du Laurens's extensive discussion of theories regarding the motion and function of the heart, which may well have been the starting point of his own *De motu cordis*.

Traditional physiology perceived the function of the human body as tripartite, based upon the principal cavities (''venters'' or ''bellies'') observed in dissection—the abdomen, the thorax, and the head. Each venter had its primary life function and dominant organ, which exerted its effects through a system of vessels containing an appropriate fluid.[6]

Anatomists like du Laurens and Bauhin believed that the abdomen served the *natural* functions of nutrition, excretion, and procreation. Its principal organ was the liver, which was the seat of the natural faculty and the root and origin of its effector vessels, the veins. Food, after it had been concocted in the stomach, passed through the intestines, where its nutritive portion, the chyle, was thought to be drawn off by the mesenteric veins. These vessels converged to form the portal vein, which then branched out through the substance of the liver. There the partially transformed chyle was believed to complete its change into venous blood. The liver expelled the perfected blood through the hepatic vein into the vena cava. The vena cava, the main distributive organ of venous blood, then ramified up and down to carry nutritive blood to all parts of the body. Food implies wastes. Thus it was believed that almost all the body's excrements were drawn off from the venous system: yellow cholor to the gall bladder, black cholor to the spleen, and urine to the kidneys. The renal veins, for example, were therefore large and inserted directly into the source of the venous blood, the vena cava, so they could draw off the serous portion of the blood.[7]

The second ''belly,'' the thorax, served the *vital* functions of maintaining and distributing heat and life throughout the body. Its principal organ was the heart, which was subserved in these purposes by the lungs. The heart was thought to be the hottest part of the body, containing an innate heat *(calidum innatum)* that was distributed out from the heart to the rest of the body, and which was moderated by the cool air in the lungs. This heat was carried by vital spirits that were created in the hot left ventricle of the heart out of air and blood. Diastole served to bring the blood and the air together, and by systole the vitalized and heated blood—now an arterial florid red—was expelled out into the aorta and thence to the rest of the body. The arteries, as the effector vessels of the heart, pulsated with the heart, and by virtue of the blood and vital spirits they contained, warmed and vivified all the parts. The arteries thereby maintained what natural heat was inherent in the body parts, and also ventilated them of their waste products.[8]

The veins and the arteries were thought to be two great vascular systems that had separate origins, contents, structures, and functions, and which had a significant interchange with each other only at the heart. The right ventricle, anatomically very different from the left, was thought to be a kind of cistern, or reservoir, for the venous system. It had a subsidiary function of perfecting a portion of the venous blood, so that this might be sent up the arterylike vein (now called the pulmonary artery) to the lungs for their nourishment. According to the most widely accepted opinion, another portion of this more perfect blood was drawn out of this cistern, the right ventricle, to pass through the intraventricular septum into the left ventricle. There it met the prepared air coming down the veinlike arteries (now called the pulmonary veins) to serve as the raw material for the creation of vital spirits and arterial blood.[9]

The third great venter was the head, whose principal part was the brain. It was the source of all *animal* functions such as motion, sense, and reason, which it carried out through its effector vessels, the nerves. The nerves were the conduits for the animal spirits, which were elaborated in the brain from the vital spirits in the arteries. These animal spirits were the instruments through which the brain received sense impressions from the outside, and by which it initiated and controlled the motions of the muscles.[10]

Within this tripartite division of the body's functions there were innumerable lesser explanations that accounted for all observed phenomena, from nutrition, heat, and respiration, to the growth of hair. Some were more important—and more controversial—than others. In summarizing only a few such detailed physiological explanations, I shall focus upon those that were explicitly contradicted by Harvey in his *De motu cordis,* and upon the broader areas of physiological explanation implicitly affected by his findings.

Harvey's key to the physiological puzzle was the motion and function of the heart. According to the traditional explanation, as reflected in Bauhin and du Laurens, the two phases of cardiac motion, systole and diastole, were equally active. In diastole the fibers of the heart dilated the ventricles, which thereby filled with blood. In systole, each ventricle compressed and expelled its contents. The arteries were anatomically continuous with the heart; therefore their motions were thought to be synchronous with that of the heart. When the heart dilated, so did the arteries; when it contracted, so did they. Active dilatation was also the cause of the visible manifestation of the heart's motion; when the heart dilated, it struck the chest wall, causing the apex beat. This dual and reciprocal motion of the heart served several kinds of functions. It transferred blood from the vena cava to the right ventricle, and thence to the lungs for their nourishment. It brought down air from the lungs to the left ventricle, where vital spirits were generated. This ventricle, in contraction, expelled waste vapors from the heart up into the lungs for expiration, and also gave the lungs the portion of arterial blood they required for life and motion. And, most importantly, the heart's motions drove blood and vital spirit out of the left ventricle into the aorta and thence into all parts of the body.[11]

The scheme made sense in many ways. Did not one feel the heart beating against the chest wall? Surely this was from the heart expanding in diastole. Could not one feel the arteries expand in every beat of the pulse? Surely this was due to the same kind of active dilatation. And since the arteries patently originated from the heart, it seemed to make sense that the whole system should dilate and contract as the same time. In this motion the heart and arteries were seen to be analogous to the lungs, where it was most certain that expansion was active, and collapse rather less so.

Several features of the traditional theory of the heart and blood are particularly noteworthy. It assumed that venous and arterial blood, although the latter derived in an indirect way from the former, were physically separated except at the heart. The theory also depended upon the ability of organs—whether heart or arteries—to dilate and contract with more or less equal activity. They could thereby alternately attract and expel—"draw and drive"—the liquids they contained. Flows in many vessels were therefore reversible, surging first in one direction and then in the other. Finally, the traditional view posited, if not always explicitly, that the sets of cardiac valves placed at the entrances and exits of the ventricles were only partially effective in maintaining a flow in only one direction.

Just as in the abdominal cavity the stomach subserved the functions of the liver, so in the thoracic cavity did the motions and functions of the lungs subserve those of the heart. The relationship between heart and lungs was perceived as mutually advantageous. The heart provided venous blood to the lungs through the artery-like vein for their nourishment, and arterial blood via the veinlike arteries for their vivification. The lungs returned the favor by providing the air from which the heart generated vital spirits, by cooling and fanning the heat of the heart, and by accepting and disposing of its wastes.

Providing raw material for vital spirits was the most important function of the lungs, one specifically associated with the motion of inspiration. The thorax dilated actively, and to avoid creating the vacuum that nature abhorred, air flowed into the lungs, down the trachea, and into every one of its branches. At their termini, the bronchi opened into, or "innosculated," the fine branches of the veinlike arteries arising out of the left ventricle of the heart. Air could thus be drawn into the ventricle by its dilatation. There it joined the vital spirit and blood already resident in the cardiac sinus, and had added to it the thin portion of blood that had sweated across the intraventricular septum from the venous system. The indwelling heat of the left ventricle thereupon joined with the continual motion of the heart to elevate this mixture of blood and respired air into a higher form, that of vital spirit.[12]

The same heat that helped create the vital spirit was also believed to give the lungs their second function. Just as a fire burns best when fanned, cherished, and moderated by air, so would the heat of the heart be stifled, and flag, unless it were ventilated by the lungs. Thus, in a certain sense, respiration served to cool and moderate the heat of the heart, which was the hottest of all the parts of the body, and the reservoir and vivifier of the natural heat resident elsewhere in the body. This refrigerative function could be confirmed by numerous observations. When a man exercised and heated himself, he breathed more rapidly to disperse the excess heat. Cold-blooded animals, such as reptiles and amphibians, respired less frequently than warm-blooded ones because they had less heat to dissipate.[13]

A third, and decidedly minor, function of respiration was to rid the heart, particularly the left ventricle, of the "smokey," "sooty," or "fuliginous" waste vapors created by the generation of vital spirits. These wastes were expelled up the veinlike arteries into the lungs via their innosculations with the bronchi, and thrust out of the body in expiration.[14]

Just as the Galenic theory of the heart and arteries seemed an admirable mixture of anatomical fact and common sense, so did the traditional theory of respiration. Air obviously contributed something to the body, and given the intimate connection of the heart and lungs, it seemed reasonable that this "something" taken into the lungs should serve a purpose in the heart. It seemed equally clear that the lungs gave off wastes that, if allowed to accumulate in a confined space, would eventually kill an animal or a man. Both phenomena could be explained very well, and very economically, by relating them to the same process of generating vital spirits in the heart. Respiration and heat could similarly be joined in a way that appealed to common sense. The hotter the animal, the more it respired. This linkage could be explained by assuming either that air cooled heat, or that air generated heat. Since the *calidum innatum* was a cardinal principle of Galenic medicine, as well as an obvious characteristic of man and all living things, anatomists generally chose the first alternative. Such a choice between mirror-image explanations was not uncommon in the premodern sciences. Ptolemaic astronomers had to assume either the earth stood still and the heavens moved, or the heavens stood still, and the earth moved. In opting for an innate heat cooled by the air, anatomists were only following their common sense in a rather less dramatic way than the astronomers.

Moreover, these three functions of respiration, as performed by the lungs, seemed confirmed by the fact that a more general process of respiration was maintained throughout the body by all reciprocating organs. The peripheral arteries, for example, also "respired," or rather, transpired. In dilating actively, they brought in air through pores in the skin, using it to moderate and cherish the natural heat of the parts. In contracting, they expelled through their coats the vital spirits necessary to vivify the surrounding parenchyma, and blew off the waste products produced by the action of those parts. In this sense, many cold-blooded animals, such as fish and the invertebrates, could "respire," even though they had no lungs. Their innate heat was so moderate that it could be ventilated merely by the beating of their arteries, without respiration properly so called.[15]

The concept of heat was obviously central to the explanatory framework of traditional physiology. Several kinds were distinguished. First, every part of the body had its own innate heat proper to it as a living being. But this heat, in the arms or legs for example, was continuously being dissipated into the surrounding environment. It therefore had to be replenished from some origin and source in the body—the heart. There, in the thick flesh surrounding the left ventricle, dwelt the innate heat of the organism. This heat, conjointly with the heart's beat, fabricated its instrument out of air and blood—the vital spirit. The vital spirit accepted and embodied the heat of the heart. Vital spirit was, in turn, carried by, and activated, arterial blood. Arterial blood and vital spirit together brought the vivifying influence of the cardiac heat to all parts of the body. The truth of this seemed clear. If a tourniquet were tied tightly around an arm, thereby compressing the artery, stopping its pulsation, and impeding its distribution of vital spirit, the arm would soon grow pale and cold. But let that tourniquet be loosened so as to allow the arm's arteries to pulsate, and the arm soon became exceedingly warm, red and tumid.[16]

The key intermediary was clearly vital spirit. What was it? In general, a spirit was conceived to be a form of subtle matter—fiery, airy, and very swift—that had the capacity to dwell in both liquids and solids, and which was the instrument of a faculty within the body. The animal faculty had animal spirits, distributed through the nerves. Similarly, the vital faculty, seated in the heart, heated and vivified through the vital spirits in arterial blood. The natural faculty in the liver, it was thought, needed no such natural spirit in the blood, since venous blood was in itself the common material of nutrition.[17]

Yet despite this subtle nature and quasi-metaphysical status, spirits were still thought to be subject to certain kinds of physical necessity. They could be driven and drawn by contraction or dilatation. They could be contained—even if with some difficulty—within vessels. Thus the reason for the thick coat of arteries: to hold in the exceedingly active and penetrating vital spirits. Conversely, when it was intended that vital spirits should diffuse out into the parenchyma, as in the limbs or internal organs, the walls of the arterioles were thinner, to allow this active matter to go about its duty.[18]

Because both nutrition and vivification were equally necessary for all organs, every part of the body had its own venous and arterial provision. The light and active nature of the lungs, for example, demanded much nourishment of a highly refined kind; therefore the vessel supplying them, the arterylike vein, was large and had a thick coat, in order to hold a large quantity of subtle nutriments.[19]

Great differences in function thus explained the observed differences between veins and arteries, and between the fluids they contained. Venous blood was relatively heavy, thick, and dark, providing the matter for replacement and growth in body parts. It needed only the thinnest of vessels to contain it. Arterial blood, on the other hand, was animated by active and subtle vital spirits. It was therefore lighter in color and texture, thinner and more penetrating than venous blood, and demanded thicker containers to keep it from being dispersed through the walls of its vessels. Color—the visible difference between bright red arterial blood and dark red venous blood—was merely one of a number of natural and essential differences between the contents and functions of the two systems. These great vascular systems were almost completely separate, like two trees growing into each other, but remaining entirely autonomous. If blood from an incision ran bright red, and pulsated because of its reciprocally contracting and dilating coat, one's lancet had inadvertently cut into the arterial system, whose blood originated in the left ventricle. If dark blood flowed slowly from one's incision, one had properly cut into the nutritive, venous system, whose blood originated in the liver.[20]

Fetal physiology, although more complicated and rather more controversial, was an extension of the same concepts of heat, spirits, and ventilation, with due allowance made for the differing anatomical arrangements of fetal vessels. Early seventeenth-century anatomists knew the following facts well. In the fetus the lungs do not move. A fistula—the foramen ovale—connects the right and left auricles, thus partially bypassing the right ventricle and lungs. Another fetal anatomical feature, the ductus arteriosus, connects the pulmonary artery directly to the aorta, thus again bypassing the lungs. Umbilical arteries run from the fetus to the placenta, and an umbilical vein runs from the placenta to insert, via the ductus venosus, into the inferior vena cava.

What distinguished the traditional physiology of the fetus from the modern was not a difference in knowledge of gross anatomy, but the way in which adult physiology was held to be reflected in the fetus. Du Laurens and Bauhin believed that the fetus, having only a small amount of its own heat, did not need the use of its lungs to moderate and cool its heart; the lungs therefore did not move. The umbilical vein, it was thought, anastomosed with the mother's veins in the placenta. Through it nourishment passed into the fetal vena cava, and thence was distributed to all parts of the fetus. The lungs were nourished with venous blood from the vena cava, via the foramen ovale and pulmonary veins. The fetus similarly drew its vital spirits from the mother through its umbilical arteries into the aorta, whence they were distributed. The nonfunctioning lungs got what little vital spirit they needed from the aorta via the ductus arteriosus and the pulmonary artery. Thus the fetus was believed to *transpire,* because its arteries beat with a motion communicated to them from the mother, but not to *respire,* moving its lungs, creating its own vital spirits, and moderating its own heat.[21] Thus did the traditional physiological scheme account for the same facts we now explain by fetal circulation, doing so with an even more clever attention to gross anatomical structure than characterized the system of adult physiology.

Such were the kinds of physiological opinions Harvey found when he turned to the textbooks of the early seventeenth century. It was a collection of explanations whose cardinal virtue was *system*. Every organ had its function, and every function had its organ. The triad was preeminent: three essential functions, three faculties, three venters, three principal parts, three sets of vessels, three fluids. Every major physiological explanation articulated closely into its conceptual neighbor. Heart, lungs, arteries, veins, blood, spirits, heat—all were bound together in an intellectual Gordian knot.

To the modern mind there is another prominent feature of traditional physiology: it lacked completely any concept of biological substructure. The body's solids were described simply by words, such as bone, muscle, parenchyma, tendon, membrane. But what were the structures of such solids? That question was nowhere asked. Indeed, in a very distinct way, it was nonsensical. Solids were thought to act either because of their nature, or because they were inhabited by some higher metaphysical entity, such as a faculty. Thus the parenchyma of the kidney attracted urine from the blood because it was its nature to do so. The parenchyma of the liver transformed chyle into nutritive venous blood because it was the seat of the natural faculty. Liquids were seen in a similar way. A fluid in a vessel might be described as a mixture of blood and air, or of blood and urine, or of blood and vital spirit, but no attempt was made, either to determine the proportions of the putative components, or even to determine in what sense the fluid was a mixture. Were its component parts present as distinct and separate particles? Or did they unite in some way when put together? For physiology, as taught in the anatomical tradition, these were nonexistent questions.

Indeed, even if one could separate out the "components" of muscle or blood, one would not thereby illuminate how these parts performed their functions. Rather, one would have simply a series of decomposition products, bearing no more necessary relationship to the essential functioning of that part than a pile of bricks bore to a house. Traditional anatomy and physiology usually rejected the idea that functions were in some kind of "mechanical" way performed by the parts of a solid or a fluid, and were thus not understandable in those terms. Could one understand a house by looking at a pile of bricks? Thus, not only did early seventeenth-century anatomists lack such instruments as microscopes to reveal structure at a level undiscernible by the naked eye, but the very nature of Galenic physiology precluded the idea that such knowledge, should it ever be gained, would have any relation to function.

THE DISCOVERY OF THE CIRCULATION

Scholarship of the last two decades, especially that of Pagel, Whitteridge, and Bylebyl, has made it clear just how much Harvey's innovations arose as a critical reaction to the foregoing melange of descriptions, explanations, and assumptions. We can now trace out, with some degree of assurance, the path Harvey followed in coming to his new insights, and how he came to argue them with unprecedented originality, precision, and clarity.

The discovery was clearly foreshadowed in the Lumleian lectures delivered, from 1616 on, before the Royal College of Physicians. When Harvey came to speak of the heart and its motion he advised his audience that they approached a subject of the greatest complexity. He first reiterated the commonly accepted description that diastole was an active phase, and that the apex beat was caused by the heart striking the breast in its dilatation. But this did not agree, he pointed out, either with his own observations or with the arguments of some other anatomists, chiefly Realdo Colombo. Rather, vivisections show a different sequence of events. First the auricles contract, thrusting blood into the ventricles and causing their diastole. The ventricles then are roused, which contract in systole, causing the heart to rise up and strike the breast, while at the same time expelling the contained blood. This blood is thrown into the arteries, expanding them, which motion is perceived as a beat of the pulse. Thus not only does the *systole* of the heart correspond to the apex beat, it is synchronous with, and the cause of, the *diastole* of the arteries.[22] Harvey had solved the long-standing problem of the arterial pulse.

This vivisectional discovery had another important consequence—the pulmonary transit. Harvey pointed out that if one assumed the cardiac valves were competent, then active systole of the right ventricle must thrust blood out the arterylike vein (pulmonary artery) into the lungs, whence it can return to the heart only through the veinlike arteries (pulmonary veins). Indeed, only such a flow could explain the thick and arterial character of the coat of the arterylike vein; it served to absorb the impact of the blood thrust into it.[23]

As each was expressed in the Lumleian lectures of 1616, none of Harvey's three main arguments—that ventricular systole caused the apex beat, that ventricular systole corresponded to arterial diastole, and that blood flowed across the lungs from the right ventricle to the left—was entirely new. Each had been argued by a number of sixteenth-century anatomists with whom Harvey was conversant. Harvey's novelty lay not so much in his conclusions as in his approach. He used *vivisection* to argue opinions that had previously been, to a large extent, argued on the basis of anatomical analogy or physiological necessity. Du Laurens had discussed at length whether the heart struck the breast in systole or diastole, and settled on diastole. He similarly had considered the cause of the arterial motion, and concluded that the arteries dilated actively because of their continuity with the heart; this active dilatation served the same respiratory purposes as the similar motion of the lungs.[24] Colombo's argument for the pulmonary transit, the most well known in the sixteenth and early seventeenth century, was based, not so much on his description of the heart's motion (which was marred by some terminological confusion), as on his firm belief that the vital spirits were generated in the lungs, not in the heart; therefore, the blood *had* to flow through the lungs.[25]

Thus, when Harvey came to write a short book based on the conclusions he had sketched out in the Lumleian lectures, he quite rightly emphasized what was new about his findings—their vivisectional foundation. In the *Prooemium* he delineated the weaknesses of previous accounts of the motion of the heart and arteries: they assumed that pulse and respiration had the same function and therefore the same kind of motion; they posited that the right and left ventricles had radically different functions; and they assumed that blood passed through the thick and dense intraventricular septum.[26] Then, in the first seven chapters, Harvey laid out for his readers, much as he had for his Lumleian auditors, his own theory of the motions of the heart and arteries, and of the consequences of those motions. The central descriptive chapters, II-IV, bore a subtitle indicative of their origins— *ex vivorum dissectione*. In Chapter V he summarized his new theory. In Chapters VI and VII he showed by anatomical arguments and vivisectional experiments that his theory of cardiac and arterial motion necessitated a pulmonary transit.[27]

The proem and first seven chapters of *De motu cordis* contained no mention of the circulation of the blood. It was first introduced in Chapter VIII, and discussed extensively in Chapters IX-XVI. This division—the first half on the motion of the heart (Chapters I-VII), and the second on the circulation of the blood (Chapters VIII-XVI)—has often been praised by scholars as an effective logical and rhetorical device. Only recently has Bylebyl taken this analysis an important step further. He has argued quite convincingly, on the basis of textual evidence, that the proem, Chapters I-VII, and Chapter XVII, were probably written first, before Harvey had any clear idea of the circulation. After he came to understand the circular motion of the blood, Harvey wrote up his arguments for it in eight more chapters, grafted the new section onto the old, and published the whole in 1628.[28] Bylebyl's interpretation, although open to some objections,[29] makes good sense, not only of the structure and content of *De motu cordis,* but of Harvey's intellectual development in the 1610s and 1620s.

The added chapters, VIII-XVI, although of a radically different shape than the first seven chapters, were cut from the same cloth. Experiment, especially vivisection, was the theme. A circulation of blood may be proved, Harvey argued, by confirming three suppositions: (1) that the mass of blood transmitted from veins to arteries through the heart was too great to be supplied by food alone; (2) that blood was driven in great quantities to all parts through the arteries, in an amount greater than was needed for the nutrition of those parts; and (3) that the veins constantly returned blood from the extremities to the heart. Harvey then proceeded to prove these three suppositions by observation and experiment, especially vivisection.[30]

These experiments, and their articulation into a brilliant structure of argument, are too well known to need recapitulation here. I wish to focus, not on what Harvey argued explicitly, but on what he saw as the implications of that argument for the survival of traditional Galenic physiology. There was much more to that explanatory tradition than the narrow aspects of heart, arteries, and blood with which *De motu cordis* was concerned. What were Harvey's early attitudes toward the comprehensive structure of Galenic explanation? How did he see it changed by his new discovery? One may see Harvey's answers to these questions in three stages, contained in his lectures, and in the two relatively independent portions of the *De motu cordis*.

HARVEY ON HEAT, RESPIRATION, VITAL SPIRITS, AND BLOOD COLOR

Heat presented, at least at the outset, few problems. In the "Prelectiones" Harvey accepted almost all of the classical, and especially Aristotelian, conceptions of animal heat. Innate heat was the author of all concoctions of food.[31] It moved fluids in the body.[32] Parts of the body were more or less noble according to their heat.[33] The heat of certain organs, such as the heart, created the need for other parts, such as the pericardial fluid or the brain, to be cool.[34] Most of all, Harvey saw heat as centering upon, and intimately connected with, the heart. The heart was the font of heat in the body. By virtue of this heat it was the chief part in the body. The heart imparted heat to all parts, and received it from none. The blood was the instrument of the heart in this office, and hence parts that contained much blood, such as the right ventricle, were especially warm.[35] This is not to say that Harvey's allegiance to the heart as a source of heat was complete and absolute. On more than one occasion he asserted that, contrary to Aristotle's belief, the heart had its embryological origins in the blood, and not vice versa; hence in terms of "primacy," the blood was of greater antiquity than the heart. Harvey also intimated that the heart was the origin of heat as much because it contained blood and spirit, as because the solid parts of the heart were in themselves hot.[36] It was the heart as a unit—flesh, fiber, blood, and spirit—that was the source of all heat in the body.

Discovering the true motion of the heart, and the resultant circulation, led Harvey to change these ideas remarkably little. In the precirculatory portions of *De motu cordis* he was at pains only to reject the idea that the pulsation of the arteries served to moderate the body's heat; he had, after all, proved that this pulse was of a purely hydraulic origin. In the new portions arguing the circulation, Harvey neatly made his new concept serve an old function. What better than to make the circulating blood a vehicle by which the heart imparted hotter and more perfect blood to all the parts? In the periphery the blood was cooled, coagulated, and worn out, only to be returned again for renewal to the center of the bodily economy, the heart. There, by the "natural, powerful, fiery heat, a store of life," it was reliquified and again sent out by the heart's pulsative movement.[37] Thus a tight tourniquet impeded the movement of blood, and hence, heat, into a limb, while loosing the ligature resulted in a sensible rush of cold blood into the center.[38] Harvey expanded on this theme in the chapter arguing that "the circuit of the blood is confirmed by probable reasons." All living things, Harvey said, following Aristotle, needed "a site and source of warmth," without which they would die.

This was without doubt the heart. Therefore the blood had to have constant recourse to the heart to renew its warmth. Chilled hands, limbs, noses, and cheeks were thus restored to warmth and liveliness, Harvey said, by the arrival of a fresh supply of heat from the center.[39] The very same theme was expressed concisely, if rather less picturesquely, in the famous note added by Harvey to his "Prelectiones" in the late 1620s:

> And for this reason Δ [Harvey's sign for "demonstratio"] it is certain that the perpetual movement of the blood in a circle is caused by the heart beat. Why? Is it for the sake of nutrition, or is it rather for the preservation of the blood and of the limbs by means of the infused heat? And the blood by turns heating the limbs and when it is made cold is warmed by the heart.[40]

Heat was, as Harvey had reminded his Lumleian auditors, intimately connected with the lungs in many ways: animals that breathed infrequently, such as frogs, snakes, or tortoises, were cold; the lungs of viviparous, hotter animals, were larger and contained more blood than those of the colder ovipara; the hotter the animal, the more filled with blood the lungs; therefore, the size of the containing thorax was proportional to the abundance of heat in an animal; and even among men, those of a hotter disposition had a wider chest and a greater abundance of blood in the thorax.[41] Most of these observations had classical, and especially Aristotelian, precedents. But Harvey's early ideas on respiration, while they linked heat and the lungs, connected them in a way at variance with both the Aristotelian and Galenic traditions. The source of Harvey's apostasy was his belief in the pulmonary transit. Since Harvey knew from his dissections that the lungs sent blood *to* the left ventricle of the heart, they could not relieve it of its excess heat, nor send it air for the generation of vital spirits, nor accept from it the fuliginous wastes believed to ascend up the veinlike arteries (pulmonary veins). What then were the lungs to do?

As early as the Lumleian lectures, Harvey solved this problem by having the lungs serve the traditional functions, but by having them do so with reference, not to the heart but to the blood.

The foremost of the classical functions was refrigeration. At certain points in the lectures Harvey intimated that the lungs actually cooled the heart itself.[42] But when he came to discuss them in detail, he spoke as though the lungs cooled instead the instrument of the heart, the blood. Many facts, Harvey believed, supported this view. The quantity of blood was always proportional to both the warmth and the size of the lungs.[43] The lungs were very abundantly supplied with veins and vessels, out of which the heat was collected.[44] By this cooling and ventilation, but chiefly by ventilation, the innate heat was conserved and maintained.[45] Harvey also said—and here he followed Colombo—that the lungs were the storehouse of spirit.[46] They were, after all, large organs that must have been made for some general or "public" function. It would seem that this general use was to concoct the blood and spirit as it passed through (subject of course to

Harvey's own special conception of spirit, to be discussed below). In fact, insofar as the lungs made spirit, they were second in importance only to the heart, and nobler than the liver.[47] Harvey accepted, although very much in passing, the excrementary function of the lungs; expiration expelled the fuliginous vapors, which were of a consistency between water and air.[48] As a last function, Harvey hinted that the lungs in subsidence propelled the blood across from the pulmonary arteries to the pulmonary veins, and thence into the left auricle.[49]

Some of these same ideas of respiratory function were developed at length in *De motu cordis,* overwhelmingly in the precirculatory portions that argued the pulmonary transit. From comparative evidence Harvey pointed out that the presence of lungs implied that of a right ventricle, and vice versa. Animals with only one ventricle, such as fish, did not have lungs.[50] The lungs were therefore a detour found only in a minority of animals; the normal path of the blood was from the veins into the arteries, directly through the heart. Indeed, the fetus was like these more primitive animals, in that the foramen ovale and ductus arteriosus were contrived to shunt blood past the lungs, creating, in effect, a two-chambered heart.[51] Why then in the adult should all the blood be moved through the lungs by the right ventricle?

> The answer may be as follows, or at least lie along some such lines. The larger and more perfect animals are naturally warmer and, when full-grown, can reasonably be described as over-heated and hard put to it to get rid of the excess. So the hot blood is carried to and through the lungs to be tempered by the inspired air and to be freed from bubbling to excess.[52]

The lungs were stuffed with this blood because they, with the heart, were the storehouse, source, and treasury of the blood, the workshop where it was brought to perfection.[53] Harvey said that he had made, moreover, numerous observations on the lungs, their functions, movements, and on the need for air, but he forebore mentioning them because they departed too greatly from his theme of the motion of the heart.[54] But it was beyond doubt, Harvey said, that the whole mass of the blood passed through the fine and spongy texture of the lungs, propelled by the contraction of the right ventricle. Harvey could even, and did, cite passages from Galen which, suitably tortured, would seem to confirm that the grand physician suspected a pulmonary transit.[55]

The passage of the blood across the lungs was in turn affected by their motions. As the lungs rose, their porosities and vessels opened, and when they fell, these openings constricted. Therefore the blood bypassed the fetal lungs, which had no motion to facilitate its transit.[56] These considerations led Harvey to suggest, in passing, another function for the mechanical motion of the lungs. Since all of the nutriment of the body passed through the lungs, perhaps this was to effect some kind of filtering, and concocting by heat, appropriate to the more perfect food demanded by more perfect animals.[57]

Such were Harvey's early ideas on heat and respiration. But what of the function that, in the Galenic tradition, had so closely linked heart and lungs—that the heart received prepared air from the lungs and used it to generate vital spirit? From the beginning of his scientific career, Harvey was suspicious of these ideas. On three separate occasions in the "Prelectiones" he asserted that there was no spirit independent of the blood. Against both the "physicians," and the vulgar masses, who believed in spirits separate and distinct from the humors that carried them, he championed the Aristotelian view that spirit was innate to the blood, part of its very definition.[58] "As light is to the candle, so is the spirit to the blood and it has actuality in being made, like flames from fire existing in continual generation and flux."[59] Moreover, Harvey clearly felt that spirits were definitely not generated in the left ventricle. Although often content to follow Bauhin's anatomical descriptions in his Lumleian lectures, Harvey pointedly omitted the Swiss anatomist's description of the ventricular generation of vital spirits.[60] If spirits were concocted anywhere from air, Harvey hinted, it was in their "storehouse," the lungs.[61]

By the late 1620s, and the publication of *De motu cordis*, Harvey even came to suspect the putative role of the air in the generation of spirits of any kind. He publicly recorded his belief that all blood, whether venous or arterial in character, had inseparable spirit.[62] Nor was this spirit derived from air. He was particularly critical of the traditional idea that the pulmonary veins carried gases—air to the heart and sooty vapors to the lungs. The walls of the pulmonary veins were thin, Harvey pointed out; why didn't they collapse if the vessels were not filled with fluid? How could the two streams move in opposite directions and not mix? How could the cardiac valves hinder some flows and not others? The question was determinable by experiment as well as by rhetorical logic. If one cut open the pulmonary veins, one always found blood, not air or sooty vapors.[63] Harvey could make his point even more forcefully by giving an old experiment a new twist:

> If one performed Galen's experiment, and incised the trachea of a still living dog, forcibly filled its lungs with air by means of bellows, and ligated them strongly in the distended position, one would find, on rapidly opening up the chest, a great deal of air in the lungs right out to their outermost coat, but no trace of such in the vein-like artery or in the left ventricle of the heart. If in the living dog either the heart drew air from the lungs, or the lungs transmitted air to the heart, they should do so much more in this experiment.[64]

By the late 1620s Harvey was so convinced that air did not enter the body, and had no part in the origin of spirits, that he added a flat denial as a marginal note to the discussion of spirits in his lectures: *spiritus non ex aere*.[65]

But if air did not generate vital spirits, why was arterial blood so bright red? This was no problem for Harvey in his lectures; he never mentioned the subject. One might think the question would certainly have arisen in *De motu cordis*. If Harvey's argument were correct, arterial blood was continuously being transformed into venous blood, and vice versa. Yet this serious implication of the circulation was addressed nowhere in the book. The only clue we have to Harvey's early speculations occurs as a later addition to the "Prelectiones." Harvey had written in 1616 that "through the lungs passes incessantly all the nutriment of the body and the whole mass of the blood." To which he added, probably sometime in the late 1620s: "which explains why arterial blood is redder."[66] Harvey seemed to hint, albeit indirectly, that the bright red color of arterial blood was due to some kind of physical straining process that took place in the lungs, producing a finer, and therefore redder, blood.

HARVEY'S "TRADITION": APPROACHES, TECHNIQUES, AND NATURAL PHILOSOPHY

Such were Harvey's early attitudes towards the problems—heat, respiration, spirits, blood color—raised by his discovery of the circulation. He extended and modified these ideas during the 1630s and 1640s and they became in turn the starting point for the investigations of others from the 1650s onward. But there was more to Harvey's legacy than explicit observations, experiments, and physiological concepts. There was also a distinct body of Harveian approaches, or "concerns" that directed his own investigations and those of his successors. Harvey used certain techniques that were taken over, sometimes in modified form, by his followers. Harvey's oeuvre was also informed by a natural philosophy that was congruent with his own training and inclinations. This, by way of contrast with Harvey's approach and his techniques, was perceived by many of his disciples, especially in the 1650s, to be outdated, and an impediment to further development.

The foundation of the Harveian approach was dissection. "Anatomy," Harvey said in the opening words of his "Prelectiones," "teaches the uses and actions of the parts of the body by ocular inspection and by dissection."[67] It was a theme that echoed through Harvey's life and writings. *De motu cordis* was written "not from books but from dissections, not from the tenets of philosophers but from the fabric of nature."[68] In *De generatione* (1651) he remarked that he had never dissected any animal without discovering something unexpected.[69]

Nor was dissection to be limited to man. Harvey began his anatomical career by counseling in the "Prelectiones" that knowledge of parts came only by comparing animals of different species, whether winged, terrestrial, aquatic, viviparous or oviparous;[70] he ended that career by reiterating the same point in *De generatione*.[71] Harvey most certainly followed his own exhortations. He had a well-developed comparative method, dissected scores of species, and in his work cited observations on hundreds of animals.[72]

But Harvey was most original in his insistence, not merely on the dissection of different animals, but on their vivisection as well. The very first sentence of the main text of *De motu cordis* told the reader that Harvey's innovations began in "many dissections" upon "living animals."[73] He reported observations and experiments on live prawns, shellfish, lobsters, snails, small fish, eels, frogs, toads, snakes, lizards, doves, pigs, and dogs; the very essence of the book was vivisection. His opponents recognized this. Harvey noted when he came to defend his discoveries in *De circulatione* (1649) that his detractors had accused him of following the "empty glory of vivisections," and had ridiculed his experiments on "frogs, snakes, flies, and other lower animals."[74]

Harvey also believed firmly that embryological evidence illuminated physiological problems. In the "Prelectiones" he reported observations of fetal lungs before and after they first respired.[75] He used embryological evidence in *De motu cordis* on at least eight different occasions, including the important passages cited above in which he reinterpreted the fetal cardiovascular system to argue for the circulation.[76] Physiological points abounded in the *De generatione*, and, conversely, Harvey marshalled embryological evidence to defend his ideas in *De circulatione*.[77] He clearly felt that origin illuminated function, and vice versa.

Harvey's experimental approach was buttressed by his techniques. Although ligature, inflation, injection, and quantification did not originate with Harvey, he used them with such ingenuity and skill that they became part of the standard armamentarium of the experimental physiologist.

Ligature had been used by physiologists from Galen onward, but by no one with such success as Harvey. In his Lumleian lectures he used ligatures to argue that the arterial pulse resulted from the thrust of blood.[78] In *De motu cordis* he used ligatures in his famous experiments demonstrating the functions of the venous valves, and that blood entered the limbs by the arteries and left by the veins.[79] Characteristically enough, he returned to this same technique in defending the circulation twenty years later: he ligated the pulmonary artery to stop the aortic pulse; he reiterated a variation of his arm-ligature experiment; and he used ligatures to carry out Galen's well-known experiment of substituting a piece of reed for a length of artery—all to confirm his ideas on the origin of the pulse and the circulation of the blood.[80]

Inflation was similarly an ancient technique that Harvey turned to good account. He referred in his lectures to experiments of inflating the intestines.[81] As noted above, in *De motu cordis* he used Galen's experiment of inflating the lungs as part of his proof that no air entered the pulmonary veins or the left ventricle.[82] In his studies on reproduction, Harvey often used inflation to reveal anatomical features: the structure of the cloaca, uterus, and ovaries, as well as the capillary connections between arteries and veins in the gravid doe. He also noted that the lungs of birds have perforations leading into the abdomen, and demonstrated them by blowing into the lungs of birds.[83]

Although it was merely inflation with a different fluid, Harvey used injection rather less. In the lectures he recounted some experiments using water injections of the intestines,[84] but his best experiment using this technique was not published in his lifetime. In 1651 Harvey reported in a letter to Paul Marquand Schlegel an experiment recently performed in the presence of colleagues. He ligated the appropriate vessels, and found that water injected into the right ventricle would not pass through the septum, but could only enter the left ventricle by passing through the lungs.[85]

Quantification was perhaps the most well known—and controversial—of the Harveian techniques.[86] In the whole of his oeuvre, he used it relatively little; most of his arguments were from observation, structural necessity, vivisection, and comparative anatomy. But it appeared so prominently in the pivotal chapters of *De motu cordis* that no subsequent anatomist could be unaware that Harvey had, for one of his arguments, deduced the necessity of a circulation from the output of the left ventricle, estimated by multiplying the stroke volume by the pulse rate.[87] Scholars disagree about the importance of this quantification in the process of discovery, but Harvey's followers were unanimous in recognizing its conclusiveness as a demonstration.

These methodological concerns and manipulative techniques were embedded in a natural philosophy that was largely traditional. Harvey was an innovator in physiology, but a conservator in scientific world-view. That world-view, as Harveian scholarship has continued to emphasize, was solidly Aristotelian.[88]

The metaphysical hallmarks of classical natural philosophy are particularly clear in Harvey's Lumleian lectures: his belief in a hierarchy of nobility among animals and organs; the equation of warmth with perfection; adherence to the doctrine that all matter tended toward its proper place; a belief in the attractive power of parts of the organism; a lack of concern for the composition of anatomical structures; the acceptance of faculties as guiding entities within the organism; and the emphasis on various parts of the body producing a balance of heat and cold.[89] Even in his later works, Harvey continued to reflect, to a very large extent, the natural philosophy with which he began his scientific career.

He generally rejected a "mechanical" approach to biological structures and processes.[90] The complex changes observable in embryogenesis especially, led Harvey to affirm repeatedly the goal-oriented action of the entities he designated variously as faculties, plastic powers, principles, or souls. The very idea of epigenesis, for which Harvey argued so strongly in the *De generatione,* was founded upon a vitalistic and teleological conception of biological processes. No "mechanical" approach to embryology, such as that of his teacher Fabricius, could properly explain how development of the fetus was ordered with regard to its end, the formed organism.[91]

Harvey therefore found the atomical philosophy, whether ancient or modern, inadequate in biology. He believed that those men erred who, with Democritus, composed all things of atoms, or, with Empedocles, of elements. Neither generation nor nutrition was simply the process of the accretion of small particles of similar type to form a larger whole. In general, he felt it was a philosophical error to seek the cause of diversity of organic parts in the diversity of the matter whence they arose.[92] Although on occasion he did speak passingly of "invisible atoms" carried in the wind,[93] Harvey clearly believed that the divisibility of matter—should it even be granted—had no bearing upon the biological world.

It was for these very reasons that Harvey remained aloof from Aristotelianism's nascent competitor in matter theory, chemistry. As early as the Lumleian lectures he expressed scepticism about the notion of a chemical fermentation in the intestines, and, although he used several metaphors that indicated his familiarity with the alchemist's art, he seems from the very beginning to have doubted the applicability of chemistry to biology.[94] His objection to chemistry, as voiced in *De generatione,* was its methodological assertion that a thing was what it could be broken down into. There were, he believed, "both mixed and compound bodies prior to any of the so-called elements." Therefore these elements, whether they be Empedoclean, Democritean, or Paracelsian, existed more in reason than in fact. Harvey's attitude was expressed in one succinct summary: "It is, however, an argument of no great cogency to say that natural bodies are primarily produced or composed of those things into which they are ultimately resolved."[95]

This is not to say that Harvey would not use quasi-chemical concepts when they seemed appropriate. He agreed with Aristotle that an ebullition, or fermentation, of the spirits in the embryonic blood may have been responsible for the origin of the heart's diastole.[96] He would not, however, grant Descartes's contention that a similar kind of fermentation was responsible for the continuation of the heartbeat beyond embryogenesis; the heartbeat was simply too quick and vigorous to be attributed to the gentle and slow effervescence of which the blood might be capable.[97] A kind of chemistry, especially if with an Aristotelian precedent, might be invoked in the beginning, but not in the completed organism.

It may therefore not be necessary to conclude from John Aubrey's comment— "He did not care for Chymistry and was wont to speake against them with an undervalue"[98]—that Harvey was completely against the new science of matter. He might grant the appropriateness of some of its concepts and processes, but he found its protagonists wearisome, and its metaphysical assumptions questionable.

Given such a conservative temperament, it is perhaps understandable that Harvey was slow in drawing out the implications of the circulation. That it had many implications was clear. The closely articulated system of traditional physiology made it impossible to transform only one element. Harvey responded initially by changing as little as possible. In the early 1630s he still felt that the heart distributed heat to all parts of the body through the blood, a process now accomplished by the circulation. Respiration still cooled, but since it could anatomically no longer cool the heart, Harvey moved toward the belief that it cooled the blood. He had doubts that respiration served to generate spirits—indeed, that such spirits even existed distinguishable from the blood.

But the discovery of the circulation, and Harvey's initial explorations into its implications, had revealed two very large questions with which Harvey was not yet prepared to deal. If air did not enter the heart or blood to generate vital spirits, why were men and animals constantly in need of a fresh supply? And if venous blood was constantly becoming arterial, and vice versa, how did one explain the differences between the two without recourse to vital spirits? Clearly with the publication of *De motu cordis,* Harvey's work had only just begun.

HARVEY'S LATER WORK AND ENGLISH DISCIPLES

2

As Harvey concluded the *De motu cordis*, he noted with no false modesty that in every part of medicine—pathology and therapeutics as well as physiology—innumerable questions could be settled, doubts resolved, and obscurities illuminated, by his new truth. A lifetime's work would perhaps not suffice to explore them all, he said, much less to treat them in this small work.[1] It must have disheartened him to find that most of the early responses to his discovery, both at home and abroad, saw only one implication: the destruction of traditional Galenic medicine. James Primrose (1630), Ole Worm (1632), and Emilio Parigiano (1635) published book-length refutations. Favorable notices by Robert Fludd (1631) and Marin Mersenne (1634) were very much in passing, while René Descartes's *Discours de la méthode* (1637) was the only major publication before 1640 to discuss Harvey's doctrine at any length.[2]

Although Harvey continued, because of his clinical skills and his position as one of the royal physicians-extraordinary, to occupy a place of high esteem at the College of Physicians during the 1630s, his colleagues did not come to his defense. This seems the more peculiar since Harvey had claimed in *De motu cordis* that he had demonstrated his ideas of cardiac motion before the College on a number of prior occasions, had freed them from the objections of skillful anatomists, and could call many of his colleagues as witnesses to his experiments.[3] If so, they were very silent witnesses. Harvey later told Aubrey that after *De motu cordis* was published, "he fell mightily in his practize, and that 'twas beleeved by the vulgar that he was crack-brained; and all the physitians were against his opinion and envyed him."[4] This was true. With the exception of Fludd, no English physician of Harvey's generation or older is known to have supported the circulation. Fortunately, Harvey lived not only to see his work accepted and extended by men of younger generations but also to contribute himself, both at Oxford and in London, to the unfolding of its implications.

HARVEY'S EARLY DISCIPLES IN ENGLAND

Harvey met two of his earliest adherents, George Ent and John Greaves, quite by chance. In the summer of 1636 Harvey went to Germany as part of a diplomatic mission led by his good friend, the Earl of Arundel, and took the opportunity to make a side trip down into Italy.[5] There in Venice, probably through a mutual friend in the Earl's entourage, Harvey met Ent and Greaves. Ent was a Cambridge man who, for the previous five years, had been studying medicine at Padua.[6] There he became friends with Greaves, an Oxford mathematician, oriental linguist, and, since 1630, the Gresham Professor of Geometry.[7] Both must have known of Harvey, for when Ent graduated M.D. at Padua in April 1636 his friends published a pamphlet of verses in his honor, including a poem by Greaves and another by one Johannes Rhode, containing extensive references to Harvey's discovery of the circulation.[8] After meeting the author of the doctrine, Ent and Greaves seem to have traveled with Harvey to Rome in October 1636[9] and possibly even as far south as Naples, before Harvey rejoined Arundel's party in early November.[10]

Ent returned to England by late 1638, was admitted to the College of Physicians in 1639, and in 1641 gave proof of his friendship by publishing a major defense of Harvey's ideas, *Apologia pro circulatione sanguinis*. Ent dedicated the work to Harvey, and prefaced it with a poem by their mutual friend, John Greaves.[11] In the *Apologia*, Ent drew out the physiological and medical implications of the circulation by defending it against one of Harvey's detractors, the Venetian Emilio Parigiano.[12] Ent emphasized the importance of experiment in disproving *a priori* judgments. He held up Harvey's attitude and methodology as a model for other physicians to emulate. Only those lacking in philosophical acumen, Ent said in his dedicatory epistle to Harvey, would assert that we cannot surpass the wisdom of our ancestors. Just because an opinion deviates from that of the ancients, we need not assume that Galen must be correct. Contrary to what Harvey's enemies had said, *De motu cordis* was truly a golden work of medicine.[13]

In succeeding years, Ent became a prime exponent of the Harveian approach at the College of Physicians. In 1642 he delivered the College's Gulstonian Lectures. In 1645 he was elected a Censor, a position he filled annually with only three exceptions until 1669. One can scarcely imagine that, in examining for admission to the College, he would pass a candidate whose knowledge of recent anatomical findings, and especially of the circulation, was defective.

Ent's fellow Harveian there was Francis Glisson.[14] Both had been at Cambridge together in the late 1620s, and when Glisson came to deliver anatomical and pathological lectures at the College in 1639−1641, he relied upon Ent's advice and help.[15] Little wonder that the surviving manuscripts of these lectures show that Glisson had not only accepted Harvey's findings, but had begun a program of anatomical research on the liver, whose function now needed to be reinterpreted.

Glisson also became a Harveian exponent at Cambridge, where, since 1636, he had been Regius Professor of Medicine. Among his many surviving lectures, orations, and disputations—most of them dating from the 1640s and 1650s—are items straight from the Harveian corpus: "On the Circuit of the Blood," "The Arteries are Moved According to the Motion of the Heart," "The Primitive Motion of the Blood is Intrinsic," "The Ideas of Descartes Concerning the Motion of the Heart are not Consonant with the Truth," as well as a number of untitled lectures on the motion of the blood that included answers to objections.[16] The origin and properties of the blood interested him no less than its movement. He lectured on how "There are No Vital Spirits Separable from the Blood," how "In every Fever There Is a Fermentation of the Blood," that "The Blood is the Cause of Sanguification," and that "Blood Alone is the Innate Heat."[17] Examples could be multiplied manyfold. There were lectures on embryology, on animal heat, on the functions of the spleen, liver, stomach, lacteals, and lymphatics. One volume of Glisson's manuscripts contains nothing but notes of anatomical dissections: of humans, pigs, chicks and chickens, fish, rats, oxen, and even a dolphin.[18]

The same innovative spirit may be seen in the topics disputed by Glisson's students, most especially in the 1650s. One young doctor recapitulated how "The Blood is Moved in a Circle."[19] In 1653 Robert Brady, who was later Glisson's deputy and successor as Regius, set forth the evidence why "The Pulsation of the Arteries is Caused by the Impulsion of Blood from the Heart."[20] Other disputations were based upon the implications of the circulation: on the spleen as a source of a sanguinary ferment, on the blood as a principal part, on its role in nutrition, and on the function of respiration. Glisson's own interest in the liver found reflection in student theses. In 1652 George Joyliffe argued the proposition that "The Separation of Bile is the Proper Function of the Liver."[21] In 1654 Henry Power debated the complementary thesis, that "The Liver is Not the Organ of Sanguification," as the Galenic tradition had held.[22] Just two years before, Power had composed, based upon numerous vivisections, a small tract entitled "Circulatio Sanguinis," in which he related no less than fifty experiments—including some of Glisson's—that supported "our Reverend and Worthy Dr. Harvey," and "that incomparable Invention of his, the Circulation of the Blood."[23]

Clearly both Glisson and Ent were, by the mid-1640s, proponents of new ideas in physiology. It is therefore not surprising that one finds them among the "divers worthy Persons"—described by John Wallis—who met in London beginning in 1645 to discuss the "*New Philosophy* or *Experimental Philosophy*."[24] The "1645 Group" has most often been treated as a precursor of the Royal Society; it may also be considered in part as arising from the desires of investigative physicians to broaden their scientific interests by meeting with such mathematicians as Wallis, John Wilkins, and Samuel Foster.

As iterated by Wallis, the physical sciences dominated the club's agenda of discussion and experimentation; geometry, astronomy, navigation, statics, mechanics, magnetism and "Natural Experiments" were the order of the day. But within these, medicine, anatomy, and chemistry claimed places as well. In addition to such topics as the nature of comets and the possibility of a vacuum, the group also considered the circulation of the blood, the valves in the veins and the lacteals.[25] Wallis, although a mathematician, was quite capable of judging such questions. As Glisson's student at Cambridge in the late 1630s, he had interested himself in what he called "the Speculative part of Physick and Anatomy" and about 1641 was the first of Glisson's "Sons, who (in a publick Disputation) maintain'd the Circulation of the Bloud, (which was then a new Doctrine)."[26]

Harvey's influence expressed itself also in the group's membership. Five of the nine participants named by Wallis were physicians, all of whom had contacts with the king's physician and his ideas. Glisson and Ent had known Harvey since the 1630s; Christopher Merrett and Charles Scarburgh had met Harvey at Oxford in the 1640s; Jonathan Goddard, at whose lodgings in Wood Street the first meetings were held, was a quondam Oxford man who had studied medicine at Cambridge with Glisson in the late 1630s and, in the 1640s, is known to have defended recent anatomical innovations in his lectures at Barber-Surgeons Hall and at the College of Physicians.[27] Although not named by Wallis, John Greaves seems also to have been a participant from 1647 on.[28]

Unfortunately, the London club described by Wallis did not long continue as a common meeting ground for medical men and physical scientists. By 1649 it had lost Wilkins, Wallis, and Goddard to Oxford, and it seems to have moved its meetings more or less permanently into Gresham College, where its interests turned much more strongly in the direction of mathematics, astronomy, and physics. Among London institutions it was rather around the College of Physicians in the late 1640s and through the 1650s that Harvey's disciples centered their research activities. Much of this work dealt with clinical and anatomical problems beyond the scope of this book,[29] but it is useful, if only for purposes of comparison with Oxford, to get some sense of the shape of that activity.

The College clearly promoted organized research in clinical subjects. The Gulstonian Lectures on pathology were given with great regularity in the 1640s and 1650s, and their incumbents included almost all the active researchers in the College.[30] Informal committees met to exchange information on the cause and cure of diseases; out of the work of one of these, dealing with rickets, arose Glisson's famous book *De rachitide* (1650).[31] Other committees considered materia medica and chemistry, on which subjects a number of fellows published works in the 1650s and 1660s.[32]

Anatomy was greatly encouraged. Glisson's *Anatomia hepatis* (1654) culminated over a decade's research on this organ. Glisson and Ent helped a younger fellow, Thomas Wharton, begin investigations of glands, which were published as *Adenographia* (1656).[33] The College's members, according to Walter Charleton, constantly employed themselves in dissecting animals of all kinds and almost daily investigated "arguments to confirm and advance that incomparable invention of Doctor *Harvey*, the Circulation of the Blood."[34] George Joyliffe continued his research on the lymphatics.[35] Scarburgh did splenectomies,[36] and both he and Charleton investigated the application of statics to muscular motion.[37]

During the 1650s much activity was encouraged, if at a distance, by Harvey himself.[38] In 1651–1654 he built for the College, out of his own resources, a magnificent "Museum," consisting of library above and meeting room below. He furnished it sumptuously and gave it many of his books and dissecting instruments.[39] In 1656 Harvey endowed an annual feast—the Harveian Oration—at which benefactors were to be praised, and the fellows exhorted "to search and studdy out the secrett of Nature by way of Experiment."[40] The College during the 1650s must indeed have seemed, in all facets of its activity, to justify Charleton's praise as the true embodiment of Bacon's vision of Solomon's House.[41]

HARVEY'S CIRCLE AT OXFORD

While Ent, Glisson, and others were upholding Harvey's cause in London and Cambridge during the 1640s, the physiologist himself was rather more concerned with his official duties, now complicated by the outbreak of the Civil War. From late 1638 on, Charles I traveled almost incessantly; Harvey, promoted in 1639 to the position of the king's senior physician-in-ordinary, accompanied him. Harvey seldom appeared at St. Bartholomew's and from 1638 to 1641 attended only seven meetings at the College of Physicians. No respite came his way until late October 1642, when Charles and his retinue retired to Oxford after the battle of Edgehill.[42] There Harvey lived for almost four years, until the summer of 1646. Some of the circumstances of his sojourn at Oxford have been discussed by biographers;[43] I wish here to focus primarily on Harvey's scientific life and contacts there, and to assess the continuing effects they had.

Oxford was, of the two ancient English universities, certainly the one most institutionally congenial to a scientifically-inclined Royalist. In addition to the Regius chair in medicine, there were two Savilian professorships (in astronomy and geometry) and the Sedleian endowment to support a lecturer in natural philosophy. John Greaves was elected Savilian Professor of Astronomy in 1643, shortly after the court's arrival in Oxford. The "Physick Garden" had been organized under the supervision of Jacob Bobart, the German botanist. The Tomlins Readership in anatomy provided support for annual dissections and lectures on anatomy, and was vested in the Regius Professor. Recently, a room in the Schools Quadrangle had been refurbished as an anatomical museum with dissection facilities, and the Tomlins endowment had been invested to produce a regular income. Cambridge, by way of contrast, had no endowed positions in science beyond the Regius Professorship of Medicine, no botanical garden, and no institutional support for anatomy.

Harvey's academic counterpart at Oxford was the Regius Professor of Medicine, Thomas Clayton. He was a man trained under the old Galenic traditions, being in 1642 three more than Harvey's sixty-four years. Much less brilliant and scientifically productive than the Cambridge Regius, Glisson, Clayton was nonetheless a man receptive to change. In 1631 he had accepted a work dedicated to him by

Harvey's arch-opponent, James Primrose.[44] Yet two years later, at the Oxford Act of July 1633, one of Clayton's students, Edward Dawson, discussed for his D. M. degree the question "An circulatio sanguinis sit probabilis?" Dawson answered the question in the affirmative.[45] In the same year Clayton seems to have arranged for an Oxford bookseller to publish a students' edition of the best standard anatomical text then available, Caspar Bartholin's *Anatomicae institutiones corporis humani* (Oxford, 1633), the very work Bartholin's son updated a few years later to reflect Harvey's discovery of the circulation. Altogether, it seems that Clayton, like Oxford, was mildly congenial to innovation.

Harvey's life there was most likely a rather cloistered and uneventful one, despite the presence of the court and army. His wife was probably not with him, and he would have had no duties except to oversee the king's health and to tend members of the court as Charles directed him. We have, therefore, little precise information about his comings and goings. In December 1642 he "incorporated" on his Padua M. D., a formality that gave him academic status in the university. In that and the following month he corresponded with Baldwin Hamey, who was tending Harvey's brothers, Matthew and Michael, in London; both subsequently died. Harvey stirred little outside the town, going once in October 1643 to tend Prince Maurice at the siege of Plymouth, and attending the negotiating party at Uxbridge in January 1646.[46] Through Greaves's good office, Harvey was elected Warden of Merton College in January 1645, although because of some obstructionist appeals, he was not admitted to that place until April. He held the position until the surrender of the city in June 1646.[47] In all, it seems Harvey had little to do in Oxford except socialize with old friends and new, and carry on scientific work as he was inclined.

Embryology was clearly the major topic of interest, one which Harvey investigated with friends both at Trinity and at Merton. John Aubrey, while he was in residence at Trinity from February to April 1643, first encountered Harvey there, "but was then too young to be acquainted with so great a Doctor." Aubrey saw Harvey come several times to Trinity to visit one of the fellows, George Bathurst, "who had a hen to hatch egges in his chamber, which they dayly opening to discerne the progres and way of generation."[48] Harvey also seems to have worked on embryology with another Trinity man, the physician Nathaniel Highmore.[49] This is evidenced by the fact that when Highmore published his book on generation in 1651, he referred to Harvey's as yet unpublished observations from the 1630s on the apparent latent period in vivipara, particularly deer, between coitus and the appearance of the fetus.[50] They almost certainly discussed this phenomenon at Oxford. The point was clearly important to Highmore, for after Harvey published these findings in his own *De generatione* of the same year, Highmore wrote a letter to Harvey about them.[51]

John Greaves, at Merton, also helped out in a rather unusual way. In the course of his travels in the eastern Mediterranean in 1637–1639, Greaves had observed the Egyptians' method of hatching eggs in incubating ovens. He evidently told

Harvey about this, and Harvey was impressed enough to mention it twice in *De generatione*.[52] The exchange did not end there. Harvey later gave Greaves a presentation copy of *De generatione*, complete with an unpublished portrait.[53] At some time in the 1640s Greaves wrote up the Egyptian method in a short memoir, which was published posthumously, by George Ent, thirty-five years later.[54]

But Harvey's closest collaborator in embryology was another political refugee, Charles Scarburgh.[55] He had been educated at Harvey's and Glisson's college, Caius, where he developed a reputation as a mathematician. But Puritan Cambridge was an uncomfortable place for a Royalist, and about 1644 he was ejected, after which he came to Oxford to enlist under the king's banner. Harvey, according to Aubrey, turned Scarburgh's attention from military affairs to medicine: "Prithee leave off thy gunning and stay here," Harvey is reported to have said, "I will bring thee into practice."[56] Harvey also brought Scarburgh into research. By June 1645 if not before, Scarburgh and Harvey were friends together at Merton where, according to Anthony Wood, a Mertonian himself, they worked on embryology:

> While he abode in Mert. coll. [Scarburgh] did help the said Dr. Harvey then warden of that house (in his chamber at the end of the said library there) in the writing his book *De generatione Animalium*, which was afterwards published by the said Harvey.[57]

The room, a well-lit square at the north end of the old library, survives today as a reminder of Harvey's work carried on in the midst of rebellion.

Scattered bits of evidence show that Harvey was involved in other kinds of anatomical activity, of which the most influential were his contacts with Highmore. The Trinity man had begun to practice medicine in the late 1630s and, by the early 1640s, had become interested in anatomy. Highmore began to conduct autopsies and dissections at Oxford, most likely in collaboration with Clayton.[58] He filled a notebook with anatomical extracts, a few from Galen and Fernel, but most from the works of anatomists in the Paduan tradition, especially Colombo and Fallopio.[59] Caspar Bauhin's monumental *Theatrum anatomicum* (1605), upon which Harvey had based his "Prelectiones," served Highmore in much the same way.[60] He also read Adriaan van den Spiegel's *De humani corporis fabrica* (Venice, 1627) especially closely, and copied out extracts *de corde* and *de pulmonibus*.[61] These in turn launched Highmore into a thirty-five page essay on respiration.[62]

Dissections and reading led Highmore most naturally to Harvey's work. He went on to fill pages of his notebook with a series of essays on the circulation, and on the function of the pulmonary artery and veins, including numerous references to Harvey's arguments and experiments.[63] Thus, when Harvey arrived in Oxford in late 1642, it was natural that their common anatomical interests should give rise to discussions. Perhaps they had even met before, since there is some tangential evidence that Highmore treated Prince Charles when the king's son fell ill with measles at Reading in 1641 or 1642.[64] In any case, about 1643 Highmore showed some writings to Harvey, who urged him to publish.[65] It was probably as a result both of Highmore's anatomical skills and his contacts with the Oxford court, that he was created D.M. on 31 January 1643.

Similar kinds of physiological questions came up in Harvey's discussions with John Greaves.[66] Greaves, in his travels through the Middle East, had noted in exploring monuments that he had less difficulty breathing in a confined space than he had expected. Greaves and his friends had, for example, spent nearly three hours in a small chamber at the center of the pyramid of Cheops. This intrigued Harvey, who pointed out that "we never breathe the same aire twice, but still new aire is required to a new respiration (the Succus alibilis of it being spent in every expiration)." He was amazed that the travelers should not "have spent the aliment of that small stock of aire within, and have been stifled." Harvey speculated that there might be "some secret tunnels" that conveyed the spent air out and brought fresh in. Greaves, for his part, doubted Harvey's assumption that air could only be breathed once and thereby depleted of its nutritive *succus*. Did not sponge and pearl divers rebreathe the same air? To this Harvey gave Greaves "an ingenious answer": the divers "did it by help of spunges filled with oile," which "corrected, and fed this aire: the which oile being once evaporated, they were able to live no longer, but must ascend up, or dye." Harvey and Greaves then went on to discuss Harvey's suggestion that small *tubuli* in the walls of the pyramid might allow fresh air in and the "fuliginous aire" out. Harvey defended the idea that because "aire [was] a thin, and subtile body," these passages could be very small and indiscernible, while Greaves pointed out that if they were so small, they might easily be blocked by drifting sand.

Beyond these items of anatomy and physiology directly connected with Harvey, other scattered bits of evidence point both to scientific activity in the garrison town and to cognizance of Harvey's findings. John Greaves's younger brother Edward practiced among the Royalists, wrote a tract on the Oxford epidemic of 1643, and later published an oration in praise of Harvey.[67] Daniel Whistler, when he left Merton in 1645 to take his M. D. at Leyden, did so with a dissertation on rickets in which he discussed the disease in light of his own recent dissections, explained and defended the circulation, and even considered the possibility that rickets, as a nutritional disease, might in part be due to an impeded motion of the blood.[68] George Joyliffe, an anatomical protégé of Clayton, is reported by a colleague to have discovered the lymphatics by 1642 and to have demonstrated them in dissections at Oxford.[69] Christopher Merrett, a friend of Joyliffe and later a protégé of Harvey, knew Jacob Bobart and botanized in the countryside around Oxford;[70] he too received his D. M. on 31 January 1643, the same day as Highmore. Walter Charleton was created D. M. a month later and soon thereafter became a physician to the King. This clearly brought him into contact with Harvey, whose ideas he espoused, and whose innovations he praised in many later publications.[71] Thomas Willis also began medical practice at Oxford in the early

1640s, although from his later descriptions of the 1643 epidemic, it seems that his practice was among rather lesser Royalists than that of Harvey.[72] Willis's friend Ralph Bathurst, the younger brother of George and also a fellow of Trinity, turned to medicine about 1642 as a complement to a career in divinity; books purchased from the 1640s on, including a first edition of *De generatione*, testify to the development of his scientific interests.[73] Timothy Clarke, later an active anatomist, was next door at Balliol.

Harvey also had contacts among poets and churchmen around the court. Abraham Cowley, another Cambridge refugee, began in Oxford a friendship with Harvey that culminated more than a decade later with the poet's famous "Ode upon Dr. Harvey."[74] William Cartwright, John Berkenhead, and Martin Llewellyn all seem to have known each other, and Harvey, at Oxford.[75] Cartwright and Berkenhead later referred to the circulation in their poetry,[76] while Llewellyn became a physician and praised Harvey in a long poem prefixed to the English translation of *De generatione*.[77] Bishops James Ussher and Brian Duppa knew Harvey, and the latter remained in contact with him until the very end of his life.[78] The courtier Thomas Henshaw later recollected some of Harvey's observations about swallows, which most likely had come up in discussions at Oxford.[79] Noblemen like William—later Viscount—Brouncker, and Henry Pierrepont, Marquis of Dorchester, were interested in medicine. Brouncker took his D. M. at Oxford in 1647 (although he never practiced) and became well known later as a mathematician and President of the Royal Society. Dorchester was a friend of Charleton, as well as Harvey's patient. He read widely in medicine, and in the late 1650s maintained a private physic garden and laboratory.[80]

In sum, Harvey's life and contacts were quite the opposite of what one might expect in a garrison town. Harvey had always been a lone worker, but at Oxford, despite political turmoil, he found a circle of congenial associates. From Harvey they sought guidance and encouragement; in return, members of this group assisted him in his own work and extended his personal influence and scientific reputation. Perhaps it was something like this investigatory comradeship that Harvey hoped to promote when, as Scarburgh later revealed, he contemplated endowing a professorship of experimental philosophy at Cambridge, complete with laboratory and botanical gardens. But Harvey was deterred from this, Scarburgh said, by the ascendency of the Parliamentarians in his own beloved university.[81]

To a considerable degree, Harvey's circle did not long survive the war. Before and after the surrender of Oxford to the Parliamentarians in June 1646, the intellectuals around the court dispersed to London, to country seats, or into exile abroad. Many of the scientists did the same. George Bathurst had been killed in 1645.[82] In the same year, Highmore went into practice in Sherborne, Dorsetshire, although he continued to see Harvey until about 1648.[83] Merrett was in London by 1645, when he began meeting with the club described by Wallis. Whistler surfaced in London in 1647 to begin practice and in 1648 was elected Gresham

Professor of Geometry. Clayton died at Oxford in 1647. John Greaves left the university at about the same time for London, where he joined the "1645 group" in meetings and experiments. He retained his Gresham Professorship of Astronomy after he was ejected, in 1648, from the Savilian chair at Oxford, and died in London in 1652. By 1648 George Joyliffe was living in Sussex, carrying on his anatomical work;[84] in 1650 he went to Cambridge, obtained his M.D. under Glisson in 1652, and immediately settled in London to practice. Charleton likewise surfaced in London in 1649, although during the 1650s he made several trips abroad. Harvey's closest friend, Scarburgh, came to London about 1647 to practice, joined both the College of Physicians and the Wallis club of natural philosophers, and in 1649 became lecturer on anatomy at Surgeons' Hall. He continued throughout his life to be closely associated with Harvey. In 1648 Samuel Hartlib, the London intelligencer, noted in his diary that Scarburgh was a very ingenious man, and that "Hee hath lived with Dr. Harveigh."[85] Two years later Scarburgh visited Hartlib and "observed that Gilbert with his Philos[ophia] Magn[etica] and Dr. Harveigh De circulatione Sanguinis etc. had more advanced real learning than many other nations." Hartlib further noted that Scarburgh's "excellency is chiefly mathematical and anatomical."[86]

Although they were to be important in Oxford science in the 1650s, only a few of the younger men stayed on. Willis qualified B. M. in 1646 and continued his practice among the Royalists in the university and town. Ralph Bathurst joined him as a junior associate and continued to see Highmore during the Dorset physician's visits to Oxford. Edward Greaves stayed on at All Souls for some years before moving to London in 1654. Timothy Clarke's movements are uncertain, although it does seem that he spent at least part of the late 1640s and early 1650s in Oxford before taking his degree there in 1654.

As for Harvey himself, the surrender of Oxford and the capture of the king changed his life greatly. Save for a short period of time spent tending the king at Newcastle in December 1646 and January 1647, he lived until 1651 in country retirement, at the homes of various of his brothers. His movements were restricted by the Parliament, and he did not attend meetings at the College; he was even rumored on the Continent to be dead.[87] It was a time in which, stripped of both his responsibilities and his privileges, separated from the conviviality he had known at the College of Physicians and more recently in Oxford, he occupied himself with ordering the work he had completed over the previous two decades.

HARVEY'S LATER WRITINGS—LOST AND SURVIVING

It is clear that, as early as the publication of *De motu cordis*, Harvey conceived of the circulation not simply as an isolated discovery, but as the beginning of a reconstruction of physiology. For years he carried out investigations in many fields of biology, recorded his findings, and wrote upon his results. The essays published in *De circulatione sanguinis* (1649) and *De generatione animalium* (1651) represent only a part of that body of research. There were, Harvey said on numerous occasions, many other works still in manuscript.

His printed books contain many scattered hints about the nature and subjects of these unpublished studies, especially in areas of physiology. In *De motu cordis* he twice mentioned a "tractatus de respiratione," concerned with the motion and use of the lungs, their role in cooling, and the necessity of the air, all derived from very numerous personal observations.[88] In *De generatione* he again twice mentioned his "essays" on the "causes, instruments, and uses of respiration."[89] Harvey's published work neglected the question of nutrition; in *De generatione* he mentioned that he had written much about how this was effected.[90] As early as 1628 he had collected observations on the spleen.[91] He wrote on the brain.[92] And *De circulatione* mentions a treatise on the motor organs of animals,[93] most probably the *De motu locali animalium* (1627), now in the British Library.[94]

In addition to these specialized studies, Harvey intended to write, or had by the 1640s already written, a general physiology encompassing all the functions of the animal economy. Speaking in the *De generatione* about the question of whether blood, or some part of blood, nourished, he promised to discuss it in his "Physiology."[95] A few pages later he forbore discussing in what way fire decomposed living bodies, since such questions belonged more properly to "the part of Physiology, which concerns the elements and temperaments, where we will discourse of them at length."[96]

Areas peripheral to physiology were also represented among Harvey's manuscript works. In embryology he had collected observations on the generation of insects to parallel those on the development of vertebrates.[97] He wrote a treatise on "Love, Passion, and Coition of Animals."[98] In pathology he kept records of autopsies conducted on his deceased patients.[99] His reasoning was impeccable: if dissections of normal animals and men revealed physiology, so would the dissections of diseased bodies, and the description of the affected sites, further pathology, and thence, therapeutics.[100] In *De generatione* he recounted cases of an enlarged scrotum and an ulcerated uterus, both transcribed from his "medical Observations." These and other cases he intended to publish at length, he said, should God grant him life.[101] Harvey's protégé Scarburgh said that Harvey had even written on the senses and on the nature of vegetation and life.[102]

It thus seems that certainly the scope, and perhaps even the volume, of Harvey's unpublished papers exceeded that which was printed in his lifetime. In the unpublished writings one may discern two themes. The first is also found in his printed work: that conclusions about living things must be founded on firsthand observations and dissections, preferably on a great diversity of animals, including man. The second theme is only implicit in his three books: that the investigative anatomical approach, coupled with the new discovery of the circulation, necessitated a complete reworking of the traditional physiology. Harvey had started a revolution and intended, even though advanced in years, to attempt to carry it through.

In this he was partially thwarted, ironically not by his scientific adversaries, but by political circumstances. Harvey's appointment in 1639 as Physician-in-Ordinary to the king entitled him to lodgings in the palace of Whitehall. He most probably occupied the set of four rooms in the northwest corner of Scotland Yard labelled in a map of c. 1669, "The Dr. of ye Household."[103] In the few years after 1639, Harvey seems to have moved some of his belongings, including manuscripts, into the lodgings. Then came the debacle. In January 1642 the king and his family fled Whitehall to Hampton Court; Harvey no doubt joined them soon thereafter to begin almost five years of travels and exile. In the king's absence the palace lay deserted and largely unguarded—"A Pallace without a Presence," lamented a pamphlet of late 1642, where in the entrance once thronged by gallants watching the ladies come and go, "you may pisse in the Porters Lodge, and never feare the losse of your hat."[104] In 1643 the palace was fortified by the Parliament.[105] Thus, although there is no record of the palace being systematically plundered, it offered an easy mark for casual looters. Sometime between 1642 and 1647 Harvey's rooms were broken into and his possessions—including many unpublished manuscripts—stolen.

Their loss was a blow to Harvey. He complained bitterly in the late 1640s that while he was attending the king, "rapacious hands" had "not only plundered my apartment of all its furnishings, but also (what I more lament) abstracted from my study my notebooks, the fruits of many years toil; by which means many observations, especially concerning the generation of insects, have perished, to the detriment, I dare say, of the republic of letters."[106] John Aubrey recorded his grief, expressed in the 1650s, at the loss of his work "De insectis," "which he had been many years about, and had made curious researches and anatomicall observations on them." Harvey's attempts to recover the papers were unsuccessful; he told Aubrey that their loss " 'twas the greatest crucifying to him that ever he had in all his life."[107] Hartlib noted in 1652 that "Dr. Harveigh complains to have lost amongst other things his anatomical histories of his patient sick-bodies when he was plundered. A worke of extraordinary great use of Physick and wherin that part of learning is yet very defective." Hartlib noted, however, that "Something of it hee hath yet left which perhaps he may publish."[108]

After Harvey's death, and especially after the Restoration, the loss of Harvey's papers in the king's service became a recurring item of Harveian biography. In 1662 Thomas Fuller reminded his readers that Harvey "had made a good progresse, to lay down a Practice of Physick, conformable to his *Thesis*, of the *Circulation of Blood*," but was "plundered of his Papers in our Civil War." The rebellion, moralized Fuller, "not only murdered many then alive," but by this "*mischief* or *mischance*," destroyed more "not yet born," whose "Diseases might have been either prevented or removed, if his worthy pains had come forth into the Publick."[109] In a Harveian Oration of the same year, Scarburgh condemned the havoc that had destroyed Harvey's papers on vegetation and life, on respiration, on the lungs, on the brain, on the motion and senses of animals, on insects, and on "many other wonderful secrets of nature."[110] A year later

Abraham Cowley praised Harvey for the circulation and for the reformation in medicine in a poem that concluded:

> These Usefull secrets to his Pen we owe,
> And thousands more 'twas ready to bestow;
> Of which a Barba'rous Wars unlearned Rage
>   Has robb'd the Ruin'd Age;
> O cruell loss! as if the Golden Fleece,
>   With so much cost, and labour bought,
> And from a farr by a Great *Hero* Brought,
>   Had sunk eve'n in the Ports of *Greece*.
> O Cursed Warre! who can forgive thee this?
>   Houses and Towns may rise again,
>   And ten times easier it is
> To rebuild Pauls, than any work of his.[111]

Charles Goodall, in his adulatory biography in 1684, took a page to list the "several other Learned Treatises" Harvey had intended to bestow upon the world.[112]

But these treatises never saw print. We have, as major evidence for Harvey's scientific activities and ideas in the last thirty years of his life, only *De circulatione* (1649) and *De generatione* (1651). Fortunately, these contain much on the implications of circulation for later physiology. Moreover, upon analysis, they prove not only to bridge the lacunae left by the destruction of Harvey's papers, but also to provide a guide to the evolution of those scientific opinions over more than a quarter of a century of Harvey's career.

De circulatione presents few problems. It consists of two essays directed to Jean Riolan, Dean of the Faculty of Medicine at Paris, prompted by criticisms of the circulation expressed in Riolan's *Encheiridium anatomicum et pathologicum* (Paris, 1648).[113] The first and shorter of the two essays is entirely taken up with answering, in detail, the points raised by Riolan; it must therefore have been written in 1648–1649.[114] The second essay is less straightforward.[115] Its first three paragraphs are addressed to Riolan, as is its concluding paragraph. But between these are pages of text discussing many physiological ideas and experiments, none of which refers to Riolan's criticisms. It seems, rather, as if Harvey had written this longer essay to answer many previous objections to the circulation and merely used the occasion of Riolan's attack to bring it into print.[116] Several kinds of internal evidence confirm this interpretation. He described a postmortem datable in the late 1610s[117] and an anatomical demonstration before the king, almost certainly from c. 1633–1636.[118] He compared the heart to a kind of bellows pump that came into use in London in the late 1620s and was described in a publication in 1634.[119] He criticized the conceptions of the heart and blood contained in Descartes's *Discours de la méthode* (1637).[120] And, at four important junctures in his argument, Harvey adduced experiments which, although he also seems to have performed them himself, were first published by Jan van der Wale

(Johannes Walaeus) in 1641.[121] Harvey did, however, make reference in the same essay to his "medical Observations" and to his treatise on physiology, which suggests that he was not yet aware of their loss.[122] It therefore seems highly likely that Essay II of *De circulatione* was composed c. 1641–1646, that is, predominantly at Oxford, and was intended as a continued discussion of problems raised initially in *De motu cordis*.

De generatione presents rather greater problems of dating. One important fact is beyond debate: the work, although published in 1651, was substantially finished four years earlier.[123] George Ent recalled in its preface that he had collected the manuscript from Harvey on a Christmas visit.[124] This must have been in 1647, since a Cambridge correspondent reported to Hartlib in October 1648 that Harvey's *De generatione* would soon appear.[125] But early publication proved as illusory in the seventeenth century as it is in the twentieth. In April 1649 Jaspar Needham implied in a letter to John Evelyn that the book was forthcoming,[126] and in November a friend told Hartlib that "Dr. Harveigh is shortly publishing his great worke De Generatione."[127] But printing still went slowly. The book finally came out in March 1651,[128] with Harvey complaining that spring to friends in Germany and Italy that until recently he had been much occupied with preparing it for the press.[129]

Not only was *De generatione* completed by 1647, but there is much evidence both external and internal to suggest that it was composed over many years. It consists not of a single unified narrative, but of seventy-two numbered *exercitationes*, plus three separately titled tracts. The essays vary greatly in length, are at times redundant, and occasionally even contradictory. Some mention dates, or refer to datable events, others do not. Some, as cited above, even refer in the present tense to treatises known to have been destroyed by approximately 1646. Clearly the work is a mixed bag.

Even so, it is distinguishable into several "clusters" of essays that seem to treat a common subject matter. The following table gives a structural analysis of the book's parts based upon typography and content. (The roman numerals given to each section are for the purpose of analysis and do not appear in the original work.)

The first section, in which Harvey described in detail the development of the chick embryo and compared his observations to those of Aristotle and Fabricius ab Aquapendente, is the cornerstone of the work. It also contains many datable references. Harvey recalled that Maurice, Prince of Orange, gave James I a cassowary that lived in the royal gardens for more than five years; Harvey was able, almost certainly in the early 1620s, to inspect one of its eggs.[130] He referred in passing to *De motu cordis* (1628), which was the latest of anatomical works cited.[131] He described Bass Rock, off the Scottish coast, seen in May 1633 when Harvey accompanied a royal progress.[132] He mentioned showing the embryo of a doe to the king and queen, most probably c. 1633–1636.[133] And he described a leaf shown him by Giuseppe degli Aromatari, when they met in Venice in August

TABLE 1

STRUCTURE OF *DE GENERATIONE* (1651)

	Essays	Subjects	Pages
I	1−24	Description of the developing chick-embryo	74
II	25−62	Theorems deduced from the descriptions	139
III	63−70	Description of fetal development in vivipara	30
	71−72	The innate heat and the primigenial moisture	12
IV	[Unnumbered]	Parturition	20
		Membranes and humors of the uterus	15
		Conception	9
			299

or September 1636.[134] All this evidence seems to suggest that Harvey wrote the first part of *De generatione* in the late 1630s, most probably before 1639, when he began to travel extensively in the king's retinue. This seems the more likely since this part also contains references to his "medical observations," and to manuscript treatises on respiration, on sexual behavior of animals, and on nutrition,[135] references he is not likely to have put into anything first written after the destruction of his manuscripts.

The third section is an analogous set of observations on the fetal development of vivipara, carried out largely in the autumn on the king's deer at Windsor. These observations and dissections were made over several seasons, of which Harvey mentions one, 1633, specifically by date.[136] It seems likely that these, and the adjoining essays on the innate heat and the primigenial moisture, date from the period c. 1630−1636. No book published later than 1628 (*De motu cordis*) was cited, and he referred twice to his "Physiology."[137] The only passage out of place is his well-known lamentation of the loss of his manuscripts, which appears as the final paragraph of *exercitatio* 68.[138] But this has every appearance of being an afterthought, penned in as the book was going off to press.

The three essays comprising the fourth section, *De partu, De uteri membranis et humoribus*, and *De conceptione*, also seem to date from approximately the same period. In *De partu* Harvey cited no works published later than 1633 and referred to his "medical observations" and to his work on respiration.[139] The short piece on the uterus said that both *De motu cordis* and Spiegel's *De formato foetu* (1626) were recent, while in *De conceptione* Harvey referred in passing to his own work on insects.[140]

It seems highly likely then, purely from internal evidence, that sections I, III, and IV, comprising slightly more than half of *De generatione*, were completed in some form by the late 1630s. Several pieces of external evidence argue the same point. Thomas Smith, of Christ's College, Cambridge, writing to Hartlib in 1648, passed on the news from Sir Thomas Browne that "Dr. Harveigh hath an excellent tract *de generatione* coming forth wch himself saw 10 years ago, full of admirable experiments & various learning.''[141] Other scientifically inclined Royalists had also seen something of his work. In his *Two Treatises* (Paris, 1644) Kenelm Digby described Harvey's surprise that in dissecting vivipara after mating, he could find neither semen nor, until many weeks later, an embryo.[142] These are observations Harvey recorded in Section III of *De generatione*. Digby's knowledge of Harvey's writings would date in all probability no later than the last time they were in London simultaneously, in 1639–1641.[143]

Conversely, the long second section of *De generatione*, comprising essays 25–62, shows many signs of having been composed rather later, in the 1640s. These *exercitationes* have many fewer citations to Continental works than appear in the other sections; they hint strongly that Harvey wrote them while away from his manuscripts and personal library. In them he made no reference to his other works, either past or projected, again implying that he wrote this section while away from London during the 1640s. Most convincingly, he remarked in one of these essays that he had lived to see the circulation admitted by almost all, surely not a statement he could have made in the 1630s.[144]

Several references, to which dates may be assigned, confirm this interpretation. In *exercitatio* 56 Harvey discussed the milklike ''chyle'' found in the healthy human fetus and noted that it was particularly abundant in the thymus.[145] This observation was almost certainly made in the 1640s, for writing against Riolan in 1648–1649, Harvey said he had ''recently found'' this milk in newborn animals and humans, especially in the thymus.[146]

Another event, mentioned in essay 52, may be even more precisely dated.[147] At the request of Charles I, Harvey treated Hugh Montgomery for a thoracic fistula created by the healing of an old suppurated wound. To Harvey's amazement, he found that the young man's heart was clearly visible through the hole. Not only was the heart insensible to touch, but one could observe

> that in the diastole it was retracted and withdrawn; whilst in the systole it emerged and protruded; and the systole of the heart took place at the moment the diastole or pulse in the wrist was perceived; to conclude, the heart struck the walls of the chest, and became prominent at the time it bounded upwards and underwent contraction on itself.[148]

An accident of nature had made it possible for Harvey, in late 1641 or early 1642,[149] to observe directly what he had been forced to infer in the first part of the *De motu cordis*: that systole of the heart corresponded to the apex beat and to diastole of the arteries.

It would seem that, with the aid of both internal and external evidence, it is possible to analyze both *De circulatione* and *De generatione* into their constituent parts, to distribute these over the ''silent'' years between 1628 and 1649, and to gain thereby a more exact picture of Harvey's scientific activity in these decades. These suggested datings are summarized in table 2. Many of the substantive arguments embodied in these two books have been scrutinized by scholars, most

TABLE 2

TENTATIVE DATINGS OF HARVEY'S WORKS

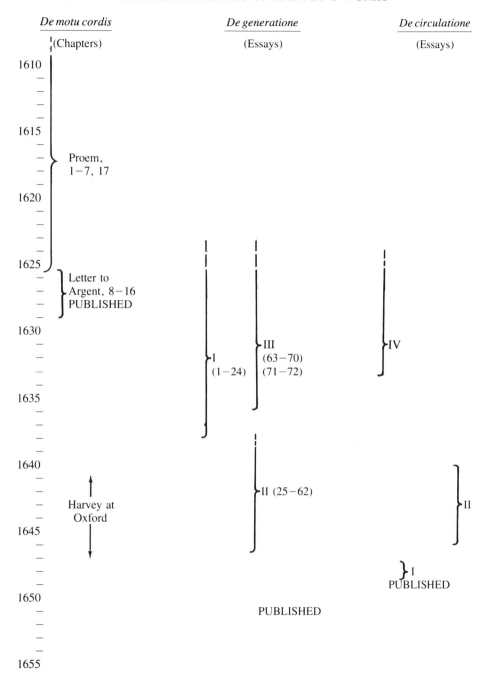

recently by Pagel and Whitteridge. But such analyses have always treated the works as finished pieces, and have emphasized the main themes of each: in *De circulatione* Harvey's treatment of cardiac action, the arterial pulse, the pulmonary transit, the existence of anastomoses, the movement of blood in the mesentery, the differences between arterial and venous blood, and the nature of spirits; in *De generatione*, his observations on the development of the chick and deer embryos, his ideas of epigenesis, fertilization, spontaneous generation, and the antiquity of the blood. I am concerned here with something rather different—not with arguments about circulation or generation *per se*, but with how Harvey used the occasion of these works to throw out hints about the broader physiological implications of his discoveries. In short, I wish to use the dating of these works to gain an insight into Harvey's program for his successors denied us by the destruction of his manuscripts.

HARVEY'S CHANGING CONCEPTS OF BLOOD, HEAT, AND RESPIRATION

Harvey's new ideas originated, as they had in the case of cardiac motion, with observation. By the late 1630s he had accumulated a large number of precise observations on the development of the chick and of the deer, and they all seemed to demonstrate one thing: the blood was the first observable part, the origin of all structure in the body. Soon after it appeared, the blood began to move. This motion was spontaneous, and even showed some signs of sensation. The blood communicated this motion to vessels, causing them to pulsate. Only after this "leaping point" of blood was discernible did one begin to see, rather later, a recognizable heart, then lungs, and after that, the liver and brain. Thus the Galenic doctrine that made the liver the origin of the blood, as well as the Aristotelian belief that attributed this function to the heart, were both wrong.[150] He was convinced, Harvey said, that "the blood exists before any particle of the body appears; that it is the first-born of all the parts of the embryo; that from it both the matter out of which the foetus is embodied, and the nutriment by which it grows are derived."[151]

Such a conclusion had been adumbrated earlier in Harvey's thought. As far back as his Lumleian lectures he had considered the question of whether the heart or the liver sanguified and settled on the heart because it had the innate heat. But, he noted, if he could show what he had seen, "yt were att an end betwene phisicians et philosophos." Neither the "physicians" (Galenists) nor "philosophers" (Aristotelians) were right: "For blood is rather the author of the viscera than they of it."[152] The same point came up obliquely in *De motu cordis*, where on two occasions Harvey noted that a throbbing point of blood could be seen in the chick embryo before the heart.[153]

By the late 1630s Harvey could thus, in the chapter "On Innate Heat," use his discovery of the blood's primacy to equate the traditional *calidum innatum* with the blood and to deny the independent existence of spirits.[154] Why, Harvey said, should anyone posit a separate heat to guide and cherish bodily processes, when there was nothing more ancient and excellent than the blood? Borrowing a distinction from Aristotle, he argued that blood was by its very essence hot; when it ceased to be hot, it ceased to be blood. Nor was the blood, the *calidum innatum*, of the nature of fire, or derived from fire. Rather, its heat was of a more divine substance, the direct expression of the soul. What could be said of heat could also be said of spirits. Why should one search after some kind of aerial or celestial spirits as motive powers within the body, when there was blood? No one had ever demonstrated the existence of such spirits. Nor were there any cavities in the heart or brain for the production and preservation of such spirits. "We, for our own parts, who use our simple senses in studying natural things, have been unable to find anything of the sort."[155] For page after page, Harvey catalogued the supposed attributes of spirits, only to argue that such concepts were unprovable, unnecessary, or absurd. True enough, blood outside the body would seem to be nothing but a fluid with several kinds of parts. But within its own vessels, it was animated by the soul and took on powers superior to the forces of the elements, thereby enabling it to warm, nourish, cherish, and preserve the entire body. Most of all, it was the formation of the embryo that demonstrated that "the remarkable virtues which the learned attribute to the spirits and the innate heat, belong to the blood alone."[156]

Identical themes appeared, *a fortiori*, in Harvey's writings during the 1640s. In the central *exercitationes* of *De generatione* he reiterated that the blood, the *punctum saliens*, was the first part to be formed and the origin of all other organs. He explained how the blood was the origin of the pulse. He rejected with even greater vigor the Aristotelian and Galenic notions that the heart and liver, respectively, were the authors of the blood. He maintained against his revered Aristotle that the blood was the prime part engendered and that the heart was the mere organ destined for its circulation. Blood was the fountain of life; the first to live and the last to die. It lived of itself; it was its own heat and spirit.[157]

The physiological implications of his strengthening beliefs came through clearly in *De circulatione*, largely in the second essay composed in the mid-1640s. The blood was the hottest part in the body, he said, not the heart; the blood imparted heat to the heart, rather than receiving it. All parts were thus warmer according to the extent that they contained blood. In embryological development this innate heat of the blood was, by its first swelling motion, the origin of the pulse. Many observations in generation confirmed this; although it would not be proper to recite them on this occasion, Harvey said, he would perhaps publish them soon.[158]

Spirits came in for particularly violent attack. Such aerial substances, separate and distinct from blood, were nothing but a subterfuge of ignorance, a *deus ex machina* to solve all dilemmas. Spirits were multiplied needlessly by bad philosophers. They were clearly not made in the heart, nor could they be found in the arteries. If there were any such things, they were in no way distinct from blood. Spirit was simply what made blood alive. Blood without spirit was nothing but gore. In fact, a separate spirit in blood was no more reasonable than a separate *calidum innatum* in the heart. One had been invented as the instrument of the other, and both were dispensable because their supposed tasks could be done by the blood.[159]

In traditional physiology, one of the fundamental functions of aerial spirits had been to account for man's obvious dependence on respiration. By denying spirits so vehemently, Harvey had cut man off from his environment. What then was the function of respiration? In *De circulatione* Harvey simply followed the logic implied by placing the innate heat in the blood. Inspiration served, he said, to temper the blood as it passed across the lungs, while expiration expelled the suffocating fumes, the waste products of the blood. In fact, Harvey said, if there were any kinds of aerial spirits in the blood, they were these waste products blown out when the blood was aired and purified, rather than some kind of separate subtle and active matter.[160] Moreover, he asserted, this tempering and purifying was tied to the physical motion of the lungs; blood could cross them only when their vessels were opened in inspiration, since such vessels were closed by the collapse of the lungs in expiration.[161]

Thus it would seem that Harvey, when he came to explicate the pulmonary transit, had settled upon three functions for respiration, of which the primary was cooling. Paradoxically, his embryological studies pushed him in a different direction. In the essay "On Parturition," he noted the Galenic idea that labor began because the fetus felt the need to cool itself more perfectly by respiration. Harvey promised to treat the question more fully in his unpublished treatise on respiration, but in the meantime voiced some doubts. Why should a seven-month fetus be expelled to respire the air, when it could very well stay *in utero* two months longer? Why could an expelled fetus live for some hours if the membranes were intact, while another that had once respired could not live for a moment without breathing? Surely this was not due to lack of "cooling"?[162]

> If any one will carefully attend to these circumstances, and consider a little more closely the nature of air, he will, I think, allow that air is given neither for the "cooling" nor the nutrition of animals; for it is an established fact, that if the foetus has once respired, it may be more quickly suffocated than if it had been entirely excluded from the air: it is as if heat were rather enkindled within the foetus than repressed by the influence of the air.[163]

Here, as in his later discussions with Greaves, Harvey had some misgivings about cutting the organism off from the air.

Denying the existence of spirits had implications for the blood as well as for the lungs. What then was the difference between arterial and venous blood? Only in the 1640s did Harvey develop an answer, or rather several answers, to this question. In the first place, Harvey said in *De circulatione*, the two bloods had no inherent physical differences. If you filled two measures of equal quantity with arterial and venous blood, you would observe that the arterial kind was in no way more blown up, frothy, or inflated with spirits. The bloods in the two receptacles would clot with practically the same consistency and coloration, fractionate their serums similarly, and occupy the same volume while both warm and cool. Why then the florid color of blood drawn in arteriotomy? That came about, Harvey said, because blood, "whenever and wherever emerging through a narrow opening, is as it were, strained off, and the thinner and lighter part, which tends to float on top and is more penetrative, is forced out." Conversely, blood flowing through a large opening, as in venesection, was thicker, had more body, and was of darker color. The statements could be reversed: if venous blood flowed out a drop at a time, it looked arterial; "blood flowing freely from a cut artery, when received into a suitable vessel, will look venous." Blood color was thus an accidental property caused, as it were, by straining. Thus, Harvey said, a "much more florid blood is found in the lungs and is expressed from them than is found in arteries."[164]

This theory of blood color as a physical artifact is illuminated by a rather longer discussion of blood in *De generatione*, written at about the same time in the 1640s. Blood, when contained naturally in the body, was everywhere of the same constitution. But when extravasated, blood's innate heat and spirit escaped; it was corrupted, and resolved itself into its dead parts. The more florid of these parts floated to the top, while the darker ones stayed on the bottom. But just as these two kinds of parts had no independent existence in the body, so did arterial and venous blood have no essential differences, only apparent ones caused by abnormal circumstances.[165]

Harvey's discussion of blood color is especially fascinating because he happened to mention, *en passant*, two observations that were to prove crucial in the later development of physiology: that blood in the lungs is highly florid, and that a florid arterial layer forms on the top of venous blood left to coagulate. But what were phenomena for Harvey's successors were for him merely artifacts.

By the early 1650s Harvey could survey an English scientific world very different from the one into which he had introduced his *De motu cordis* a quarter-century earlier. The College of Physicians was dominated by his supporters. Some of them, like Ent and Glisson, he had persuaded in the 1630s. Others, like Goddard and Wharton, were in turn students of his first converts. Still others, such as Scarburgh, Merrett, Joyliffe, and Charleton, he had met at Oxford. Some Oxonian disciples, such as Highmore, had gone out into the provinces, while others, such as Bathurst, Willis, and Clarke, had stayed on in the university town.

While acquiring disciples, Harvey had also developed the implications of the circulation, a process whose full import is denied us by the destruction of his manuscripts, but which may be reconstructed by examining his remaining writings from the 1630s and 1640s. He centralized physiological functions strongly in the blood. It was there, and not in the heart, that the innate heat was to be found. The blood acted through its vital properties, not through some kind of spirits generated in the left ventricle of the heart. Having abolished spirits, Harvey needed no air to enter the body as their raw material, and thus emphasized the function of respiration in cooling the innate heat in the blood, depurating it of its wastes, and allowing it to cross from right to left through the mechanical motion of the lungs. The two colors of the blood were due, not to spirits, but to artifacts created by their observation. It was an admirably coherent set of explanations, marred only by the few doubts introduced into Harvey's mind by fetal "respiration." The details of this process seemed to suggest a respiratory function—apart from the mythical generation of vital spirits, and unlike cooling and depuration—that required air to enter the animal body. These doubts, although veiled for fear of having to readmit spirits, were to be the starting points for the next generation of English physiologists.

THE SCIENTIFIC COMMUNITY IN COMMONWEALTH AND RESTORATION OXFORD 3

Although Harvey had disciples in many places throughout England, and, by the 1650s, firm adherents throughout Europe, it was only in one place and within one scientific community—Oxford—that his successors developed a set of successful new explanations for the central physiological functions of blood, heat, and respiration. These explanations were based, not on the Aristotelian natural philosophy that had been so congenial to Harvey, and which he had shared in varying degrees with Ent and Glisson, but on the new atomistic and mechanical philosophy then coming out of France. Chemistry, both theoretical and highly practical, was used to explore the properties of corpuscular organic units. New experimental apparatus and techniques were developed to argue these reconstructed physiological explanations. In short, during the 1650s, 1660s, and 1670s, Oxford natural philosophers solved many outstanding Harveian problems because they redefined, in their own particular way, the nature of physiological explanation.

Such a transformation of scientific problematics is not readily achieved. It seems to have occurred at Oxford in these decades because anatomists and physiologists not only individually cultivated an interest in chemistry and the physical sciences but also were associated in scientific activity with men more knowledgeable in these sciences than the physicians who had traditionally advanced anatomy and physiology. Such cooperative scientific activity was possible because Commonwealth and Restoration Oxford provided institutional support, intellectual resources, and a congenial environment for a core of scientific leaders, such as William Petty, John Wilkins, Thomas Willis, and Robert Boyle, who in turn recruited large numbers of Oxonians to an interest in the new experimental philosophy. Oxford thereby became, in those years, the center of a large, diverse, and active scientific community.

Its size is particularly noteworthy. During the period from the early 1640s to the late 1670s there were at Oxford, for varying lengths of time, over 110 individuals for whom there is evidence of scientific interest and activity. Perhaps more impressive, such a figure does not include physicians who merely took a medical degree there, without corroborative evidence of additional scientific activity. These Oxford natural philosophers and virtuosi, along with details about their careers and scientific interests, are listed in table 3 at the end of this chapter.

As a community of intellectuals and scientists, it was phenomenally productive. Its members wrote, while at the university or after leaving Oxford, over 580 books, of which over 240 were concerned with scientific, medical, or technological subjects. Some men contributed far more than their share of scientific works: Boyle (49), Wallis (23), Charleton (17), Petty (15), and Hooke (13). But a surprisingly large number of men—twenty-eight in addition to those mentioned above—wrote two or more scientific works, thereby giving evidence of much more than passing interest in science.

These Oxford scientists fall into three well-defined classes. At the top is a small group of approximately a dozen "major" scientists, men of noteworthy scientific contributions and literary productivity, such as Robert Boyle, Walter Charleton, Nathaniel Highmore, Robert Hooke, Richard Lower, John Mayow, Walter Needham, William Petty, John Wallis, Seth Ward, John Wilkins, Thomas Willis, and Christopher Wren. Below this class is a much larger group of some forty-six "minor" scientists, men of rather more limited scientific activity and insight, although many of these did write one or more books of merit. This group includes such men as Ralph Bathurst, Timothy Clarke, Jonathan Goddard, John Greaves, Nathaniel Hodges, George Joyliffe, John Locke, Christopher Merrett, Thomas Millington, Henry Oldenburg, Lawrence Rooke, Charles Scarburgh, Robert Sharrock, Henry Stubbe, Daniel Whistler, and Robert Wood. The apparent predominance of physicians is not illusory; over three-quarters of the "minor" scientists took up such medical topics as anatomy, physiology, chemistry, and materia medica. The third class of approximately fifty comprise the "virtuosi"— men who did little original scientific work themselves, but were eager participants in, and followers of, the activities of their more gifted peers. About a third of the virtuosi were divines, such as Joseph Glanvill, William Holder, Gerard Langbaine, Samuel Parker, or Thomas Sprat. Another third were physicians, such as Edward Greaves, Richard Lydall, and Walter Pope. Civil servants like Peter Pett, Robert Southwell, and Joseph Williamson, make up the remainder of the "virtuosi."

Not all these men were in Oxford at the same time. Career choices, political exigencies, and personal inclinations took men away from Oxford, and sometimes returned them to it. Relatively few stayed in Oxford even for the greatest portion of their lives: Wallis's fifty-four years and Bathurst's seventy years of university residence were very much the exception. Of fifty-seven scientists, seven ended their careers at Oxford, thirty-two in London, and the remainder in the provinces or abroad. Yet despite this tendency of Oxford to disperse its scientific progeny, many men spent a significant amount of time in the university town, much of it after the inception of their scientific interests. The average scientist spent almost fourteen years in the university, almost seven of which were after the beginning of active scientific interests. Moreover, this avocation arose, in the great majority of the cases, during the individual scientist's education: three-quarters of the "major" and "minor" scientists in the Oxford scientific community were educated there. Even after leaving Oxford, many members of the community, whether scientists or virtuosi, continued their scientific interest; an even fifty became Fellows of the Royal Society.

THE UNIVERSITY BACKGROUND OF OXFORD SCIENCE

Such a large and complex community of scientists did not arise *de novo*. It was made possible, even if not inevitable, by a number of changes that overtook Oxford in the early seventeenth century and remade a stodgy medieval university into a congenial place for scientists and scientific activity.

The engine of these changes was growth—of numbers, of buildings, and of endowments. In the late sixteenth and early seventeenth centuries the merchant and gentry classes had discovered that a university education, previously reserved for the poorer sort intent upon a church career, was desirable background for personal improvement and public advancement.[1] In the 1540s Oxford had admitted about 40 B.A.s and about 20 M.A.s each year. These numbers climbed slowly in the late sixteenth century, and swelled even higher in the first four decades of the seventeenth century, such that by the 1620s, Oxford was taking in an average of more than 400 students each year, admitting over 240 of those to B.A.s, and almost 150 each year to the M.A. The population of junior members had grown more than sixfold in less than a century.[2]

This influx of students called forth a vigorous program of construction. New colleges were founded—Trinity, St. John's, Jesus, Wadham, and Pembroke—and old ones rebuilt. The boom reached its height in the early seventeenth century. Wadham was built in 1610–1613, Jesus in 1620–1643, and Pembroke in 1624–1626. Merton and Exeter rebuilt their quadrangles, while Lincoln, Brasenose, and St. John's added completely new ones. Trinity added Kettel Hall, and Oriel, a fourteenth-century foundation, was completely reconstructed in the decades before the Civil War.[3]

The growth was not just one of stone and mortar, undergraduates and fellow-commoners. Although the evidence is harder to trace, it seems that the Jacobean period also saw an increase in foundation revenues. New scholarships and exhibitions were contributed; lands were left to augment the incomes of the fellows.[4]

Science had shared in this largesse. Before the seventeenth century, Oxford had only a few endowed teaching positions in the sciences: the Regius Professorship of Medicine and the Superior and Inferior Linacre Lectureships, vested in Merton College. These were more than doubled by 1630. Professorships of Geometry and

Astronomy were founded in 1619 by Henry Savile, the Warden of Merton. In 1621 Henry Danvers, the Earl of Danby, had given the university £250 to found a "Physick Garden" for growing botanical specimens and teaching the use of medicinal simples. In the same year Sir William Sedley founded a professorship in natural philosophy. The university would have received yet another professorship of natural philosophy in 1626 had Francis Bacon's estate been sufficient to carry out his will. The last large gift came in 1624, when Richard Tomlins founded a Readership in Anatomy, to be vested in the Regius Professor of Medicine. All of these benefactions had come to maturation by the late 1630s. Landed endowments were secured for the Savilian and Sedleian chairs. Jacob Bobart, a German botanist, had organized and planted the physic garden. Rooms in the Schools Quadrangle had been refurbished as the Anatomy School, and the endowment of the Readership invested to produce a regular income. By the outbreak of the Civil War, Oxford thus had seven endowed positions in the sciences (compared to two at Cambridge), with stipends ranging from modest (£6 to £12 per annum for the Linacre Lecturers) to handsome (£160 per annum for the Savilian Professor).[5]

Library facilities were similarly improved in the Jacobean period. Neglect and religious strife during the sixteenth century had practically destroyed Oxford's medieval library, such that Sir Thomas Bodley began the task almost anew in 1598. By 1620 he had rebuilt the fifteenth-century library of Duke Humfrey, added the Arts End, and persuaded the university to consolidate their scattered lecture rooms into a new two-storied Schools Quadrangle, which he crowned with a third story to house the more than 16,000 books donated by him and his friends. Bodley also negotiated an agreement with London's Company of Stationers, whereby a copy of every book printed by its freemen was to be presented to the Bodleian, making it the first (and for some time the only) deposit library in England.[6] By the 1670s the Bodleian had grown to over 40,000 titles, including almost every book that a physician or natural philosopher might wish to consult: thirty works of Bacon, in four languages; fifteen of Pierre Gassendi; ten of Descartes; almost all of the Continental writers on anatomy and chemistry, as also such English authors as Harvey, Ent, Highmore, Boyle, Willis, Lower, and Hooke.[7]

The book trade had expanded with the university. From a handful in the sixteenth century, the number of publishing booksellers had grown such that by the 1630s there were twenty-five publishing booksellers in Oxford, almost twice the number at Cambridge.[8] The largest, Richard Davis, had a stock of well over 30,000 volumes, including over 2,000 in mathematics and philosophy, and over 4,000 in medicine and the medical sciences.[9] In shops like these, scholars gathered to socialize, to examine the books coming out of a dozen major Continental publishing centers, and to see their own productions through the press. Oxford scientists took advantage of the local facilities: Austen, Bohun, Boyle, Dickenson, Highmore, Lovell, Mayow, Mundy, Parker, Sharrock, Wallis, Ward, and Willis all published books—in some cases several—through Oxford booksellers. Indeed, it seems that during the Commonwealth and Restoration periods certain booksellers, such as Davis, Thomas Robinson, and Amos Curteyne, were early commercial supporters of the new philosophy. They not only published and stocked scientific works, but also distributed such scientific periodicals as the *Philosophical Transactions*.

These institutional supports for learning and science created during the first four decades of the century were only temporarily affected by the cataclysmic events of the 1640s. During the siege the number of students admitted to degrees fell precipitously, but then recovered by the early 1650s to the late Elizabethan level of about 140 B.A.s and 75 M.A.s per annum.[10] Revenues and benefactions were similarly interrupted during the War, but were recovered and consolidated by the early 1650s. Scholars at Commonwealth and Restoration Oxford found themselves in the enviable position of enjoying restored and augmented incomes, new and commodious buildings, an atmosphere largely detached from political and religious strife, and far fewer pressures to leave the confines of the Thames and Cherwell.

This orderly recovery was in large part due to the progressive policies adopted by the Visitors appointed by Parliament in 1647 to purge the university of intransigent Royalists, and to purify it of the detested Laudian religious "innovations" of the 1620s and 1630s.[11] The Visitation of 1648 ejected many students, fellows, professors, and heads of house who refused to recognize Parliament's authority, thereby creating vacancies into which others, either Puritan or merely moderate, were appointed. On the whole the sciences benefited from these events. Many scientists and virtuosi—such as Bathurst, Whistler, Langbaine, Dickenson, and Richard Lydall—simply submitted, although many retained High-Church sympathies. Others, such as Willis, Clarke, Clerke, Coles, Smith, Barlow, Conyers, John Lydall and Edward Greaves, were continued in their places despite nonsubmission.

In making new appointments, a few key proponents of the new philosophy, such as the religious moderate John Wilkins, used the opportunity to advance their confreres. The newly appointed Warden of Wadham was an important member of the committee set up by the Visitors on 5 July 1648 to examine applicants for vacated positions. Others included the physician John Palmer, Warden of All Souls, and the Baconian sympathizer, Henry Langley, Master of Pembroke.[12] Thus in 1649 Wilkins's colleague in the London club, John Wallis, was appointed to the Savilian Professorship of Geometry in the place of a stubborn Royalist. In the same year Seth Ward, although a staunch High-Churchman, was appointed as John Greaves's successor to the other Savilian chair. Greaves and Charles Scarburgh conspired, almost certainly with the approval of their fellow clubman, Wilkins, to appoint Ward, who was known to Scarburgh as an able mathematician from their common days at Cambridge in the early 1640s.[13] Other appointments of many younger men were obviously made without knowledge of the scientific interests they were later to exhibit. But even in these, the Visitors' committee based its selection on ability and recommendations, rather than simply on puritan religious conformity. Those few who came from Cambridge, such as Millington, Hodges, Pett, and Pope, seem to have moved for the sake of friendship or advancement, not to Puritanize Oxford. Other appointees, such as Dickenson, Sharrock, Stephens, Wood, Barksdale, Crosse, Gorges, and Owen, were of diverse religious persuasions, but all had been educated at Oxford. In any case, the Visitors' interference in the internal affairs of the university was almost surgically brief. Within a short time the leading colleges—Wadham, Merton, Christ Church, and Trinity—were granted permission to elect their own fellows and govern their own affairs. By the early 1650s, when Oxford had returned to normal academic life, the Visitors and their friends functioned more to protect the university from the demands of outside radicals, than to impose Parliament's will upon it.[14]

MEDICAL EDUCATION AT COMMONWEALTH AND RESTORATION OXFORD

Oxford medical education, and consequently the medical sciences of anatomy and physiology, were among the beneficiaries of institutional innovation and political upheaval. In the 1650s and 1660s the number of Oxford physicians grew rapidly and the scope of their scientific activity expanded greatly, so that by the Restoration and for some years after, one may truly speak of an Oxford medical community.

Such was not always the case. In the first four decades of the seventeenth century, while the rising numbers of resident undergraduates and bachelors swelled colleges to the bursting point, the faculty of medicine remained small, admitting fewer than ten D.M.s per decade. The statutory course was long—seven years to the M.A., and another seven to the D.M.—and consisted largely of the three elements inherited from medieval tradition: residence, lectures, and disputations.[15] These continued in force after the Civil War, but were increasingly supplemented by more interesting kinds of informal teaching in anatomy, chemistry, botany, and clinical medicine, all of which made medicine a more attractive career, especially at a time when one of the alternatives, the clerical life, was fraught with political and ideological difficulties. The medical faculty almost tripled in size, admitting 33 D.M.s in the 1650s, not to speak of the additional medical students who left to practice without degrees, or who were laureated on the Continent.

The formal elements of medical training improved greatly in the 1650s. Prerequisite lectures in astronomy, geometry, and natural philosophy, taken by medical students for the M.A., were given by neoterics (Ward, Wallis, and Crosse) who, according to Wilkins and Ward, taught Copernicus and Kepler, defended the "Atomicall and Magneticall" hypotheses, promoted the doctrine of arithmetic indivisibles, and in general felt free to "discent" from Aristotle, "and to declare against him.''[16] When Thomas Willis became Sedleian Professor of Natural Philosophy in 1660, he neglected the statutory injunction to lecture on Aristotle, and instead spoke on the senses as interpreted by the corpuscular philosophy.[17] Thomas Clayton, Jr., who in 1647 had succeeded his father as Regius Professor of Medicine, was more an astute university politician than a physician, but he did manage to secure talented deputies as Tomlins Reader in Anatomy: William Petty and Henry Clerke.[18] When medical disputations began again in 1651 after a decade's intermission, Clayton supervised a series of exercises that were as *au courant* as those before had been outdated.[19]

Medical students supplemented lectures with reading. Two such students, John Ward and John Locke, both of Christ Church, kept sets of commonplace books in the 1650s and early 1660s that show the way a medical student was expected to educate himself.[20] While still an undergraduate, Locke filled a notebook with medical prescriptions, or receipts, interspersed with notes drawn from his reading. He read, not primarily the ancients, but such very recent authors as Jan Baptista van Helmont, Pierre Borel, Francis Glisson, George Ent, and Ole Rudbeck.[21] He wrote out little essays on the heart, sanguification, fermentation, heat, and even a longer piece on the diseases of women.[22] As bachelors, Locke took lengthy notes on physiology from Harvey and on herbs from Fuchs, while Ward extracted Johnson on plants, de Renou on pharmacy, Sala on chemistry, Glisson on anatomy, and Biggs and Deverti on medicine.[23] Once past the M.A., Locke read extensively in recent and more specialized literature: in Moebius, Sylvius, Sydenham and especially Sennert on general and clinical medicine; Descartes and the medical Cartesians such as Hooghelande and Velthusius; Valentine, Boyle, and le Febure on chemistry; Conring, Deusing, van der Heyden, Vesling, Bartholin, Wepfer, and Swammerdam on anatomy and physiology, as well as Locke's own countrymen Ent, Wharton, and Charleton, and his Oxford colleagues Highmore, Castle, and especially Willis.[24] Medical training was in no way in tutelage to Galen.

Although Oxford did not acquire a professor of botany until 1669, botanical instruction in no way suffered; the medical student had only to walk down the High Street to the Physic Garden to examine hundreds of plants. There Bobart taught informal classes, led herbarizing expeditions, and in 1648 published a *Catalogus plantarum Horti Medici Oxoniensis* as a guide to the Garden's use. John Ward, for example, learned the herbalist's art in 1659–1661. Under Bobart's direction, he went on simpling expeditions to nearby Shotover Hill and botanized the top of New College wall to find "Ruta muraria." Bobart recommended authors to Ward, discussed the new curiosity of the "sensitive plant," passed on scientific gossip, and handed down dicta, which Ward duly recorded, on everything from chestnuts to grapes.[25] A few years later Locke followed the same course, collecting a two-volume herbarium pressed neatly between leaves of his tutees' Latin exercises, and indexed by an annotated copy of Phillip Stephens's and William Brown's guide to the Garden, *Catalogus horti botanici Oxoniensis* (1658), which succeeded Bobart's.[26]

The teaching of anatomy similarly supplemented the official lectures of the Tomlins Reader with informal learning. Ward noted that for five or six years in the 1650s, Stephens "went over a bodie of Anatomie" with students, charging £5 for the course; he used Johann Vesling's *Syntagma anatomicum,* one of the few students' textbooks that accepted the circulation.[27] In the 1660s Willis and Lower taught informal anatomy classes which, according to a French visitor, were even more popular than the official instruction.[28]

Groups of physicians and medical students also gathered to conduct anatomies on executed criminals, a practice that could sometimes bring spectacular results. On 14 December 1650 William Petty and Thomas Willis met at Petty's lodgings in Buckley Hall to dissect the body of Anne Greene, a young woman hanged for infanticide, only to discover that the proposed cadaver was still very much alive. With the help of Bathurst and Clerke, they revived her and nursed her back to health, thus moving a commentator to note that "whilst they missed the opportunity of improving their knowledge in the dissection of a Dead body, they advanced their fame by restoring to the world a Living one."[29] Petty, in describing the resurrection in a letter to Hartlib, echoed the same theme: "My Endeavours in this businesse have bettered my reputation"[30]—a conclusion justified two weeks later when Petty was formally appointed Tomlins Reader in Anatomy.[31]

Additional anatomical experience could be gained by participation in autopsies. Conyers did frequent post-mortems in the 1650s and early 1660s, whose results were duly recorded by both Anthony Wood and by Ward, who occasionally assisted.[32] In late 1659 Ward looked on at the dissection of Edmund Gwynn, a *serviens* at Christ Church recently deceased, and gave descriptions of the post-mortem state of the spleen, heart, kidneys, lungs and liver; Boyle, also present, compared the state of the lungs to those of "sound men yt hee had opened."[33] Lower, in letters to Boyle in 1663 and 1664, similarly reported post-mortem findings.[34] Willis was perhaps the most active pathologist: in 1667 he reported post-mortems on a newborn child, a hydrocephalic, an hysteric, and several who had died of convulsions, scurvy, whooping cough, and complications of a spontaneous abortion.[35] When cadavers were not available, dogs took their place. Ward did a number of animal experiments with the apothecary Stephen Toone, including splenectomy and one "dissection of a dog" which "ended not finding what wee lookd for to wit ye passages of ye chyle."[36]

For those interested, chemistry could also be learned informally and cooperatively. In early 1649 John Lydall reported to Aubrey that he, Bathurst, and "our Chymist" Willis had been at work confecting the explosive *aurum fulminans.*[37] Anthony Wood recorded that his fellow Mertonian, Dickenson, "spent much labour and money in the art of chymistry, kept an operator," and intended to write a book on the subject; Ward recorded many of Dickenson's chemical dicta.[38]

Willis, after he moved out of Christ Church and into Beam Hall in the late 1640s, set up a laboratory there and took on Robert Hooke as his assistant.[39] After Boyle arrived in late 1655, his chemical work was at first assisted by Lawrence Rooke, and later, on Willis's recommendation, by Hooke. Conyers was another devotee of the chemical arts. By the late 1650s he had set up a furnace to concoct medicaments, which Ward recorded by the dozen.[40] In late 1659 Boyle brought in the Alsatian chemist Peter Stahl to act as one of his operators and to teach chemistry to all comers. Many took advantage of the opportunity, and, over the next few years, Stahl taught chemistry to numerous groups that included the physicians Millington, Bathurst, Yerbury, Jeanes, Lower, Locke, Levinz, and Griffith, the mathematicians Christopher Wren, Wallis, and Brancker, and the virtuosi Williamson, Crew, and Anthony Wood.[41] John Ward too seems to have been an informal pupil of Stahl, recording a large number of his opinions in 1661.[42] In 1667 Ward, in the company of Robert Plot and John Mayow, learned laboratory protocols in a more formal *cursus chymicus* taught at Oxford by William Wildan.[43] John Locke and David Thomas conducted their own course in 1666–1667, clubbing together to buy supplies and cooperating in carrying out the experiments.[44]

Clinical medicine was learned through the same network. Medical students attached themselves to local practitioners as informal apprentices. Willis seems to have been successively doyen to Bathurst, Hodges, Stubbe, Lower, and Masters. Conyers was Ward's mentor. Students like Ward would exchange professional information with local surgeons, such as Francis Smith, William Day, and John Gill, and spend time discussing drugs in the shops of such apothecaries as Stephen Toone, John Crosse, and Arthur Tillyard, who also operated hospices for nursing the sick. Most of all, would-be physicians traded receipts and advice. From their very inception, the notebooks of Ward and Locke record the unusual cases, the modes of treatment, and the favorite prescriptions of a score of Oxford practitioners, including Willis, Bathurst, Conyers, Stephens, Dickenson, Clerke, Lydall, Lamphire, Lovell, Millington, Boyle, Lower, Needham, Jeamson, and even a tale about a case of "colick" treated by Harvey, a story still circulating from the physiologist's sojourn in Oxford.[45]

THE OXFORD EXPERIMENTAL PHILOSOPHY CLUB

Mathematics and the physical sciences, although generally more solitary avocations than chemistry and the medical sciences, could similarly be pursued cooperatively in the university milieu. In the early 1650s Wallis discussed with Seth Ward, Lawrence Rooke, Richard Rawlinson, Robert Wood, and Christopher Wren the ideas later embodied in his *Arithmetica infinitorum*.[46] Wren and William Neile worked together on problems of rectifying various families of curves; their results were published by Wallis and later exposited by Wren in London as his Gresham lectures.[47] The two divines, Wallis and Ward, explored together the notion that the three dimensions in one cube might be a useful metaphor for the Trinity.[48] Observational astronomy was a more intrinsically cooperative activity. In 1654 Rawlinson and Wren helped Wallis observe a solar eclipse.[49] A year later Hartlib noted that Wilkins and Wren were building a telescope eighty feet long to observe the moon.[50] By 1656 Wren, William Neile, and his brother Sir Paul Neile were carrying out detailed observations of Saturn and making a model to account for its unusual shape.[51] Wallis's treatment of the distance of comets first arose in discussions with Wren in 1661–1662.[52] In 1664 the same two were joined by Boyle and the visiting John Evelyn in their unsuccessful attempt to observe a transit of Mercury using a telescope mounted on the tower of the Schools Quadrangle.[53] Experimentation in physics offered other opportunities for joint work. Wilkins and others came to Boyle's lodgings c. 1657 to pommel water sealed in a pewter globe, in order to find out whether it was compressible.[54] At about the same time, Boyle and Wren were doing experiments on the variation of the compass needle.[55] The development of the vacuum pump by Boyle and Hooke in 1659 offered unparalleled opportunities for cooperative research. That year Ward, Wren, and most especially Wallis, helped Boyle and Hooke perform diverse pneumatic experiments with the new device.[56] Again in the mid-1660s, Wallis, Wren, and Millington helped Boyle do measurements of the highest column of water supportable by atmospheric pressure; Wallis also helped Boyle produce maximum expansions and compressions of air.[57]

Given this rich and complex pattern of small-scale cooperative investigation in both the physical and medical sciences, it is hardly surprising that, under the aegis of suitable leaders, a larger and more formal club of experimental philosophers prospered at Oxford in the 1650s and 1660s. One of its incarnations, the meetings at Wadham presided over by Wilkins in the 1650s, has frequently been mentioned by generations of writers tracing the origins of the Royal Society.[58] But scholarship in the last decade has made it increasingly clear that organized science in the university town existed before the Wadham meetings, continued for some time after Wilkins left, and exhibited a diversity and influence that is obscured by seeing Oxford science and the "Experimental Philosophy Club" merely as a provincial way-station between the "1645 Group" in London, and the foundation of the Royal Society at Gresham College there in 1660.[59] Webster, especially, has brilliantly summarized, for the Commonwealth period, much of the scattered evidence bearing upon these "Academies within the Academies"—their evolution, composition, political and philosophical inclinations, and general subject interests.[60] I wish here to delineate more clearly the changing loci of Oxford cooperative science in the 1650s and 1660s, to assess the participation of physicians in these proceedings, and to indicate the patterns of scientific interaction that proved most important in shaping the Oxford approach to Harveian physiological problems.

Scientific gatherings were held in different locales in Oxford, all within a few hundred yards of each other.[61] At the club's inception, most likely in late autumn

1649, it met in William Petty's lodgings in Buckley Hall (107 High Street), opposite the entrance to Brasenose College.[62] It was there that Petty, Willis, Bathurst and Clerke gathered to dissect Anne Greene. Several of those that contributed witty poems on this event, such as Henry Berkenhead, Joseph Williamson, Robert Sharrock, Christopher Wren, and Walter Pope, were probably participants in the Buckley Hall club. In late 1651 or early 1652, upon Petty's departure, the club moved to Wilkins's lodgings in Wadham College, where it continued, probably until late 1656 or early 1657, lapsing no doubt after Wilkins's marriage in 1656 to Oliver Cromwell's sister involved him increasingly in London politics. Gatherings began again, probably in late 1657, at Boyle's lodgings in Deep Hall (88 High Street), next to University College and opposite the entrance to All Souls; in February 1658 Petty wrote to Boyle from Ireland that he had not, amongst all his intelligence, "heard better news, than that the club is restored at *Oxford.*"[63] The meetings continued there, whenever Boyle was in town, until his final departure to London in April 1668. The Boyle group seems to have been particularly active in 1664–1668, such that Sharrock entertained the hope "that wee shall have some Endowemt for a Society to Study Experimentall Philosophy att Oxford," which would not only promote "Knowledge in Generall," but also increase the reputation of the university.[64] When Boyle was absent, as he was for the most part from December 1659 to July 1664, the virtuosi seem to have congregated at Thomas Willis's house, Beam Hall, in St. John Street opposite Merton College.

In each case, both the leaders and the venue were natural. Although only twenty-six when he began his club, in 1645–1646 Petty had participated in the Parisian scientific circle *chez* Marin Mersenne, and from 1646 to 1649 was part of the "invisible college" organized by Hartlib, Boyle, and Benjamin Worsley to promote Baconian schemes in London and Ireland.[65] Buckley Hall was a suitable meeting place, Wallis recalled, "because of the conveniences we had there (being the house of an Apothecary) to view, and make use, of Drugs and other like matters, as there was occasion."[66] The apothecary was John Clarke, whom Hartlib knew as a "very great chymist," an experimenter on bees, who had formerly been the only apothecary to William Harvey.[67] Wadham was appropriate because of Wilkins's great stature in university affairs, his previous participation in the "1645 Group" in London, the number of scientists (Ward, Rooke, Wren, Morton, Neile, Pope, and Sprat) who by c. 1652 were living at Wadham, and most of all because of the large collection of instruments and mechanical devices that Wilkins had built up there.[68] Boyle's aristocratic background, wealth, extensive scientific connections, talented assistants, and large store of chemical and physical apparatus made him a logical successor to Wilkins. Deep Hall also had its apothecary shop, that of John Crosse, a knowledgeable botanist and well-connected virtuoso in his own right.[69] Boyle made Deep Hall attractive even in his absence, when Stahl taught chemical classes there in 1660–1663. Although seemingly not as well stocked, Beam Hall did have a chemical laboratory and dissecting facilities which, together with Willis's status as Sedleian Professor and Oxford's leading practitioner, made it attractive in the 1660s, especially to young physicians.

Predictably for meetings continued over nearly twenty years, the size and composition of the group, as well as the intensity of its activity, varied greatly. At one juncture it would be well organized and forgather regularly; at another the "club" would be closer to the seventeenth-century meaning of the word—a circle of friends and associates, sharing common interests, and common activities, as the occasion arose. Through it all there was a certain continuity of personnel. The Buckley Hall club brought together scientists already resident in Oxford—Willis, Bathurst, Rawlinson, Richard and John Lydall, and Robert Wood—with those newly arrived—Petty, Wallis, Wilkins, and Seth Ward. It grew to its largest and most active in late 1651. Ward said in early 1652 that there were about thirty in the club, and Wallis recalled its meetings as "very numerous, and very considerable."[70] One cooperative venture, launched in mid-1651, attempted to survey the Bodleian "to make medullas of all authors in reference to experimental learning."[71] The twenty-four participants in this project, neatly organized into three teams by Gerard Langbaine, probably reflect quite well the membership of the Buckley Hall club.[72] They include Wilkins, Wallis, Bathurst, Willis, Rawlinson, Wood, Conyers, Henry Clerke, Joshua Crosse, Seth and John Ward, the two Lydalls, and Langbaine himself. The men were of varying religious persuasions, ranged in age between 26 and 41, and included three heads of house and five out of the seven incumbents of university-endowed scientific positions; half of the twenty-four were physicians.

Of the Petty club, Willis, Bathurst, Rawlinson, Wood, Wilkins, Ward, Wallis, and Christopher Wren were among those who continued into the Wadham phase. Although meetings were, according to Wallis, less regular,[73] the club attracted new members, among them Christopher Wren's cousins Matthew and Thomas, Jonathan Goddard (the newly appointed Warden of Merton), Lawrence Rooke, Edmund Dickenson, Robert Boyle, Robert Hooke, George Castle, William Holder, John Lamphire, Peter Pett, Walter Pope, and Thomas Sprat. Robert Boyle, in particular, had been attracted to Oxford, rather than being recruited from among university scholars. He had known Dickenson in the late 1640s, had visited Wilkins several times in the early 1650s, and finally, attracted by the "knot of such ingenious and free philosophers,"[74] had settled there in late 1655.

Boyle's catholic interests and engaging personality drew others into intellectual discourse once the club had moved to Deep Hall, including Henry Stubbe, Robert Sharrock, John Locke, David Thomas, Ralph Austen, as well as members of Willis's Beam Hall circle—Thomas Millington, Richard Lower, Nathaniel Hodges, Walter Needham, and Thomas Jeamson. In short, the Oxford club circa 1665 was no less distinguished than it had been fifteen years before, although it was somewhat smaller, less formally organized, and had only two individuals, Willis and Wallis, who had been participants since the very beginning.

As an informal gathering of friends, lacking corporate existence, the club kept no records to tell us of its proceedings and interests. As Hooke later lamented,

> divers Experiments were suggested, discours'd and try'd with various successes, tho' no other account was taken of them but what particular Persons perhaps did for the help of their own Memories; so that many excellent things have been lost, some few only by the kindness of the Authors have been since made publick . . . [75]

One may however recover some sense of the club's aspirations from a draft set of eight rules for governing the meetings that were drawn up by Langbaine in October 1651, although they suggest a greater formality of organization than was regularly achieved.[76] Admission to these "Philosophical Meetings," as Hooke called them, was by majority vote in a secret ballot, with a quorum of eleven, and with provision for dropping resident members lax in their attendance. Meetings were to be held every Thursday "before two of the clock." The member appointed for that day was to perform an "exercise" or "bring in such experimentes as shall be appoynted for that day"—a requirement the club may have had some difficulty enforcing strictly, since several rules stipulate penalties for failure. Experiments were clearly the company's *raison d'être*, for each new member was to pay a *pro rata* share "for the instruments in stock."

Scattered bits of evidence reflect these club enterprises. In December 1650 Petty described to Hartlib the club's experiments in using inflated bladders to lift great weights.[77] These experiments had a curious longevity in the minds of the participants. Willis used the example in a university lecture c. 1661 to illustrate the motion of muscles.[78] Wilkins repeated the experiments for the Royal Society in the same year.[79] Wallis entertained the Society in 1663 with an account of the statics involved, which he published in 1671.[80] Another communal effort, the Bodleian project headed by Langbaine, was described in a letter written by Seth Ward in February 1652 and was echoed in later comments made to Hartlib by two of its participants, Wallis and Robert Wood.[81] In the same letter Ward noted that, in addition to the "greate clubb" involved in cataloging the library, there was a smaller group of eight "who have joyned together for the furnishing an elaboratory and for making chymicall experiments wch we doe constantly every one of us in course undertaking to manage the worke."[82] Willis, one of the eight, duly recorded in a clinical notebook his expenditures "laid out at Wadham Coll.": to the carpenter, smith, and mason for alterations, for glassware, stills, mortar and pestle, drugs, reagents, and even the payment of one shilling "For carriage of the Glasses and things from my chamber."[83] In 1654 Ward cited exactly this "conjunction" both of "Purses and endeavours" to prove to a critic that Oxford was not neglecting chemistry.[84] Even some years later, in the summer and autumn of 1665, Boyle's letters to Oldenburg trace out the club's activities, this time in experiments on injections of poisons into animals.[85]

One continuing concern of the Oxford club, microscopy, had special applicability to anatomy and physiology. The "1645 Group" in London had met at Goddard's lodgings "on occasion of his keeping an Operator in his house, for grinding Glasses for Telescopes and Microscopes."[86] By 1648 Petty had an excellent and highly prized microscope that had been given to him by Thomas Hobbes.[87] Wilkins also had one by then, which led him to proclaim the "excellent beauty" that might by "the use of *Micrescopes*" be seen "in the parts of the most minute creatures."[88] When men such as Goddard, Petty, and Wilkins came to Oxford in the 1650s, they gathered around them a circle that actively improved and used these microscopes. In September 1655 Boyle reported Wilkins's belief that the Oxford group, especially Christopher Wren, had brought microscopes to "that certainty and exactnes that it's impossible to adde more unto it." Wren's microscopes could now "multiply exceedingly," and were capable of measuring "geometrically" the objects under observation. Wren was so pleased that he told Hartlib "of a book hee is preparing with pictures of observ[ations] microscop[ical]."[89] This work of the Wadham circle was well enough known by the late 1650s for James Harrington, in attacking John Wilkins and Matthew Wren for their political opinions, to accuse these "University Wits" of being "good at two Things, at diminishing a Commonwealth and at Multiplying a Louse."[90] Matthew Wren responded that, as practiced by the assembly at Wadham, microscopy inquired into "the Figure and positions of those smaller parts of which all Bodies are composed," and was thus a branch of mathematics. The "Company" observed the structure of "little Animals, as a Flea, a Louse, or a Mite," as a way of showing "that the Wisedome of the great Architect of Nature is not more conspicuous in the larger Bulks of an Elephant or Camel, then in these little Creatures."[91]

Onlookers and visitors alike remarked on the freedom, vigor, and inventiveness with which the Oxford club and its members carried on their inquiries. Aubrey, who kept in touch with his Trinity friends by correspondence and visits, felt that "Till about the year 1649, when Experimental Philosophy was first cultivated by a Club at Oxford, it was held a strange Presumption to attempt an Innovation in Learning."[92] All through the 1650s Hartlib peppered his diary with notes about Oxford activities gleaned from his correspondence and conversations with Petty, Wilkins, Wallis, Ward, Austen, Wren, Boyle, and Wood. The entries tended, as was his wont, to concern largely the practical side of club projects: malleable glass, the chemistry of gold; machines for double writing and perspective drawing; grasses, fruit trees, grafting and irrigation; hygroscopes, burning glasses, and compasses; telescopes of twenty-four, forty-five, and even eighty feet long; even a universal grammar and real character appropriate to communicating such solid learning.[93] Evelyn was likewise taken with the contrivances in Wilkins's garden, as well as by his collections of magnets, balances, thermometers and perspective glasses.[94] Walter Charleton, who had been Wilkins's tutee at

Magdalen Hall in the late 1630s, visited in the mid-1650s. He was impressed with the work of the Savilian professors in mathematics and in elliptical astronomy, and especially with the advancements there in optics, telescopes, and microscopes, with which latter instruments "the eye ere long may be enabled" to behold "the smallest *Moleculae,* or first collections of Atoms concurring to determinate the Figures of Concretions."[95]

One particularly appreciative sojourner was Henry Oldenburg. A German educated in theology at Bremen, he had come to England in 1653 as an envoy, where he became acquainted with Robert Boyle's sister, Katherine Jones, Countess Ranelagh.[96] He was engaged as tutor to her son, Richard, and arrived at Oxford in April 1656. Through Boyle he rapidly became acquainted with the "men who bend their minds to solid studies," especially Wilkins, and his circle at Wadham.[97] Oldenburg spent a happy year there, in a place that had, as he reported to a correspondent, all the advantages:

> Oxford is indeed a city very well furnished with all the things needed for the grounding and cultivation of learning: among which rich libraries, fat revenues, convenient houses, and healthy air easily hold first place.[98]

Oldenburg carried the friendships formed there to the end of his life, and always looked back fondly to his time with "the *Oxonian* sparkles."[99]

The importance of the Oxford club lay, in fact, not in the results of its formal cooperative projects, but in the way in which its meetings both epitomized the smaller scale collaborative research that became common at Oxford in these years, and served as a forum for intellectual exchange between the two major groups of Oxford scientists—the "physicians" and the "mathematicians." Out of countless gatherings, formal and informal, private and more public, arose a complex network of scientific friendships that crossed the boundaries of discipline and faculty. Again and again one finds physical scientists like Wilkins, Wallis, Wren, and Hooke or the chemist Boyle, doing physiological experiments with physicians such as Willis, Lower, Needham, and Clarke—or all meeting on the common ground of chemistry. Borrowings, then as ever since, were largely by the medical sciences from the physical sciences, but this made the exchange nonetheless enjoyable to both sides.

THE BONDS OF A SCIENTIFIC COMMUNITY

The small and close environment of a university town bound together Oxford scientists in a host of ways other than strictly scientific.

The college was the natural focus of a scholar's life. Some houses at Commonwealth and Restoration Oxford not only supported scientists with endowed places and congenial surroundings, but became the reservoirs out of which more were recruited. Of Oxford's twenty-five colleges and halls, six—Christ Church, Wadham, Merton, All Souls, Queen's, and Trinity—supplied a greatly dispro-portionate number of scientists. Collectively, these six were distinctive because, in contrast to the other houses oriented largely toward the education of under-graduates, they provided places for large numbers of M.A.s (over 40 percent of the resident masters), men in their twenties and thirties who were more likely to be inclined toward science.[100] The constitutions of these six colleges, especially in the case of Wadham, Merton, and All Souls, provided either nonclerical fellow-ships, or, in the case of Christ Church, a lengthy period of support before one had to decide whether to take orders.[101] Most of all, these colleges had heads who either actively encouraged science or, at the very least, saw it as an acceptable form of intellectual activity. At Christ Church, John Owen (Dean, 1651–1659) was a school chum of Wilkins, a close friend of Langbaine, a participant in the Bodleian catalogue project, and a virtuoso whose library contained the works of such Continental neoterics as Galileo, Gassendi, Mersenne, and Descartes, as well as those of his fellow Englishmen like Bacon, Hartlib, Wilkins, Wallis, Ward, Highmore, Willis, and Harvey.[102] Owen was succeeded as Dean by John Fell, Thomas Willis's brother-in-law and a man well-known among Oxford virtuosi. Wadham had Wilkins as Warden from 1648 to 1659, while Goddard's tenure as Warden of Merton from 1651 to 1660 was an appropriate scientific successorship to Savile and Harvey. The Warden of All Souls was John Palmer, an Oxford D.M. and son of a Taunton apothecary; after his death in 1660, his widow married Ralph Bathurst.[103] At Queen's, Langbaine's tenure as Provost was followed by that of Thomas Barlow, a virtuoso and good friend of Robert Boyle. Trinity's Common-wealth President is not known to have been scientifically inclined, but Bathurst seems to have been almost as powerful in college affairs; Bathurst got Ward in as President of Trinity in 1659–1660 and rose to that office himself in 1664.

Scientific friendships and associations coalesced along collegial lines. At Wadham, Pope was Wilkins's half brother, wrote Rooke's biography, and became Ward's client and confidant. Sprat's historical account of the Royal Society emphasized its Wadham origins. During the 1650s yet another Wadhamite, Christopher Wren, was associated closely with Wilkins and Ward in numerous researches noted in Hartlib's diary. At Queen's, Langbaine, Barlow, Rawlinson, Williamson, and Southwell all knew each other well. When Wren moved in 1654 from Wadham to a fellowship at All Souls, he joined another group consisting of Millington, Castle, Pett, and latterly John Mayow. This circle at All Souls, in addition to meeting with Boyle and others at Deep Hall, gathered socially in the coffee rooms run by the apothecary Arthur Tillyard at 90 High Street, two doors west of John Crosse's and opposite the main gate of All Souls. The Christ Church crowd was the largest of collegial science groups, including Willis, his assistant Hooke, the botanist Robert Lovell, and the physicians Lower, Locke, Stubbe, Hodges, Masters, and John Ward.

Christ Church exemplifies another very strong bond among one influential set of

Oxford scientists—their common origin at the seventeenth-century's most famous school, Westminster, in London. This may in turn be traced to Westminster's redoubtable headmaster, Richard Busby, who protected his royal foundation from parliamentary sequestrations in the 1640s, and who bred up during the Commonwealth a brilliant, disciplined, and loyal generation of pupils, many of whom were elected to the closed Studentships reserved for them at Christ Church.[104] Although Busby was known largely for his teaching of Latin and Greek, he had broad interests in mathematics and natural philosophy. He was a good friend of the mathematician John Pell, whose papers Busby collected, as well as those of other mathematicians, such as Hariot, Thorndyke, and Warner.[105] As a young Westminster student, Hooke learned his Euclid while lodging in Busby's house.[106] Hartlib recorded Busby's interest in fermentation. Busby and another former student, Walter Needham, corresponded in the 1650s about the new philosophies of Descartes and Gassendi.[107] Among the benefactions set up in his will, Busby included a mathematical lectureship at Christ Church.[108] Little wonder that some of his students not only followed an avocation in science, but often pursued it in the company of those wearing the old school tie. In the decade from the mid-1640s to the mid-1650s Busby sent Hodges, Lower, Stubbe, Locke, Hooke, Vernon, and Woodroffe to Christ Church; Pope and Wren to Wadham; Millington and Castle to All Souls; and Williamson to Queen's.[109] Needham, when he came from Trinity College, Cambridge, to study with Willis c. 1662–1664, was likewise merely learning his science through the old boy network.

Ties among scientists in different colleges were reinforced by associations in the many functions of general university life. Members of the same age cohort disputed against one another, sat on examining boards with one another and served on the same committees of university governance. Lower and Jeanes examined students together in 1653; they were both students of Stahl almost a decade later. Lower examined Castle in 1657. Jeamson examined Glanvill in 1658; he in turn was examined by Locke in 1660, and examined Guidott in 1664.[110] Again and again one notes friendships between scientific virtuosi of different colleges which seem to correspond to their mutual places in the university hierarchy. The cohort admitted to regency in the five-year period 1641–1645 included Bathurst, Lamphire, Willis, Joyliffe, and Clerke; that in the period 1654–1658 included Wren, Hodges, Lower, Stubbe, Castle, Sprat, Locke, Thomas, Crew, Glanvill, Brancker, and Williamson.[111] Much collaborative investigation in each case mirrored the age cohort. University functions seem in some cases to have even provided an opportunity to introduce scientific questions into more general intellectual discourse. Was it purely by accident that Pope, as Proctor, was in the moderator's chair in 1658 when the eighty or so incepting Masters discussed "Whether the world is composed of atoms?"[112]

These scientific links, both within and without colleges, not only found expression in associations and in correspondence when one party left Oxford, but in the numerous books Oxford scientists dedicated to one another. Highmore dedicated one anatomical work to Harvey and another to Boyle. Willis wrote a *dissertatio epistolica* on urine to Bathurst. Wallis dedicated mathematical books to Langbaine, Wilkins, Goddard, Rooke, Ward, and Boyle; he even wrote against Hobbes a long defense of Oxford pneumatical experiments in the form of a letter to Boyle. Ward prefaced an astronomical work with a dedicatory epistle praising Goddard; Dickenson did the same in an historical work. As befitted a powerful, wealthy, and influential man, Boyle received numerous dedications: mechanical works by Petty and Merrett; botanical works by Austen and Sharrock; a tract in physics by Hooke; medical and anatomical works by Lower, Stubbe, and Needham. Stubbe also dedicated a medical work to Willis, and Millington received homage from both Lower and Castle. Literary gift-giving was clearly a natural outcome of shared friendship and research.

The final, and in some senses the most immediate, bond of the Oxford scientific community was simple propinquity. Seventeenth-century Oxford was a small city whose very dimensions brought men together in a way not possible in such a metropolis as London. Only by walking around it can one get some idea of how really small it was on a human scale. Let us take a stroll through scientific Oxford as it was in the late spring of 1659.

All good walking-tours of Oxford begin at Christ Church; ours should be no exception. Owen presides over a vast and rich college whose common room includes Nathaniel Hodges, John Ward, Robert Lovell, Richard Lower, Robert Hooke, Henry Stubbe, John Locke, and more than a dozen lesser philosophers and medical men. A few minutes' walk takes us across Christ Church's great quadrangle, through Peckwater, and out by way of Canterbury quad, where Willis had his rooms, into Oriel Square. Looking east 120 yards or so down St. John Street (now Merton Street), we can see Merton, where Dickenson has his laboratory and Richard Lydall, his rooms. Richard Trevor is on leave studying medicine in Padua, but Goddard might possibly be in the Warden's lodgings on one of his periodic visits. On the north side of St. John Street, opposite Merton, is Beam Hall. There Thomas Willis lives and practices with his younger associates, such as Lower, and maintains the chemical laboratory in which Hooke served his apprenticeship.

A few minutes' walk north on Oriel Street brings us to the High Street, in the center of intellectual and scientific Oxford. Twenty-five yards to our left, on the south side of the High, is Buckley Hall, where Harvey's apothecary, John Clarke, has his shop, and where the "clubb" met in Petty's lodgings on the second floor; Matthew Wren lived in the same building. In the next block, also on the south side, are the bookshops of Richard Davis and Thomas Robinson. But we turn right, to the east, and walk along the south side of the High Street the hundred yards to Arthur Tillyard's, where we check his coffee rooms for friends. A few steps farther takes us past the Three Tuns to Deep Hall. Here, opposite John Palmer's lodgings, John Crosse has his apothecary shop and, more importantly, Robert Boyle his lodgings. If we had arrived in late 1659, we might have seen

Peter Stahl setting up to give classes in chemistry. As it is, there is no meeting of the club today, and Boyle, Hooke, and Daniel Cox are conducting experiments with the vacuum pump.

If we were to continue for a quarter of a mile down the High Street, we would come to the Physic Garden, where Jacob Bobart presides and gives instruction in botany to medical students such as John Ward. Across the street from him is Magdalen College, where Petty's Tomlins deputy, Henry Clerke, has a fellowship. Joshua Crosse, the Sedleian Professor, lives nearby; Bobart is his next door neighbor.

But we decide to cross the High and enter All Souls. Millington, Castle, and Pett are in their rooms, as is Christopher Wren, who is the college Bursar. Things are unsettled in London, and Wren is not reading his lectures there as Gresham Professor of Astronomy. Leaving All Souls and walking the 150 yards up Cat Street to the Schools Quadrangle, we pass John Wallis's house, and then on our right, Hart Hall, where the Principal, Phillip Stephens, is giving his anatomy class. Peeking up New College Lane, we can see the college of the same name, where Boyle's friend Robert Sharrock is writing up his experiments on plants as Boyle suggested he do. Sharrock will soon be editing Boyle's *New experiments physico-mechanicall touching the spring of the air* for publication by Thomas Robinson.

It is early afternoon as we enter the Schools, so there are no lectures or disputations. Tacked to the door are notices announcing the times and places of the lectures Millington and Hodges will read for D.M. degrees. To the east and south are the scientific schools. Natural Philosophy is in the far left corner on the first floor, and directly above it is the Anatomy School with its collection of bones and exotic specimens. Mornings, Seth Ward reads in the Astronomy School on the second floor to the right, behind us, while Wallis lectures in the sister room on the same floor to our left. Immediately ahead of us is the entrance to Bodley's great library, where Stubbe is the Under-Librarian.

Leaving the Schools, we walk the few hundred yards up to Wadham. Wilkins is still Warden, although he will soon resign to become Master of Trinity College, Cambridge, and has recently been spending much time out of town. Seth Ward is in residence; his friend Ralph Bathurst will exercise his good offices to have Ward elected into the vacant presidency of Trinity. Some of the scientists who had lived in Wadham are gone: Lawrence Rooke to London, Francis Crosse to study medicine at Leyden, Charles Morton to a parish living in Cornwall, and William Neile to his country estate at White Waltham in nearby Berkshire. But there are still a number of virtuosi, such as the fellows Thomas Sprat and Walter Pope, in residence. Thomas Jeamson is a scholar there, and John Mayow and Thomas Guidott are still undergraduates. Mayow will soon be elected scholar, a position he will hold only briefly before going on to a fellowship at All Souls.

The last few hundred yards west through the garden brings us to Trinity, where Ralph Bathurst is senior fellow. His good friend and fellow chemist, John Lydall, has been dead for two years, and some of the virtuosi who made the college an attraction for Harvey—men such as Nathaniel Highmore and George Bathurst—are also gone. The scientific tradition at Trinity has long since merged with other circles.

We have walked slightly more than two-thirds of a mile, and it has been less than half an hour since we left Christ Church gate. For the scientists in mid-century Oxford, it was a small town indeed. The university and its institutions, the circumstances of the times—each contributed their part in creating a local community rich in scientific friendships. This community in turn shaped the physiological ideas elaborated by its members in the 1650s, 1660s, and 1670s. It exposed medical scientists to new ideas in chemistry and matter-theory, which made possible a redefinition of the key problems of post-Harveian physiology. It made them privy to new instruments, apparatus, and techniques. It provided an audience both to applaud and to criticize new ideas and experimental results. It held up the ideal of collaborative research as a way of solving physiological problems. In sum, the Oxford community provided the perceptual lens to focus physiological inquiry in a new way, the environment and tools to carry on that inquiry, and the lines of dissemination along which the results of such inquiry could be diffused.

TABLE 3

OXFORD SCIENTISTS AND VIRTUOSI, 1640–1675

The following table gives an overview of the participants in Oxford science from the eve of the Civil War to the mid-1670s. I have used the following abbreviations.

"Origins"	=	Family background, where raised and educated.
F	=	Father's occupation.
"Club"	=	Citation in memoirs recording the names of men active in various cooperative scientific projects in Oxford.
B	=	Participant in cataloging of the Bodleian Library, c. 1652: Bodleian Library, MS. donat. Wood 1, pp. 1-3.
W1	=	John Wallis, *A defence of the Royal Society* (London, 1678), pp. 5–9.
W2	=	John Wallis, autobiography (1697), in Thomas Hearne, ed., *Peter Langtoft's chronicle* (Oxford, 1725), I, clxi-clxiv.
W3	=	Wood, *Life and times*, I, 201, 290, 466, 472–474.
W4	=	Wood, *History and antiquities,* II, 632–634.
E	=	John Evelyn to William Wotton, 12 September 1703, in *The diary and correspondence of John Evelyn, F.R.S.,* ed. William Bray (London: Henry G. Bohn, 1863), III, 390–398.
S	=	Thomas Sprat, *The history of the Royal-Society of London, for the improving of natural knowledge* (London, 1667), pp. 52–58.
St	=	Student of Peter Stahl: Wood, *Life and times,* I, 290, 472–475.
A	=	Aubrey, *Brief lives.*
G	=	Participant or laudatory poet in the "resurrection" of Anne Greene: [Richard Watkins], *Newes from the dead* (Oxford, 1651).
P	=	Walter Pope, *The life of the right reverend father in God Seth, Lord Bishop of Salisbury* (London, 1697), pp. 18–29, 110.

MAJOR SCIENTISTS

Name	Birth Date	Origins	Oxford Period & College Affiliation	Degrees	First Scientific Interest & Clubs
BOYLE, Robert	1627	F: Earl Waterford, Ireland & Stal-bridge, Dorset Eton, 1635-1639	1655-1659 1664-1668 With John Crosse, "Deep Hall", 88 High St	Hon DM 1665	1646 W1, E, W3, W4, S, P
CHARLETON, Walter	1620	F: Rector Shepton Mallet, Somerset	1635-1646 Magd Hall 1635-1646	DM 1643	Late 1630s
HARVEY, William	1578	F: Yeoman Folkstone, Kent Canterbury Sch	1642-1646 Merton Warden 1645-1646	(Camb BA 1597 Padua MD 1602)	1599
HIGHMORE, Nathaniel	1613	F: Rector Fordingbridge, Hants Sherborne Sch	1631-1646 Trinity 1632-1646	BA 1635 MA 1638 BM 1641 DM 1643	c.1640
HOOKE, Robert	1635	F: Vicar Fresh Water, Isle of Wight Westminster Sch c.1649-1653	1653-1660 Christ Church 1653-1660	MA 1663	c.1651
LOWER, Richard	1631	F: Gentleman Tremeer, St Tudy, Cornwall Westminster Sch c.1643-1649	1649-1666 Christ Church 1649-1662	BA 1653 MA 1655 DM 1665	c.1654 St
MAYOW, John	1641	F: Gentleman Bray, Morval, Cornwall	1658-1674 Wadham 1658-1660 All Souls 1660-1678	BCL 1665 DCL 1670	Mid 1660s
NEEDHAM, Walter	c.1631	Salop Westminster Sch c.1645-1650	c.1662-1664 (Trinity, Camb) Lived in Town	(Camb BA 1654 MA 1657 MD 1664)	1655

Scientific Subject Interests	Closest Oxford Scientific Friends	Occupation & Location	Joined Royal Society	Scientific Books	Death
Chemistry, Pneumatics, Physiology, Anatomy	Dickenson, Wood, Rooke, Wilkins, Wallis, Ward, Wren, Stubbe, Lower, Willis, Hooke, Bathurst, Needham, Petty, Bohun, etc.	Aristocrat, Oxford & London	1660	1660, 1661(2), 1662, 1663, 1664, 1665, 1666(2), 1669, etc.	London 30 Dec 1691
Natural Philosophy, Anatomy, Physiology	Wilkins, Harvey, Evelyn, Brouncker, E. Greaves	Physician, Oxford & London	1661	1650(3), 1654, 1659, 1661, 1665, 1668, 1669, 1672	London 24 Apr 1707
Anatomy, Physiology, Medicine	J. Greaves, Charleton, G. Bathurst, Highmore, Scarburgh	Physician, London		1628, 1649, 1651	London 3 Jun 1657
Anatomy, Physiology, Medicine	Harvey, Bathurst, Boyle, Willis, Petty	Physician, Sherborne, Dorset		1651(2), 1660, 1670	Sherborne, Dorset 21 Mar 1685
Mechanics, Astronomy, Optics, Chemistry, Physiology	Boyle, Willis, Wilkins, Wallis, Ward, Wren, Lower, Locke, Hodges, Sharrock, Stubbe, Oldenburg	Gresham Prof Geom London 1664-1704	1663	1661, 1665(2), 1674(2), 1677, 1678(2)	London 3 Mar 1703
Anatomy, Physiology, Chemistry	Willis, Millington, J. Ward, Needham, Boyle, Hooke, Wallis, Locke, Stubbe, Bathurst	Physician, Oxford & London	1667	1665, 1669	London 17 Jan 1691
Physiology, Chemistry, Natural Philosophy	Boyle, Millington, Willis, Lower, J. Ward, Plot, (Wren, Castle?)	Physician, Oxford, Bath & London	1678	1668, 1674	London Sep 1679
Anatomy, Physiology, Medicine	Lower, Boyle, Willis, Millington	Physician, Shrewsbury & London	1667	1667	London 5 Apr 1691

MAJOR SCIENTISTS *(Continued)*

Name	Birth Date	Origins	Oxford Period & College Affiliation	Degrees	First Scientific Interest & Clubs
PETTY, William	1623	F: Clothier Romsey, Hants	1649-1652 Brasenose Lived in "Bulkeley Hall", 106 High St	DM 1650	1644 W1, W2, W4, S, A, G
WALLIS, John	1616	F: Vicar Ashford, Kent Felsted Sch Essex	1649-1703 Exeter Lived in Catte St	(Camb BA 1636 MA 1640) DD 1654	1640 B, W1, W2, St, W4, S
WARD, Seth	1617	F: Attorney Buntingford, Herts Buntingford Sch	1649-1660 Wadham 1650-1660	(Camb BA 1637 MA 1640) DD 1654	1637 B, W1, W2, E, W4, S, A, P
WILKINS, John	1614	F: Goldsmith Oxford Sylvester's Sch Oxford	1627-1637 1648-1658 Magd Hall 1627-1637 Wadham Warden, 1648-1659	BA 1631 MA 1634 BD 1648 DD 1649	Mid 1630s B, W1, W2, E, W4, S, A, P
WILLIS, Thomas	1621	F: Yeoman North Hinksey, Berks Sylvester's Sch Oxford	1637-1667 Christ Church 1637-1648 "Beam Hall" 1648-1667	BA 1639 MA 1642 BM 1646 DM 1660	Mid 1640s B, W1, W2, W4, S, A, G, P
WREN, Christopher	1632	F: Dean Windsor & Bletchingdon, Oxon Westminster Sch 1641-1646	1649-1657 1661-1665 Wadham 1649-1653 All Souls 1653-1662	BA 1651 MA 1653 DCL 1661	1647 E, W3, St, W4, S, G

Scientific Subject Interests	Closest Oxford Scientific Friends	Occupation & Location	Joined Royal Society	Scientific Books	Death
Mechanics, Anatomy, Chemistry	Wilkins, Clerke, Willis, Bathurst, Wood, Boyle, Ward, Wallis	Physician & Civil Servant, Ireland & London Tomlins Reader c.1650-1652	1660	1647, 1648, 1674	London 16 Dec 1687
Mathematics, Mechanics, Astronomy, Chemistry	Wilkins, Ward, Wren, Boyle, Goddard, Wood, Millington, Rawlinson, Willis, Lower, Oldenburg	Divine Savilian Prof Geom 1649-1703	1661	1655, 1656(2) 1657, 1658, 1659, 1662, 1669, 1670, etc.	Oxford 28 Oct 1703
Astronomy, Mathematics, Chemistry	Scarburgh, Wilkins, Rooke, Boyle, Wren, Hooke, Bathurst	Divine Savilian Prof Astron 1649-1661	1661	1653(2), 1654(2), 1656(2)	Knightsbridge 6 Jan 1689
Mechanics, Mathematics, Astronomy, Microscopy	Wren, Ward, Boyle, Wallis, Charleton, Petty, Willis, Hooke, Goddard, Bathurst, Morton, Neile, Oldenburg, Rooke	Divine	1660	1638, 1640, 1641, 1648, 1668	London 19 Nov 1672
Chemistry, Anatomy, Physiology, Medicine	Bathurst, Highmore, Petty, R. & J. Lydall, Clerke, Wilkins, Boyle, Wren, Millington, Lower, Needham, Hooke, Dickenson, etc.	Physician, Oxford & London Sedleian Prof Nat Phil 1660-1675	1660	1659, 1664, 1667, 1670, 1672, 1674, 1675	London 11 Nov 1675
Mathematics, Astronomy, Physics, Anatomy, Chemistry	Wilkins, Scarburgh, Willis, Holder, Wallis, Ward, Wood, Boyle, Hooke, Petty, Bathurst, Goddard, Jeamson, Neile, Oldenburg, J. Lydall, Pett, Rawlinson, M. Wren	Savilian Prof Astron 1661-1673 Architect, London	1660		London 25 Feb 1723

MINOR SCIENTISTS

Name	Birth Date	Origins	Oxford Period & College Affiliation	Degrees	First Scientific Interest & Clubs
AUSTEN, Ralph	c.1600	Staffs	1647-1676 Lived in Town		1652
BATHURST, Ralph	1620	F. Gentleman Hothorpe, Northants Coventry Sch	1634-1704 Trinity 1634-1704 President 1664-1704	BA 1638 MA 1641 DM 1654	Late 1630s B, W1, W2, E, St, W4, S, A, G, P
BOBART, Jacob, Sr.	1599	Brunswick, Germany	1632-1680 Lived in Town		c.1630
BOHUN, Ralph	c. 1641	F: Gentleman Counden, Warwicks Winchester Sch c.1652-1658	1658-1664 1667 New Coll 1658-1676	BCL 1665 DCL 1685	Early 1660s
BRANCKER, Thomas	1633	F: Headmaster Barnstaple, Devon	1652-1663 Exeter 1652-1663	BA 1655 MA 1658	Late 1650s St
BROUNCKER, William Viscount	c.1620	F: Viscount	c.1636-1646	DM 1646	Late 1630s
BROWNE, Edward	1644	F: Physician & Knight Norwich Norwich Sch	1665-1667 Merton 1665-1667	(Camb MB 1663) DM 1667	c.1660
CASTLE, George	c.1635	F: Physician Oxford Westminster & Thame Schs	1652-1665 Balliol All Souls 1655-1665	BA 1654 MA 1657 DM 1665	Late 1650s W3, W4
CLARKE, Timothy	c.1620		c.1642-1654 Balliol	DM 1652	Late 1640s

Scientific Subject Interests	Closest Oxford Scientific Friends	Occupation & Location	Joined Royal Society	Scientific Books	Death
Gardening, Botany	Langley, Wallis, Boyle, Joshua Crosse, Sharrock	Gardener, Oxford		1653, 1658, 1676	Oxford Oct 1676
Physiology, Chemistry, Medicine	Willis, Ward, Wilkins, Rooke, Highmore, Boyle, Bohun, Lower, Sprat, Parker, Whistler	Physician & Divine Oxford	1663		Oxford 14 June 1704
Gardening, Botany	Locke, Boyle, J. Ward, Sharrock, John Crosse, Merrett, Coles	Keeper, Physic Garden 1632-1680		1648	Oxford 4 Feb 1680
Natural Philosophy	Bathurst, Boyle	Cleric, Surrey & Wilts		1671	Salisbury 12 Jul 1716
Mathematics, Chemistry	Boyle, Stahl, Wallis	Cleric & Schoolmaster Cheshire & Lancs		1662, 1668	Macclesfield, Lancs 26 Nov 1676
Mathematics, Medicine		Aristocrat, London	1660	1653	London 5 Apr 1684
Anatomy, Medicine, Natural History	Wren, Mayow	Physician, London	1667	1673	Northfleet, Kent 28 Aug 1708
Medicine, Chemistry	Millington, Wren, Pett, Willis, Sprat, (Mayow?)	Physician, London	1669	1667	London 12 Oct 1673
Anatomy, Physiology, Chemistry	Joyliffe, Boyle, Wren, Neile, Dickenson, Lower	Physician, London	1660	1670	London 11 Feb 1672

MINOR SCIENTISTS *(Continued)*

Name	Birth Date	Origins	Oxford Period & College Affiliation	Degrees	First Scientific Interest & Clubs
CLARKE, William	c.1642	F: Gentleman Somerset	1660-c.1668 Oriel 1660-1666	BA 1661	1660s
CLERKE, Henry	c.1622	F: Bourgeois Willoughby, Warwicks Coventry Sch	1638-1687 Magdalen 1639-1667 President 1672-1687	BA 1641 MA 1644 BM 1648 DM 1652	Late 1640s B, G
COLE, William	1635	F: Cleric	1653-1666 Balliol Glos Hall	BM 1660 DM 1666	Late 1650s
COLES, William	1626	F: Cleric Adderbury, Oxon	1642-1651 New Coll 1642-1650 Merton 1650-1651	BA 1650	c.1645
COX, Daniel	1640	F: Gentleman Stoke Newington, London	1657-1659 1663-1665 (Asst to Boyle)	(Camb MD 1669)	c.1657
CROSSE, Francis	1631	F: Cleric Stogumber, Somerset	1649-1656 Wadham 1649-1656	BA 1652 MA 1655 (Leyden MD 1664)	Mid 1650s
DICKENSON, Edmund	1624	F: Rector Appleton, Berks Eton c.1638-1642	1642-1677 Merton 1642-1664	BA 1647 MA 1649 DM 1656	1654 E
ETTMULLER, Michael	1644	Leipzig, Germany	1668-1669 Lived in Town		
GODDARD, Jonathan	1616	F: Shipbuilder Greenwich Chatham Sch	1632-1637 1651-1656 Magd Hall 1632-1637 Merton Warden 1651-1660	(Camb MB 1638 MD 1643)	Late 1630s W1, W2, P, S

Scientific Subject Interests	Closest Oxford Scientific Friends	Occupation & Location	Joined Royal Society	Scientific Books	Death
Chemistry, Medicine	(Mayow?)	Physician, Bath & Middlesex		1670	Stepney, Middlesex 24 Apr 1684
Anatomy	Willis, Millington, Lower, Bathurst, Petty	Physician, Oxford Tomlins Reader c.1652-1664	1667		Lancs 24 Mar 1687
Medicine, Anatomy, Physiology		Physician, Worcester & London		1674, 1689, 1693, 1694, 1702	London 12 Jun 1716
Botany	John Crosse, Bobart, Stephens, J. Lydall	Cleric, Surrey & Winchester		1656, 1657	Winchester 1662
Chemistry, Medicine	Boyle	Physician, London	1665	1669	London 19 Jan 1730
Medicine	Wilkins, Gorges, Joshua Crosse	Cleric & Physician, Bristol		1664	Bristol 1675
Chemistry	Goddard, Boyle, Stahl, T. Clarke, Willis, J. Ward	Physician, Oxford & London	1678	1655, 1686, 1702	London 3 Apr 1707
Chemistry, Medicine	Boyle, Wallis, (Mayow?)	Physician, Leipzig Prof 1681-1683		1670 & posthumous works	Leipzig 9 Mar 1683
Chemistry, Optics, Anatomy	Wilkins, Wallis, Ward, Dickenson, Millington	Physician, Gresham Prof Med 1655-1675 London	1660	1670	London 24 Mar 1675

MINOR SCIENTISTS *(Continued)*

Name	Birth Date	Origins	Oxford Period & College Affiliation	Degrees	First Scientific Interest & Clubs
GREAVES, John	1602	F: Rector Colemore, Hants	1617-1634 1640-1647 Balliol 1617-1624 Merton 1624-1648	BA 1621 MA 1628	Late 1620s
GRIFFITH, Richard	c.1635	F: Gentleman Eton 1650-1654	1654-1663 University 1654-1663	BA 1657 MA 1660 (Caen MD 1664)	Late 1650s St
GUIDOTT, Thomas	1638	F: Gentleman Limington, Hants Dorchester Sch	1656-1667 Wadham 1657-1666	BA 1660 MA 1662 BM 1666	Mid 1660s
HODGES, Nathaniel	1629	F: Vicar Kensington, Middlesex Westminster Sch c.1641-1646	1648-1659 Christ Church 1648-1659	BA 1652 MA 1654 DM 1659	Late 1650s
JEAMSON, Thomas	1638	F: Rector Rycote, Oxon	1654-c.1671 Wadham 1654-1670	BA 1657 MA 1660 BM 1664 DM 1668	Early 1660s
JOYLIFFE, George	1621	F: Gentleman East Stower, Dorset	1637-1648 Wadham 1637-1639 Pembroke 1639-1648	BA 1640 MA 1643 (Camb MD 1652)	c. 1642
LOCKE, John	1632	F: Attorney Pensford, Somerset Westminster Sch c.1647-1652	1652-1667 Christ Church 1652-1684	BA 1656 MA 1658 BM 1675	c.1652 St
LOVELL, Robert	c.1630	F: Rector Lapworth, Warwicks	1648-1661 Christ Church 1650-1660	BA 1650 MA 1653	Mid 1650s

.

Scientific Subject Interests	Closest Oxford Scientific Friends	Occupation & Location	Joined Royal Society	Scientific Books	Death
Astronomy	Harvey, Scarburgh, Holder, Langbaine	Gresham Prof Geom, London 1630-1643 Savilian Prof Astron 1643-1648	1648 1650 1652		London 8 Oct 1652
Chemistry, Medicine	Stahl	Physician, Surrey & London	1681		London Sep 1691
Chemistry, Anatomy	Willis, Hodges, Stubbe, E. Greaves, Glanvill, Pett, (Mayow?), Highmore	Physician, Bath & London		1673, 1674 1676, 1684, 1691, 1694, 1705	Bath After 1703
Medicine, Chemistry	Willis, Millington, J. Ward, Guidott	Physician, London		1665, 1671, 1684	London 10 Jun 1688
Anatomy	Wren	Physician, Oxford & London Tomlins Reader c.1669-1671	1665		Paris Jul 1674
Anatomy, Botany	Clayton Sr, Merrett, T. Clarke	Physician, London			London 11 Nov 1658
Medicine, Chemistry, Botany, Natural Philosophy	Lower, Stubbe, Thomas, Boyle, Stahl, Willis, Cox, Bathurst	Physician & Civil Servant, Oxford & London	1668		Oates, Essex 28 Oct 1704
Botany, Natural History, Medicine	J. Ward, Hooke	Physician, Coventry		1659, 1661	Coventry Nov 1690

MINOR SCIENTISTS *(Continued)*

Name	Birth Date	Origins	Oxford Period & College Affiliation	Degrees	First Scientific Interest & Clubs
MERRETT, Christopher	1615	F: Bourgeois Winchcombe, Glos	1631-1648 Glos Hall 1631-1633 1636-1643 Oriel 1633-1635	BA 1635 BM 1636 DM 1643	Mid 1630s W1, W2
MILLINGTON, Thomas	1628	F: Gentleman Newbury, Berks Westminster Sch c.1640-1645	1649-1676 All Souls 1649-1680	(Camb BA 1649) MA 1651 DM 1659	c.1652 W3
MORRIS, Samuel	c.1643	F: Bourgeois Sussex	1658-1667 Magd Hall	BA 1662 (Leyden MD 1668)	Mid 1660s
MORTON, Charles	1627	F: Cleric Egloshayle, Cornwall	1648-1655 Wadham 1649-1655	BA 1649 MA 1652	
MUNDY, Henry	1623	F: Bourgeois Henley, Oxon	1642-1656 Merton 1642-1656	BA 1647	
NEILE, William	1637	F: Knight White Waltham, Berks	1652-1657 Wadham 1652-1657		c.1655
OLDENBURG, Henry	c.1617	F: Teacher Bremen, Germany	1656-1657 Lived in Town	(Bremen M. Theol. 1639)	1655
PLOT, Robert	1640	F: Bourgeois Borden, Kent Wye Sch	1658-1690 Magd Hall 1658-1676 University 1676-1690	BA 1661 MA 1664 DCL 1671	1668

Scientific Subject Interests	Closest Oxford Scientific Friends	Occupation & Location	Joined Royal Society	Scientific Books	Death
Botany, Natural History, Medicine	Harvey, Joyliffe, Bobart	Physician, London	1660	1660, 1665, 1666, 1669, 1670, 1682	London 19 Aug 1695
Chemistry, Anatomy, Botany	Wallis, Boyle, Goddard, Willis, Lower, Mayow, Hodges, Smith, Wren, Pett, Castle, Bathurst	Physician, Oxford & London Sedleian Prof Nat Phil 1675-1704			London 5 Jan 1704
Anatomy, Medicine	Lower	Physician, London & Petworth, Sussex	1668	?	
Mathematics, Natural Philosophy	Wilkins, Neile	Cleric & Schoolmaster Cornwall, London & New England			Charleston, New England 11 Apr 1698
Physiology, Medicine	Dickenson, Wood, Willis, (Mayow?)	Schoolmaster & Physician, Henley, Oxon	1680		Henley, Oxon 28 Jun 1682
Astronomy, Mathematics, Physiology	Wilkins, Wren, Ward	Gentleman, Berks	1663		White Waltham, Berks 24 Aug 1670
Scientific Communication	Boyle, Wilkins, Wallis, Stubbe	Tutor, Editor, Sec. R.S., London	1660		London 3 Sep 1677
Natural History, Chemistry	Willis, Wallis, Boyle, Bathurst, Millington, J. Ward, Mayow	Prof Chem 1683-1690	1677	1677, 1685, 1686	Kent 30 Apr 1696

MINOR SCIENTISTS *(Continued)*

Name	Birth Date	Origins	Oxford Period & College Affiliation	Degrees	First Scientific Interest & Clubs
POPE, Walter	c.1627	Fawsley, Northants Westminster Sch c.1640-1645	1648-1662 Wadham 1648-1662	BA 1649 MA 1651 DM 1661	1651 W4, G
ROOKE, Lawrence	1622	F: Rector Horton, Kent Eton c.1635-1640	1650-c.1657 (Kings, Camb) Wadham	(Camb BA 1644 MA 1647)	Mid 1640s S, P
SCARBURGH, Charles	1616	F: Gentleman London St. Paul's Sch	1643-1646 Merton	(Camb BA 1637 MA 1640)	c.1640 W1
SHARROCK, Robert	1630	F: Rector Aldstock, Bucks Winchester Sch 1643-1649	1648-1665 New Coll 1649-1660	BCL 1654 DCL 1661	c.1657 G
SMITH, John	1630	F: Gentleman Bucks	1647-1659 Brasenose	BA 1651 MA 1653 DM 1659	Mid 1650s B(?)
STAHL, Peter		Strassburg, Germany	1659-1665 1670-c.1674 John Crosse's 88 High St 1659 Arthur Tillyard's 90 High St 1660-1662 Ram Inn, 113-114 High St 1662-1664		
STEPHENS, Phillip	c.1620	F: Bourgeois Devizes, Wilts	1637-1643 1648-1659 New Coll 1649-1653 Hart Hall Principal 1653-1660	BA 1640 (Camb MA 1645) DM 1656	Early 1650s

Scientific Subject Interests	Closest Oxford Scientific Friends	Occupation & Location	Joined Royal Society	Scientific Books	Death
Astronomy	Wilkins (Half-brother), Ward, Boyle, Lower, Millington, Sprat, Hooke	Gresham Prof Astron 1661-1687 London	1661		London 25 June 1714
Astronomy, Mathematics, Chemistry	Ward, Boyle, Wren, Wilkins, Pope, Wallis, Bathurst	Gresham Prof Astron 1652-1657 Geom 1657-1662	1660		London 26 Jun 1662
Anatomy, Mathematics	Harvey, Ward, Wallis, Greaves, Wood, Wren, Wilkins	Physician, London	1661		London 26 Feb 1694
Botany, Chemistry	Boyle, Willis, Stahl, Bobart, John Crosse, Dickenson	Divine, Hants		1660	Winchester 11 Jul 1684
Physiology		Physician, London		1666	London Winter 1679
Chemistry	Boyle, Williamson, Willis, Sharrock, Dickenson & Numerous Students	Chemist, Oxford, London & Provinces			London 1675
Botany, Anatomy	Wm. Browne, Bobart, John Crosse	Physician, Oxford & London	1658		London c.1660

MINOR SCIENTISTS *(Continued)*

Name	Birth Date	Origins	Oxford Period & College Affiliation	Degrees	First Scientific Interest & Clubs
STUBBE, Henry	1632	F: Rector Partney, Lincs Westminster Sch c.1642-1649	1649-1653 1656-1659 Christ Church 1649-1659	BA 1653 MA 1656	1656
TRAPHAM, Thomas, Jr.	c.1638	F: Surgeon Oxford	1654-1663 Magdalen 1654-1660 Magd Hall 1660-c.1663	BA 1658 MA 1661 (Caen MD 1664)	Early 1660s
TYSON, Edward	1650	F: Civil Servant Clevedon, Somerset	1667-1679 Magd Hall	BA 1670 MA 1673 (Camb MD 1680)	Early 1670s
WHISTLER, Daniel	c.1619	F: Gentleman Goring, Oxon Thame Sch	1635-1648 Trinity 1635-1640 Merton 1640-1645	BA 1639 MA 1644 (Leyden MD 1645)	Early 1640s
WOOD, Robert	c.1622	F: Rector Pepper Harrow, Surrey	1640-1656 Merton 1642-1650 Lincoln 1650-1660	BA 1647 MA 1650 ML 1656	1647 B, A

VIRTUOSI

Name	Birth Date	Origins	Oxford Period & College Affiliation	Degrees	First Scientific Interest & Clubs
ALLESTREE, Richard	1619	F: Bourgeois Uppington, Salop Coventry Sch	1637-1648 c.1655-1681 Christ Church 1637-1648 1660-1681	BA 1640 MA 1643 DD 1660	1667
AUBREY, John	1626	F: Gentleman Broad Chalke, Wilts	1642-1643 1646-1648 Trinity		1647
BARKSDALE, Francis	1618	F: Gentlemen Newbury, Berks	1633-c.1643 1648-1656 Magd Hall Magdalen 1648-1653	BA 1636 MA 1639 DM 1649	Mid 1640s B

Scientific Subject Interests	Closest Oxford Scientific Friends	Occupation & Location	Joined Royal Society	Scientific Books	Death
Medicine, Physiology, Chemistry	Lower, Willis, Locke, Boyle, Guidott, Sprigg, Oldenburg, J. Ward	Physician, Jamaica, Warwicks		1662, 1666, 1670(4), 1671(4)	Bath 12 Jul 1676
Medicine	Willis	Physician, Jamaica		1679	? Jamaica 1692
Anatomy	Plot	Physician, London	1679	1680, 1698, 1699	London 1 Aug 1708
Medicine	Bathurst, (Harvey?)	Physician, London Gresham Prof Geom 1648-1657	1661	1645	London 11 May 1684
Mathematics, Medicine	Ward, Wren, Petty, Wallis, Boyle	Physician, Math Teacher, Civil Servant Dublin	1681	1680, 1681	Dublin 9 Apr 1685
Chemistry	Willis, (Boyle, Ward, Wallis, Lower, Parker?) Millington, Needham	Divine, Oxford			Oxford 28 Jan 1681
Mathematics	J. Lydall, Bathurst	Antiquary	1663		Oxford June 1697
Medicine	Joshua Crosse, Stephens	Physician, Oxford & Newbury			

VIRTUOSI *(Continued)*

Name	Birth Date	Origins	Oxford Period & College Affiliation	Degrees	First Scientific Interest & Clubs
BARLOW, Thomas	1607	F: Bourgeois Orton, Westmorland Appleby Sch	1625-1675 Queen's 1625-1675 Provost 1657-1675	BA 1630 MA 1633 DD 1660	Mid 1630s
BATHURST, George	1610	F: Bourgeois Hothorpe, Northants	1626-1645 Trinity 1626-1645	BA 1629 MA 1632 BD 1640	1632
BROOKES, Christopher			1649-1665 Wadham (Manciple) 1649-1665		
CONYERS, William	1622	F: Gentleman Walthamstowe, Essex Merchant Taylors 1631-1639	1639-1660 St. John's 1639-1660	BA 1643 MA 1646 DM 1653	1651 B
COWLEY, Abraham	1618	F: Stationer London Westminster Sch	1644-1646 Lived in Town	(Camb BA 1639 MA 1642) Hon DM 1657	
CREW, Nathaniel	1633	F: Baron Sterne, Notts	1653-1672 Lincoln 1653-1672 Rector 1668-1672	BA 1656 MA 1658 DCL 1664	St
CROSSE, John	c.1620		1620-1690s "Deep Hall" 88 High St	(Non Academic)	
CROSSE, Joshua	1615	F: Bourgeois Newark, Notts	1632-1642 1648-1676 Magd Hall 1632-1642 Magdalen 1648-1660	BA 1634 MA 1637 DCL 1650	Mid 1640s B
FELL, Philip	1633	F: Cleric Oxford Winchester Sch 1645-1652	1652-1670 Trinity All Souls	MA 1660 BD 1667	Mid 1650s

Scientific Subject Interests	Closest Oxford Scientific Friends	Occupation & Location	Joined Royal Society	Scientific Books	Death
	Boyle, Owen	Divine, Bodleian Librarian 1652-1660			Bugden, Hunts 8 Oct 1691
Mathematics, Embryology	Harvey, Bathurst (brother)	Tutor, Oxford			Farringdon, Berks c. Jul 1645
Mathematics	Wilkins	Mathematical Instrument Maker	1651		Oxford 1665
Medicine, Anatomy, Chemistry	Bathurst, Clerke, Willis, J. Ward, Boyle	Physician, Oxford & London			London c. Jul 1665
Botany, Medicine	Harvey	Poet	1661	1661	Chertsey, Surrey 28 Jul 1667
Chemistry		Lawyer & Divine, Oxford & London			Durham 18 Sep 1721
Chemistry	Boyle, Stahl, Hooke, Williamson, Lower, Mayow	Apothecary, Oxford			Alive in 1693
Natural Philosophy	Wallis, Wilkins, Ward, Langbaine, Boyle	Sedleian Prof Nat Phil 1648-1660			Oxford 9 May 1676
Medicine	Willis	Physician & Schoolmaster, Eton			Worcester 1682

VIRTUOSI *(Continued)*

Name	Birth Date	Origins	Oxford Period & College Affiliation	Degrees	First Scientific Interest & Clubs
GLANVILL, Joseph	1636	F: Bourgeois Halwell, Devon	1652-1660 Exeter 1652-1656 Lincoln 1656-1660	BA 1655 MA 1658	Late 1650s
GORGES, Robert	c.1625	Cheddar, Somerset	1647-c.1656 St. John's 1648-c.1656	MA 1648 (Trin Coll Dublin LLD)	1651 B
GREAVES, Edward	1608	F: Rector Colmore, Hants	c.1631-1654 Merton 1631-1634 All Souls 1634-1654	BA 1633 MA 1637 BM 1640 DM 1641	Late 1630s
HAMILTON, William	c.1625	Scotland	1648-1651 All Souls 1648-1651	(Glasgow MA)	Late 1640s
HOLDER, William	1616	F: Cleric Southall, Notts	Bletchingdon nr Oxford 1646-1662	(Camb MA 1640) DD 1660	Mid 1640s W1
JEANES, Thomas	c.1627		1652-1663 Magdalen 1652-1663	(Camb BA 1650) MA 1652 BM 1655 DM 1659	St
LADYMAN, Samuel	1626	F: Bourgeois Dinton, Bucks	1643-c.1655 Corpus Christi	BA 1647 MA 1649	1651 B
LAMPHIRE, John	1614	F: Gentleman Winchester, Hants Winchester Sch 1627-1634	1634-1688 New Coll 1634-1648 Hart Hall Principal 1663-1688	BA 1638 MA 1642 DM 1660	Mid 1650s W3

Scientific Subject Interests	Closest Oxford Scientific Friends	Occupation & Location	Joined Royal Society	Scientific Books	Death
Scientific Controversy		Divine, Bath	1664	1661, 1665, 1668, 1671(3) 1676	Bath 9 Nov 1680
	Petty, Wood	Civil Servant, Ireland			
Medicine	Charleton, Guidott, (Harvey?)	Physician, Oxford, Bath & London		1643, 1667	London 11 Nov 1680
Chemistry		Librarian, London			Alive in 1661
Mathematics, Music, Linguistics, Astronomy	Wren (Brother-in-law), Wallis, Wilkins, Hooke	Cleric, Bletchingdon & London	1661	1669, 1670, 1694(2)	Hertford 24 Jan 1698
Medicine, Chemistry		Physician Oxford & Northants			Peterborough, Northants Nov 1668
		Cleric			Limerick, Ireland 1684
Medicine	Tillyard, Bathurst, Masters, Willis, Pett, Millington, Wren, Castle	Cleric & Physician, Oxford			Oxford 30 Mar 1688

VIRTUOSI *(Continued)*

Name	Birth Date	Origins	Oxford Period & College Affiliation	Degrees	First Scientific Interest & Clubs
LANGBAINE, Gerard	1609	F: Bourgeois Barton, Westmorland	1628-1658 Queen's 1628-1658 Provost 1646-1658	BA 1630 MA 1633 DD 1646	Early 1640s
LEVINZ, William	1625	F: Gentleman Evenley, Northants Merchant Taylors Sch	1641-1698 St. John's 1641-1698 President 1673-1698	BA 1645 MA 1649 DM 1666	Early 1660s St
LYDALL, John	1623	F: Gentleman Ipsden, Oxon	1640-1657 Trinity 1640-1657	BA 1644 MA 1647	1648 B
LYDALL, Richard	1621	F: Gentleman Uxmore, Oxon	1638-1704 Oriel 1638-1641 Merton 1641-1704 Warden 1693-1704	BA 1641 MA 1647 BM 1656 DM 1657	1648 B
MASTERS, John	1637	F: Knight Cirencester, Glos	1654-1672 Christ Church 1654-1672	BA 1657 MA 1659 DM 1672	Mid 1660s
OWEN, John	1616	F: Cleric Stadham, Oxon Sylvester's Sch c.1626-1631	1631-1637 1651-1660 Queen's 1631-1637 Christ Church Dean 1651-1660	BA 1632 MA 1635 DD 1653	1630s B
PARKER, Samuel	1640	F: Lawyer Northampton	1656-1667 Wadham 1656-1660 Trinity 1660-1667	BA 1660 MA 1663	Early 1660s

Scientific Subject Interests	Closest Oxford Scientific Friends	Occupation & Location	Joined Royal Society	Scientific Books	Death
Mathematics	Southwell, Wilkins, Williamson, Bathurst, Rawlinson	Divine, Oxford			Oxford 10 Feb 1658
Chemistry	(Willis, Lower, Needham, Boyle?), Stahl	Physician & Cleric, Oxford			Oxford 3 Mar 1698
Mathematics, Anatomy, Botany, Chemistry	Bathurst, Highmore, Willis, Aubrey, Wren, Petty, Wilkins	Tutor, Oxford			Oxford 12 Oct 1657
Medicine	Willis	Physician, Oxford			Oxford 5 Mar 1704
Anatomy	Willis, Lower	Physician, Oxford & London			
Mathematics	Willis, Langbaine, Wilkins	Divine, Oxford & London			Ealing, Middlesex 24 Aug 1683
Natural Philosophy, Mathematics	Bathurst, Willis, Hooke	Divine, London & Oxford	1666		Oxford 20 Mar 1688

VIRTUOSI *(Continued)*

Name	Birth Date	Origins	Oxford Period & College Affiliation	Degrees	First Scientific Interest & Clubs
PETT, Peter	1630	F: Shipbuilder Deptford, Kent St. Pauls Sch	1647-1659 Pembroke 1647-1648 All Souls 1649-1660	(Camb BA 1648) BCL 1650	Mid 1650s W3, W4
PIERREPONT, Henry (Marquis of Dorchester)	1606	F: Earl	1642-1647 Lived in Town	MA 1642	
QUARTERMAINE, William	1618	F: Gentleman Shabbington, Bucks	1634-1657 Magd Hall 1635-? Pembroke 1657	BA 1635 MA 1638 DM 1657	Early 1650s
RAWLINSON (RALLINGSON), Richard	1618	F: Bourgeois Milnethorp, Westmorland	1636-1661 Queen's 1636-1661	BA 1641 MA 1643 BD 1657 DD 1661	1642 B
SOUTHWELL, Robert	1635	F: Gentleman Kinsale, Ireland	1653-1659 Queen's 1653-1659	BA 1655 DCL 1677	Late 1650s
SPRAT, Thomas	1635	F: Cleric Tallaton, Devon	1651-1663 Wadham 1652-1670	BA 1655 MA 1657 DD 1669	Mid 1650s
SPRIGG, William		F: Gentleman Oxford	1652-1660 Lincoln 1652-1660	BA 1652 MA 1655	Mid 1650s
THOMAS, David	c.1634	F: Gentleman Preshute, Wilts Winchester Sch 1646-1651	1651-1666 New Coll 1651-1670	BA 1655 MA 1658 ML 1666 DM 1670	Early 1660s
TILLYARD, Arthur	1615	F: Baker Oxford	1615-1693 90 High St.	(Non Academic)	Early 1650s W3

Scientific Subject Interests	Closest Oxford Scientific Friends	Occupation & Location	Joined Royal Society	Scientific Books	Death
	Millington, Barlow, Wren, Castle, Boyle	Civil Servant, London & Dublin	1661		1 Apr 1699
Medicine, Botany, Chemistry	Charleton, Scarburgh, Harvey, Rooke, (T. Clarke?)	Aristocrat	1663		London 1 Dec 1680
Chemistry	Stahl, Williamson, Clerke, T. Clarke	Physician, Navy & London	1661		London June 1667
Mathematics, Astronomy, Cartography	Wren, Wallis, Langbaine, Williamson, Whistler	Cleric Oxford London & Sussex	1660?		Pulborough, Sussex Autumn 1668
Mathematics, Astronomy	Petty, Langbaine, Williamson	Civil Servant, London & Abroad	1662		King's Weston, Glos 11 Sep 1702
Scientific Controversy	Wilkins, Wren, Pope, Ward, Bathurst	Divine, London	1663	1667	Bromley, Kent 20 May 1713
Natural Philosophy	Stubbe	Schoolmaster & Barrister, Ireland		1657	Alive in 1695
Chemistry, Anatomy	Locke, Boyle, Glanvill	Physician, Salisbury			Salisbury 1694
Chemistry	Boyle, Stahl, Pett, Millington, Castle, Wren, M. Wren	Apothecary & Coffee House Proprietor, Oxford			Oxford 14 Dec 1693

VIRTUOSI *(Continued)*

Name	Birth Date	Origins	Oxford Period & College Affiliation	Degrees	First Scientific Interest & Clubs
TREVOR, Richard	c.1628	F: Knight Bucks	1648-1655 Merton 1648-1660	BA 1648 MA 1651 (Padua MD 1658)	1651 B
VERNON, Francis	c.1637	F: Bourgeois London Westminster Sch c.1647-1654	1654-1668 Christ Church 1654-1677	BA 1658 MA 1660	1658
WALKER, Obadiah	1616	F: Bourgeois Worsborodale, Yorks	1633-1648 1660-1661 1665-1688 University Master 1676-1688	BA 1635 MA 1638	Mid 1640s
WARD, John	1629	F: Vicar Spratton, Northants	1646-1661 Magd Hall Christ Church 1649-1660	BA 1649 MA 1652	1648 B
WILLIAMSON, Joseph	1633	F: Vicar Bridekirk, Cumberland Westminster Sch c.1646-1650	1650-1655 1658-1660 Queen's 1650-1678	BA 1654 MA 1657 DCL 1674	1651 St, G
WOODROFFE, Benjamin	1638	F: Cleric Oxford Westminster Sch c.1650-1666	1656-1672 1692-1711 Christ Church 1656-1672 Glos Hall Principal 1692-1711	BA 1659 MA 1662 DD 1673	St
WREN, Matthew	1629	F: Divine Ely	1651-1657 with Francis Bowman in "Bulkeley Hall" 106 High St	MA 1661	W3, W4, S, P
YERBURY, Henry	1628	F: Gentleman Trobridge, Wilts	1642-1648 1660-1686 Magdalen	BA 1646 (Padua MD 1654)	St

Scientific Subject Interests	Closest Oxford Scientific Friends	Occupation & Location	Joined Royal Society	Scientific Books	Death
Medicine		Physician, London			London 17 Jul 1676
Mathematics, Scientific Correspondence	Hooke, Oldenburg, Lower	Civil Servant, France & Levant	1672		Ispahan, Persia Spring, 1677
Mathematics, Optics	Wood	Divine & Tutor, Oxford & Surrey		1679, 1680	London 31 Jan 1699
Medicine, Anatomy, Chemistry	Conyers, Lower, Willis, Mayow, Stahl, Bobart, Lovell, Needham, Hodges	Cleric, Stratford-on-Avon			Stratford-on-Avon 12 Sep 1681
Astronomy, Chemistry, Mathematics	Langbaine, Stahl, Boyle, Sharrock, Locke, Rawlinson, Southwell, Thomas	Civil Servant, London	1662		London 3 Oct 1701
Medicine, Chemistry	Aubrey, Ward, Boyle, Stahl	Divine, Oxford & London	1668		London 14 Aug 1711
Natural Philosophy	Wren, Wilkins, Sprat, Pope, Boyle, Pett, Willis	Civil Servant, London	1660		14 Jun 1672
Medicine	Stahl	Physician & Tutor, Oxford			Oxford 25 Mar 1686

OXONIANS AND NEW APPROACHES TO PHYSIOLOGY 4

T he redefinition of such physiological problems as heat, respiration, and the nature of the blood took place largely because the post-Harveian generation, especially at Oxford, rejected the Aristotelian natural philosophy that had been so dear to Harvey, and substituted for it a corpuscular and chemical concept of biological structure and function. Chemistry had been an adjunct to medical explanation, or at least one stream of it, since the innovations of Paracelsus and his followers in the sixteenth century.[1] Corpuscular and atomistic concepts, while as philosophical positions went back to antiquity,[2] were in their application to science much newer on the scene.[3] Such notions had been imported largely from across the English Channel.

It goes far beyond our present purposes to explore the origins and general content of the mechanical and corpuscular philosophies of René Descartes, Pierre Gassendi, Marin Mersenne, and others as they emerged in France and Holland in the 1630s and 1640s. A few temporal reference points must suffice. Descartes's particulate matter theory, although adumbrated in the essays appended to his *Discours de la méthode* (1637) and *Meditationes de prima philosophia* (1642), emerged in full clarity only in his supremely influential *Principiae philosophiae* (1644). His mechanistic explanations applied to animal and human bodies, *De l'homme*, appeared posthumously in 1664, but many of his physiological ideas had long since been expounded by his Dutch disciples Henrik de Roy (Henricus Regius) in his *Fundamenta physices* (1646) and *Fundamenta medica* (1647), and by Cornelis Hooghelande in his *Cogitationes* (1646).[4] Mersenne's brand of mechanism was most clearly delineated in *Cogitata physico-mathematica* (1644).[5] Gassendi was known largely as an astronomer and critic of both Aristotle and Descartes until his massive three-volume *Animadversiones in decimum librum Diogenis Laërtii* (1649) established him as the leading spokesman for, and interpreter of, Epicurean atomism as applied to science. The substance of the *Animadversiones* was incorporated into Gassendi's masterpiece, *Syntagma philosophicum*, published posthumously in the *Opera omnia* (1658).[6]

Gassendi's corpuscular ideas are particularly interesting, both in themselves and because of their later usefulness in underpinning changing concepts in biology and chemistry. His system was built upon the traditional two pillars of atomism, matter in motion and the void. All the matter in the universe was organized as atoms that had, Gassendi said, four general properties. The first was similarity of substance: all atoms were of one and the same nature. The second was magnitude; atoms might differ in the quantity of matter that various classes of them possessed. Figure was their third property, defined in terms of three real dimensions. Their fourth, and final, general property was weight, or gravity (*pondus*), which was the principle of their motion.[7] Gassendi then went on, in page after double-columned folio page, to develop the implications of these four properties.[8] Especially noteworthy was his claim, put forward in discussing the two properties of magnitude and figure, that microscopy demonstrated atomism. Minute animals, viewed through lenses or microscopes, could be seen to have the full complement of vital organs, a fact that was only possible because the atomic composition of aggregates allowed such small units to be functional.[9]

The origins of what the Aristotelians called qualities were, Gassendi said, equally to be understood in terms of the particulate nature of matter. Atoms of various size and figure might produce all the perceived qualities when sufficient numbers of them were agglomerated in the proper way to cause motion within the sense organs. Qualities that were not the simple result of the sizes and figures of homogeneous masses of atoms, might be produced by atoms of one type, joining with atoms of another, to form concretions whose sensible qualities would differ from those of the constituent components. Such concretions, these second-order producers of qualities, Gassendi called *moleculae*.[10]

In subsequent pages Gassendi developed in great detail how his conceptual scheme of atoms and molecules, of particles and corpuscles, might be used to account for all the common characteristics of physical description. Light, color, sound, odors, rarity, density, perspicuity, opacity, subtility, hardness, smoothness, fluidity, humidity and ductility—all could be transposed out of the Aristotelian categories of essences and substantial forms, into the matter-and-motion categories of the atomistic conceptual scheme.[11] Even heat and cold could be accounted for on a particulate basis. A body was heated by fire because it interacted with small, light, round, fast "calorific atoms." A body was cold because it possessed heavy, slow "frigorific atoms."[12] For every set of observed physical properties or changes, there was, Gassendi posited, an explanation that could be framed in terms of the properties of the discrete units of matter, atoms, and/or the properties of the corpuscles that those atoms combined to make up, the *moleculae*.

Although Descartes's philosophy differed radically from Gassendi's on issues of epistemology and religion, the two nearly converged on matter theory. Descartes too saw corpuscles as composed of a single *prima materia*, although he denied that they were true indivisible atoms. These corpuscles were of three classes: the largest particles of gross matter in various shapes, the intermediate "subtle matter" that moved the grosser particles, and extremely fine irregular particles that filled in all the spaces left by the two larger classes of particles. Descartes believed that the universe was a plenum, and that the void of the atomists was merely an illusion. Action took place, not by atoms or *moleculae* moving in a vacuum, but always by contact, the vortices of subtler particles moving the grosser and more sensible ones.

The ideas of the French mechanists were enthusiastically received by Englishmen, especially by those who had contacts with the circle of English emigrés in Paris. Petty was an especially influential intermediary, since he had both studied with Hooghelande in Holland and been part of Hobbes's coterie at Paris during the few years, 1645–1648, when Gassendi was in the capital. As he expressed himself in a letter to Henry More in 1649, Petty thought highly of Descartes as a philosopher who "hath indeed made use of sensible principles such as are Matter, Local Motion, Magnitude, figure, situation, &c." to solve ingeniously "the phaenomana of Nature." Yet, because the Cartesian system was not solidly grounded on experiments, it, like those of Aristotle, Galen, Paracelsus, and Helmont alike, was largely "witt & phancy" compared to the real knowledge gained by men "that are daily conversant in the works of nature."[13] Petty voiced no such objections to the speculations of Gassendi and was kept well informed of their progress. In 1648 Charles Cavendish wrote to Petty from Paris, thanked him for his "new discoveries in Anatomie," which had been passed on to Hobbes, and gave this news of Gassendi:

> Your worthy Friend and myne Mr. Gassendi is reasonable well and has Printed a book of ye life and Manners of Epicurus since your going from hence as I thinke. He hath now in ye Presse at Lyons, ye philosophie of Epicurus in wch I beleeve wee shall have much of his owne philosophy wch doubtlesse will be an exellēt worke.[14]

Cavendish then went on to describe a variant on the strongest tangible evidence for the existence of atoms and the void—the Torricellian experiment, with its vacuum at the top of what we now call a mercury barometer.

Boyle, who had returned in 1644 from a five-year tour of the Continent, was similarly well apprised of French developments. In 1647 he wrote Hartlib from his country seat at Stalbridge, Dorsetshire, of his reading in "a late mechanical treatise of the excellent *Mersennus*." He praised Hartlib for resuming his correspondence with the French savant and commended the astronomical observations of *Gassendus*, "a great favorite of mine."[15] A year later Hartlib passed on to Boyle an extract from Cavendish's letter to Petty telling of Gassendi's forthcoming *Animadversiones*, including details of the modified Torricellian experiment.[16] It seems clear both from Boyle's printed works and from his surviving manuscripts that he followed up these contacts by reading the *Animadversiones* almost immediately after it came out in 1649.

Highmore was similarly well placed to learn of Gassendi's work. He was a kinsman of William Petty.[17] Boyle, when he settled at Stalbridge in early 1647, was located only five miles from Sherborne, where Highmore had his practice. The two had become acquainted, probably before 1648. It is therefore not surprising that, three years later, Highmore's published work showed that he also was intimately acquainted with Gassendian atomism.

The pattern of rapid dissemination of French mechanism, and especially of continental atomism, was repeated again and again. John Wallis had both the *Animadversiones* (1649) and the *Syntagma* (1658).[18] Walter Charleton, who may well have visited Paris in 1651–1652,[19] soon sprinkled a theological tract with numerous laudatory references to Mersenne, Descartes, and to the atomistic ideas of Gassendi as expressed in the *Animadversiones*.[20] Within another two years Charleton had produced his massive *Physiologia Epicuro-Gassendo-Charltoniana* (1654), an augmented translation of the physical sections of the *Animadversiones*.[21] Even Thomas Browne, in Norwich, had read Gassendi's atomistic work within a year of its publication.[22]

A CORPUSCULAR FOUNDATION FOR BIOLOGICAL AND CHEMICAL PHENOMENA

Particulate philosophies, and Gassendian atomism especially, were important because they provided the generation coming to scientific maturity in the late 1640s and 1650s with an alternative way to explain the myriad new facts adduced in the chemical and biological sciences. One may see this, for example, in the early scientific career of Robert Boyle.

By late 1646, although only nineteen years old, Boyle had been drawn into the London circle of Hartlib's friends interested in mechanics, chemistry, and husbandry, a group he dubbed "our *invisible college*." Boyle carried these interests, especially chemistry, with him when he returned to Stalbridge in February 1647. After some mishaps, such as a furnace broken in transit, he set up a laboratory there and continued experiments. Chemical medicaments seem to have been of particular interest,[23] for he composed, and sent on to Hartlib, a lengthy essay pleading with physicians to share their receipts freely.[24] Boyle also learned anatomy, most probably with Highmore's help. Essays written in 1648–1649 refer to the circulation of the blood, to Whistler's theory of rickets as a nutritional disorder, and in general to how Boyle's "Curiosity for Dissections" had shown him the "many Bones, and Muscles, and Veins, and Arteries, and Gristles, and Ligaments, and Nerves, and Membranes, and Juices, a humane Body is made up of.''[25] Reading and microscopy were joined to chemistry and anatomy to lead Boyle to marvel at God's handiwork. When he surveyed the heavens,

> and when with excellent Microscopes I discern in otherwise invisible Objects the unimitable Subtlety of Nature's Curious Workmanship; And when, in a word, by the help of Anatomicall Knives, and the light of Chymicall Furnaces, I study the Book of Nature, and consult the Glosses of Aristotle, Epicurus, Paracelsus, Harvey, Helmont, and other learn'd Expositors of that instructive Volume; I find my self oftentimes reduc'd to exclaim with the Psalmist, *How manifold are thy works, O Lord! in wisdom hast thou made them all.*[26]

Boyle was perfectly prepared, therefore, to see the chemical and biological potential in Gassendian atomism, even though atomistic speculations had traditionally been associated with scepticism and atheism. By January 1650 he had written a significant and revealing essay "Of Atoms,"[27] which seems to correspond to one entitled "Essay Of ye Atomicall Philosophy," now among the Boyle Papers at the Royal Society.[28] Interestingly, in later life Boyle was still aware enough of the heterodox connotations of atomism to write across the top of the manuscript "These Papers are without fayle to be burn't.''[29]

Boyle began his "Essay" by praising the efforts of "the learned pens of Gassendus, Magenus, Des Cartes & his disciples our deservedly famous Countryman Sir Kenelme Digby & many other writers" to resurrect the atomism of Democritus, Leucippus, and Epicurus, after it had been long submerged by the "Peripateticke Philosophy." Contrary to Aristotelian objections, atoms did have real existence. Though they might be further divisible in imagination (thus placing no limitation on God's power), they were not so in reality. They were, Boyle said, the natural stopping points in the resolution of real bodies.[30]

Evidence from both chemistry and biology, Boyle believed, supported the atomistic hypothesis. First of all, atomism accounted for the properties of homogeneous substances. How except by being divisible into atoms could silver be dissolved in aqua fortis (nitric acid), and then be recovered again? Secondly, the posited smallness of atoms was no impediment to their credibility. Light passed through a solution of a salt showed no atoms, even though they must be there. The same was true of scents, he said—we cannot see them, but matter must be there.[31] More convincing still was, Boyle believed, the argument from organic forms. Aristotle, in his *Historia animalium*, had mentioned a small white insect as the smallest of living things. Although this "Mite" would seem to be "a moving Atome," Boyle had "several times both discovered by selfe and showne to others," through an "excellent Microscope," not only "the severall limbs of this little Animall but the very haire growing upon his legs." Let us but consider, he said, "what a multitude of Atomes must concurre to constitute the severall parts externall and internall necessary to make out this little Engine."

> How many must goe to the contexture of the eyes & other internall senses how many to ye snout which it has like a hog & the severall parts of it, how many to the

stomacke & the guts & the other inward parts addicted to ye concoction of Aliment & exclusion of Excrements, & to be short how unimaginably little must be the parts that make the haire upon the legs, & how much more subtle must be the animal spirits that run to & fro in nerves suitable to such little legs?[32]

Here is the argument encountered in Gassendi only in passing: microfunction argues microstructure. Boyle obviously thought it telling, for in another essay written c. 1649–1651 he again used the microscopic mite to argue "how strangely skilful and delicate a Workmanship" was needed to make the parts of such an animal so small and yet functional.[33]

Boyle's third point in his "Essay," one developed at great length, was that only by assuming the existence of effluvia, and their atomic character, could one explain a vast range of natural phenomena. Atomic effluvia accounted for the lingering nature of animal scents, as well as for the insensible loss of weight of human bodies that Sanctorius had observed in his statical experiments. They explained the scented nature of plants and solids, the attraction of electrified bodies, the healing efficacy of certain stones, and the magnetizing and magnetic qualities of the lodestone.[34] Boyle even confided to Hartlib in 1651 that it was "not despaired" that the "effluvia of things could bee made discernible by microscopia."[35]

Boyle's appreciation of the contrivance of biological structure, both gross and microscopic, only increased as he improved himself in anatomy during the early 1650s. He met the aged Harvey and had several discussions with him. He met Willis, Joyliffe, and Timothy Clarke—the former introduced to him at Oxford by Wilkins, and the latter two through their practice in London. Boyle wrote almost rhapsodically of the many hours he spent "conversing with dead and stinking Carkases" which, far from being a hated employment, delighted him more than time spent at court or in libraries, so great was his joy "in tracing in those forsaken Mansions, the inimitable Workmanship of the Omniscient Architect." He could "with more delight look upon a skilful Dissection, than the famous Clock at Strasburg."[36] Although Boyle's wealth enabled him to hire assistants to perform experiments, he had not "been so nice as to decline dissecting *Dogs, Wolves, Fishes,* and even *Rats* and *Mice,*" with his own hands.[37] In these dissections he had not only the help of, as Petty said, "My cousin *Highmore's* curious hand,"[38] but of Petty himself during the time from late 1652 to mid-1654, when Boyle was in Ireland. There, where he lacked "glasses and furnaces to make a chemical analysis of inanimate bodies," Boyle exercised himself "in making anatomical dissections of living animals." With Petty's assistance, Boyle satisfied himself about the circulation of the blood, dissected out the venae lactae and the newly discovered receptaculum chyli, and saw, especially in the dissection of fishes, "more of the variety and contrivances of nature, and the majesty and wisdom of her author, than all the books I ever read in my life."[39]

Many anatomical structures and physiological mechanisms, Boyle found, showed the admirable design that the microscope had revealed in the smallest mite: the spinning of a silkworm; the exact image cast upon the retina by the vertebrate eye; the perfect mechanical powers of the muscles, transmitted through the tendons; even the way in which fetal vessels carefully brought the circulating blood past the motionless lungs, "So careful is Nature not to do things in vain."[40] The development of the chick embryo, as observed by "our accurate and justly Famous Anatomist" Highmore and "that great Promoter of Anatomical Knowledge," Harvey, showed the exquisite purpose of nature in the order and fashioning of the parts.[41] Contemplation of the valves in the heart and veins, Harvey told Boyle, had led him to the discovery of the circulation.[42] Even the accidents of dissection could lead to the unveiling of contrivance. Had not Joyliffe himself told Boyle that he had discovered the *vasa lymphatica* "by an accident."?[43]

As Boyle learned more anatomy, he also considered more deeply the implications of the atomical philosophy. Aristotelianism was clearly unsatisfactory; it explained phenomena in such unintelligible terms as sympathy, antipathy, and occult qualities.[44] Even using its own concept of nature "abhoring a vacuum," Aristotelian natural philosophy could not explain the Torricellian experiment, one Boyle had done with his own hands.[45] In the organic realm there was "a multitude of Problems, especially such as belong to the use of the Parts of a human Body," for which one simply could not say that Nature did "such and such a thing, because it was fit for her so to do."[46] Rather, Boyle asserted, it made more sense to conceive that God created matter in "an innumerable multitude of very variously figur'd Corpuscles." God then connected these particles into bodies, put them into such motions and governed them by such laws, that, although remaining unconscious, these creatures would act as if they had "a Design of Self-preservation." Animals were like the famous Strassburg clock—their seemingly purposive behavior was due to the cunning of the artificer, who had carefully framed all their parts to perform the end for which the whole was contrived.[47] The marvelous activity of silkworms, or spiders, or nest-building birds was not due to souls, but to heavenly design.[48]

Boyle could—and did—reject many aspects of ancient atomism that he viewed as unproved: that matter was eternal and from eternity had been divided into atoms; that the number of atoms was infinite; that they had an infinite space in which to move; that they were endowed with an infinite variety of determinate figures; that they had been from eternity their own movers; and, most of all, that a fortuitous concourse of atoms had produced the structure of the world.[49] Boyle believed strongly that inorganic substances exhibiting structure, such as crystals, had not been produced by chance. Animals, and their parts, were even less likely to have been so created. Who could imagine the eye, or the valves in the veins, resulting from an Epicurean concourse of atoms? All of nature, although atomical in structure and mechanical in process, needed an artificer to create the parts in the proper form and to put them into motion in the proper way.[50] "The greatest part of the Favorers of the Atomical Philosophy," Boyle asserted, "do not want to do without God."[51] Boyle, in large part because he approached Gassendian atomism as a philosophy that had to account for both living and nonliving matter, could shear it of its atheistical nonessentials and emphasize its congruence with a providential world.

NATHANIEL HIGHMORE: HARVEIAN AND ATOMIST

In 1651 Nathaniel Highmore published two books uniquely reflective of the rapidly changing nature of biological explanations. The first, *Corporis humani disquisitio anatomica*, was a thick folio textbook designed to recast anatomy and physiology into a form congruent to the circulation of the blood. It had been under Highmore's hand for some years and was dedicated, appropriately enough, to William Harvey. The second, *The history of generation*, was a thinner octavo that attempted to interpret embryology, nutrition, and action at a distance in atomistic terms. It was fulsomely dedicated to Highmore's Dorsetshire friend and neighbor, Robert Boyle, and almost certainly had been written within the previous few years. The two works show, not mutual contradiction, but rather how a man searching for new physiological explanations could find in the corpuscular philosophy a powerful new conceptual framework.

As Highmore noted in his preface to Harvey, the *Disquisitio* was anatomy brought forth clothed in the circulation, the doctrine to which Harvey had devoted his life. Harvey had himself seen some of these "lucubrations" eight years before at Oxford and had urged Highmore to publish. The raging torrent of public disturbances had seemed to condemn this work to "perpetual shadow," but now, with the return of peace and the cessation of arms, Highmore was emboldened to bring forward, under Harvey's protection, this, his small token of gratitude.[52]

Although the *Disquisitio* assumed the circulation as its base, it was not until Highmore had disposed of the descriptive anatomy of the viscera, genitals, and thorax that he could get down to arguing it in earnest. Not unexpectedly, he began his chapter on the use and action of the heart with Harvey's metaphor: the heart was the king of the human realm; it was to the microcosm as the sun was to the macrocosm. It was the seat of the heat that sustained life.[53] After praising the importance of the heart in detail, he quite properly proceeded to the central anatomical evidence for the circulation—Harvey's analysis of the motion of the heart. He recapitulated how Harvey showed the erring ancients that the active phase of cardiac motion was *systole*, not diastole. This contractive motion was, in turn, the cause of the passive diastolic motion of the arteries.[54] Highmore then went on to define, as Harvey never did explicitly, the origin and function of the circulation. According to Highmore, circulation was the violent motion of the blood, caused by the contraction of the heart, by which blood was impelled impetuously into the arteries, and which not only reliquified the blood grown cold in the periphery, but also carried spirits and food to every particle of the body.[55] Although these functions were implicit in Harvey's work, he had perhaps rightly never been quite willing to hazard a precise definition.

Having launched into controversy, Highmore spent the next dozen pages defending Harvey from the strictures of James Primrose, the most vociferous of his English critics. Only then did Highmore get down to codifying his own reasons for the circulation. His mode of argument is interesting. Leaving it to his reader to peruse Harvey's experimental demonstrations, he concentrated on, as his chapter title indicated, arguments "De sanguinis circulationis necessitate." First of all, the circulation preserved the blood from coagulation. This could be seen easily from the fact that extravasated blood coagulated rapidly, not because it was outside the veins, but because it no longer moved. Second, the blood's movement preserved it from putrefaction, which Highmore defined as a union of the particles of the blood, each with its own kind. Blood, he said, consisted of many different kinds of such particles and was thus very liable to putrefy. Third, taking a page from the penultimate chapter of *De motu cordis*, Highmore argued that the necessity of the circulation was proved by the rapid diffusion of poisons and medicaments throughout the body, even when they were applied only at one place. Fourth, as Harvey also noted, the circulation was necessary to dispense nutrition to all parts of the body. This too, Highmore elaborated, was based on the particulate nature of the blood; many different kinds of particles are circulated in it, and each part of the body has an affinity for its own proper cognate particles.[56]

Yet even in attempting to systematize some of Harvey's conclusions, Highmore was at the same time subtly shifting the grounds of discussion. Whereas Harvey had been rather ambiguous about the functions of the circulation and in *De generatione* had emphasized the unitary and vital nature of the blood, Highmore plumped strongly for a particulate interpretation of the blood, a blood that carried heat and nutrition and whose very particulate constitution necessitated its constant motion. Although Harvey might have applauded Highmore's conclusion, he probably regarded such a concept of the blood as erroneous and would have viewed with dismay his disciple's attempt to render an analytical definition of the vital fluid in terms of heterogeneous particles.

The same kind of subtle shift took place in Highmore's discussion of the pulmonary transit. After recapitulating the structure and function of the pulmonary artery and veins, Highmore argued that their older names, *vena arteriosa* (arterylike vein) and *arteria venosa* (veinlike artery), had only led to confusion. In line with the concept of the circulation, he proposed the terms *arteria pulmonum* and *vena pulmonum*—the first time, to my knowledge, that these terms were used. After using Harvey's anatomical and vivisectional arguments to establish the pulmonary circulation, Highmore went on to consider in detail the putative functions of the pulmonary vein. Men like Galen and Fernel thought it served to carry air to the heart, where it was elaborated into vital spirits by virtue of the innate heat of the heart. But, he pointed out with obvious satisfaction, the ancients had been very ambiguous about whether the actual substance of the air entered the heart, or only its qualities. Aristotle implied substance in one place and qualities in another. Galen in some places implied only that the frigidity of the air entered and in others, that its substance did.[57]

But all of this equivocation, Highmore announced triumphantly, could be contradicted by both reason and experiment. First of all, the substance of the air could not be carried to the heart because there were no communicating vessels between the two. Moreover, this could be proved by experiment. If an animal were vivisected, and the lungs inflated by bellows ligatured to the trachea, it could be kept alive for the space of several hours. Yet, in all of this, it could clearly be seen that air was in no way transferred to the heart. The pulmonary vessels contained blood, as could easily be demonstrated by sectioning them.[58]

Highmore's experiment, in its technique, was not new. Vesalius had recommended this open-thorax technique as a way of keeping an experimental animal alive while the investigator examined the motion of the heart.[59] Harvey had used inflation of the lungs, although seemingly not in a living animal, to prove the same point: that air did not pass from the lungs into the pulmonary veins and left ventricle.[60] Highmore, in using Vesalius's technique, was using it for Harvey's purpose—to disprove the generation of vital spirits from the air. As a technique of examining the relationship between air and the organism, it was to recur again and again in the Oxford tradition to which Harvey's work gave rise.

Not satisfied with just one argument, Highmore went on to give other reasons—both Harvey's and his own—for rejecting the generation of vital spirits as a function of respiration. If the heart attracted the air, it had to do so either in systole or in diastole. It could not do so in systole, and in order to attract air in diastole, the movement of the heart and lungs would have to be synchronized, which they patently were not. Moreover, if air were necessary for the generation of vital spirits, how could fish or the fetus *in utero* obtain the necessary air? Finally, one knew that the spirit of the blood was an oily material, with no resemblance whatsoever to air. In this last argument Highmore clearly accepted the chemists' conclusion that distillation of blood gave a proper "anatomy," or analysis of its composition. For all these reasons, he concluded, air did not go to the heart and did not give rise there to the vital spirits.[61]

Having rejected the generation of vital spirits, Highmore then went on to accept the other classical respiratory functions—but in each case with a distinct corpuscular twist. He seems to have believed that heat was innate in the heart and that this was communicated to the blood. Respiration served to expel the excess heat particles from the blood which might otherwise accumulate and break the bonds that held the blood's moisture particles together. This accounted for the difficulty of breathing at high altitudes and for suffocation in closed spaces; in both instances there was not sufficient play for the heat particles to escape. Fuliginous waste vapors from the blood were similarly particulate and were also blown off in respiration. As in a laboratory fire, if the particles of fumes were not carried off, they impeded the union of fire particles that constituted the flame-generating process.[62] Respiration had yet another function. The mechanical motion of the lungs served to aid the transit of the blood from the right to the left ventricle. This could be proven by the pulmonary inflation experiment; when the lungs collapsed, the blood did not pass through them as freely.[63]

Highmore's physiology was true to the intentions expressed to Harvey in the preface, although more in his conclusions than in his mode of reasoning. Both had come to reject, in their books of 1651, the notion that air entered the body to generate vital spirits; but they had done so on differing assumptions. Both held to innate heat, but Highmore lodged it in the heart, while Harvey transferred it to the blood. Highmore elaborated at length Harvey's ideas of the circulation distributing heat and nutrition, but did so according to corpuscular notions that Harvey was at the same time explicitly rejecting. Highmore's ideas clearly arose from Harvey's and from their Oxford sojourn together. The authors cited by Highmore in the *Disquisitio* were those abstracted in his Oxford notebooks. He even recounted in one passage a post-mortem performed at Oxford by him and his "most esteemed friend," probably either Thomas Willis or Ralph Bathurst.[64] Yet, in taking up Harvey's work, and in elaborating more fully Harvey's hints about the blood, circulation, and respiration, Highmore had started to change the framework within which physiological problems were discussed. His chemical references were passing. His particulate matter-theory was rather ill-defined and something less than thoroughgoing. But it was a beginning.

In the *History of generation* Highmore carried corpuscularianism into explicit atomism. There were, he said, certain "universal and general Laws of Nature" by which phenomena were to be interpreted:

> First, it is absolutely true, and an unquestionable law of Nature, (if, prejudice laid aside, right reason takes her place) that all actions and motions are performed by Atomes, or small bodies, moving after a different manner, proportionable to their severall figures; and not by I know not what qualities, (which have only a notional subsistence,) acting without the bodies to which they belong, and leaping from one subject to another, without changing their forms.[65]

Other "Laws" were simply a selective restatement of the principles of Gassendian atomism: all bodies emitted effluvia composed of these atoms; atoms were not of one figure or magnitude, a condition that affected their behavior and perceptibility; these atoms desired rest, which gave rise to unions; preferential places of rest were determined by the cognate figures of atoms; and finally, these atoms followed a random motion until they found their cognate pores.[66] Thus, effects that seemed to be worked at a distance, such as healing by sympathy, were due to the action of atomic effluvia.[67]

Highmore used these laws to refute the vaguely mechanistic embryological theories of Kenelm Digby, the knight-errant Royalist and friend of Harvey. In 1644 Digby had defended Harvey's doctrine of cardiac motion and circulation and had used Harvey's then unpublished dissections of deer to support his theory that all parts of plant and animal embryos were laid down according to the necessary expansions of matter, under force of heat, when it was exposed to cold.[68] Highmore rejected this theory of embryonic development because it could not explain why progeny resembled their parents.[69] Our model, Highmore said, should be the atomic nature of nutrition, as it was carried out by the circulation. Food was broken up in digestion into the multitude of its constituent atoms and released into the blood stream. Atoms that were cognate to various parts of the organism were absorbed at the points of congruence and became part of the body.[70] Genetic continuity was maintained in the same way. Seminal atoms from the male and the female interacted to lay down the basic body plan, which was then augmented by the process of cognate attraction of atoms from either the yolk (in oviparous animals) or the maternal blood stream (in viviparous animals).[71]

Highmore's atomistic approach led him to some conclusions at variance with his friend and mentor. He, like Harvey, observed the developing chick embryo. But, because Highmore used a microscope, he could see that what Harvey had taken to be a pulsating point of blood existing before the heart, was rather a transparent heart contracting to expel its contained blood.[72] To Highmore's eye, aided by the microscope, embryology did not necessarily prove the primacy of the blood. Digby had cited Harvey's findings that in the doe, there was no evidence of seminal fluids mixing and that the conception did not appear until two months after rutting. Highmore, since his theory demanded the physical mixing of the two sets of seminal atoms, could not accept these results. He criticized the experimental controls and suggested deferentially that, although Harvey's "curious eye" seldom took anything on trust, the great investigator may have been deceived by his presuppositions.[73] Clearly, by the early 1650s, Highmore had diverged widely from the natural philosophy, if not the physiology, of his Oxford colleague and had moved forcefully in the ideological direction of his "much honoured Friend,"[74] the squire of Stalbridge.

THE NEW PHYSIOLOGY IN OXFORD MEDICAL TEACHING

Highmore and Boyle, the past and future Oxonians, were not alone in seeing new lines of inquiry opened by anatomical innovations, corpuscular philosophies, and chemical discoveries. Throughout the 1650s and early 1660s, much evidence suggests that the Oxford medical community was acutely conscious of the changing state of physiological explanations. Lectures, study groups, disputations—all of these elements of academic life reflected an awareness of the innovative currents in the medical sciences.

The lectures of William Petty, which have survived among his papers at Bowood, are probably the most strongly programmatic in content.[75] Upon being admitted D. M. at Oxford in March 1650 Petty read, in conformance to the Laudian Statutes, "six solemn lectures."[76] In the first he sketched briefly the history and aims of medicine. The second he devoted to defending the necessity and applicability of chemistry to medicine and to the medical sciences. In the remaining lectures he showed how chemical theories were applicable to the more traditional areas of therapy, such as the treatment of fevers and hysteria. Petty's appointment as Tomlins Reader in late 1650 brought him to the Schools on numerous occasions in 1651. On 4 March, he delivered his "Oratio inauguralis anatomica,"[77] and followed this with a set of five lectures on the cadaver in the early spring.[78] He graced the Act in July with a "Lectio anatomica proemialis,"[79] and in the autumn delivered the statutory "Three Osteological Lectures."[80]

The dominant characteristic of Petty's lectures was the aggressiveness with which he preached the Harveian doctrines of the heart's pumplike motion and the resultant circulation of the blood. His "Lectio" at the Act, for example, began with the recitation of a poem in praise of Harvey and his work. Major portions of the lectures on the cadaver were devoted to discussing theories of the heart's motion, the function of respiration, and the problem of the origin of the blood. Petty's views on these questions seem to have been distinctly Cartesian, especially in his analysis of the structure of matter and his emphasis on the necessity of mechanical, deterministic explanations. But this was to be expected, bearing in mind his studies under Hooghelande and van der Wale in Leyden. Hooghelande, especially, was a partisan of the Cartesian position, while van der Wale adopted an experimental approach to physiology more in line with the tenor of *De motu cordis*.

Such a combination of Descartes's mechanism and Harvey's substantive description of the circulation is not as incongruous as one might think. Historians of science have always, rightly, contrasted Descartes's explanation of the heartbeat as thermal expansion of the blood with Harvey's argument for its contractile, pumplike nature. They contrast Descartes's rejection of faculties, qualities, and innate directing forces with Harvey's very comfortable acceptance of Aristotelian presuppositions. But, at mid-seventeenth century, especially to a man of Petty's temperament and training, the choice was by no means so clear-cut. If a man has two allegiances, he is not easily forced to a decision. Petty's solution, one which was to be followed by many of his friends and colleagues over the next few years, was to explain Harveian problems and results according to mechanistic methodologies.

The chemical theme, especially in Petty's lectures for the D. M., is also understandable. In 1645 he and Hobbes had taken a seven-month course in laboratory chemistry at Paris under the Scottish chemist, William Davisson.[81] Petty's notes record more than fifty protocols that demonstrate the range of practical chemistry. He compounded medicaments. He purified substances such as mercury, sulfur, and antimony.[82] He made such inorganic acids as *spiritus nitri* (nitric acid) by distilling niter pulverized with fuller's earth.[83] He learned techniques that were by no means trivial: fractional distillation, use of sand and water baths, precipitation, filtration, and recrystallization.[84] Little wonder that Petty should have been so enthusiastic about the experimental side of natural philosophy, that his Oxford club should have an interest in "inspecting drugs," or that a medical diary he kept at Oxford from March 1650 to November 1651 should show a predilection for such inorganic preparations as those made from sulfur, mercury, borax, and coral (calcium carbonate).[85]

Petty's lectures on the cadaver would only have been of use to men who had some introduction to the rudiments of anatomy. Philip Stephens's anatomy classes, using Johannes Vesling's *Syntagma anatomicum*, provided just such an introduction—in a distinctly Harveian form. According to John Ward, Stephens taught these classes "16 termes." It is unlikely that he would teach four terms a year, so it seems probable that the period over which he tutored anatomy stretched from the early 1650s to his departure from Oxford in the summer of 1659. Stephens died in London immediately "after the restoration."[86]

The text Stephens used was a modest, straightforward introduction to human anatomy.[87] As Vesling noted in his preface, his book explained the parts of the body as they were encountered in dissection, without reference to the many controversies that had exercised academics to the detriment of true anatomical knowledge.[88] Vesling was very largely true to his promise. Excepting, that is, the pages in which he gave a clear and concise description of the motion of the heart and the circulation of the blood, as drawn from the work of Harvey. The heart, he wrote, was the source of heat and spirit in the body. The heart drew this heat from the blood, and by the cardiac motion it stirred the blood's spirit to perfection. Since the cardiac valves kept blood from returning after contraction, there must needs be a perpetual circular motion of the blood. It was to William Harvey that we owed this knowledge.[89] Vesling's argument was much less subtle than Highmore's, but it would do for beginners.

In some ways Vesling's text remained conservative. Although asserting that vital spirit originated in the motion of the heart rather than from the air, he continued to believe that the pulmonary vein carried blood mixed with air to the left ventricle. He also held that cold air cooled the vital heat and carried off the fuliginous vapors.[90] Written before 1651, Vesling's text continued to state that the liver was the "manufactory" of the blood and natural spirits.[91] Although at one point he hinted that the heart drew its heat from the blood, in most other places he followed more traditional notions in making the heart itself the source of heat, the *fons vitalis caloris*.[92] His treatment of spirits was equally inexact. He remained solidly within the Aristotelian concept of biological materials; fluids, as well as organs and their parts—chyle, blood, flesh, fat, fiber, muscle—were ultimate and self-defining entities, without further substructure, which worked their effects by their qualities and occult properties. In many ways Vesling's *Syntagma* was the perfect vade mecum for young students. It presented solid anatomical facts enlivened with Harvey's discoveries, set within the traditional natural philosophy that was still the official posture of the university.

By the end of their medical education, however, many students at Commonwealth and Restoration Oxford were clearly interested in rather more innovative anatomical and physiological ideas. This may be seen in the topics chosen for those most traditional of university exercises, the *Disputationes in Vesperiis* and *in Comitiis* held each year in June and July. These disputations are important, not so much as substantive contributions to a scientific subject—that could hardly be expected in the twenty or so minutes it took to read the declamation on each subject—but rather as indicators of the major problems that were felt to be of interest or importance. As such, they seem to be tolerably faithful. Disputants in medicine chose their own questions, as opposed to those in arts, who debated topics set for them by the university. More significantly, for a disputant proceeding D. M. in the same year, the questions debated were often propositions abstracted from the statutory lectures he read for his degree, and thus provide a clue to the contents of some more important academic exercises.

Before the Civil War, inceptors in medicine showed little inclination to take up anatomical and physiological questions. The highly consolidated state of Galenic physiology made pathology and therapeutics more attractive areas in which the disputant could demonstrate his originality and rhetorical ability. George Bate, for example, later a respected Oxford physician, affirmed in 1637 that beer was more beneficial in fevers than barley water, and that bloodletting and purging did not expel pestilential essences.[93] In 1639 John Edwards argued that dinner ought to be more abundant than luncheon, that a diversity of victuals contributed to health, and that fermented bread was more wholesome than unleavened bread.[94] A year later Bridstock Harford asserted that every pleurisy had four phases, that one should purge at the beginning of pleurisy, and that the disease could be cured without bloodletting.[95] The rare physiological thesis was either trivial, such as Tobias Garbrand's defense of human dissection in 1639, or set firmly within the classical tradition, such as Thomas Clayton, Jr.'s assertion in the same year that the veins sanguified.[96]

In contrast, when exercises began again in 1651, they expressed strong and consistent concerns for the physiological problems raised by anatomical innovations. The clinical implications seen in new discoveries were fewer, since it seems by common consent to have been agreed that "recent inventions in anatomy" did not change the classical methods of therapy—a point argued explicitly by William Quartermaine in 1657 and Thomas Jeamson in 1669.[97] Only Peter Gerard, in 1669, proposed the radical doctrine that, since the discovery of the circulation, there was no longer any need for ligatures in phlebotomy.[98]

Pecquet's discovery of the thoracic duct in 1651, coupled with the acceptance of the circulation, had robbed several of the visceral organs of their traditional functions. Alternatives were hotly debated at Oxford. Thomas Arris (1652) asserted that the chyle was not brought to the liver, and Sherrington Sheldon (1663), the corollary that the liver was not the manufactory of blood.[99] Timothy Clarke (1653) argued that sanguification did not take place in the liver, spleen, or the veins, leaving one wondering where he thought it did take place.[100] Traditional notions of chyle and its origins were challenged as well. On two occasions, 1651 and 1654, Bathurst rejected the Galenic idea of chylification by heat and suggested instead that it was done by an acid ferment in the stomach.[101] Arris (1652) seems to have agreed.[102] In 1669 Gerard argued against heat concoction, while his fellow inceptor, Nicholas Hele, took the traditional side.[103]

The spleen attracted great interest. Robert Fielding (1654) suggested that it was the organ of fermentation, and further, that rickets resulted from a debility of the fermentation so conferred on the blood, and a consequent impeding of the circulation.[104] In 1661 George Castle agreed that the spleen conferred a ferment on the blood and, moreover, that hypochondriacal distempers originated there and that genius depended on it.[105] Thomas Jeamson (1664) countered this splenic enthusiasm in asserting that the spleen was neither the cause of sanguification nor the seat of the sensitive soul.[106] In 1669 Hele argued that the spleen was neither the source of black icterus nor the seat of scurvy, while Gerard added, most likely on anatomical grounds, that it did not minister a ferment to the stomach.[107]

The nature and function of blood was a favorite topic. Traditionalists like Anthony Nourse, who argued in 1651 that blood was composed of four humors and contained spirit, were soon routed.[108] Acland (1653), Quartermaine (1657), and Sheldon (1663) all argued that the blood did not nourish; only Jones (1669) held to the classical position.[109] Quartermaine went further to assert that blood was not made up of four humors.[110] William Page (1653) defended the recent Harveian doctrine that there were no spirits distinct from blood.[111] Nor was milk made from blood, said Sheldon (1663) and Alvey (1671), in contradiction of the Galenists.[112] As to blood's origins, now that the liver was deposed, Richard Lydall (1654) suggested that heat itself sanguified, while Edward Stubbe (1657) believed that the sanguifying faculty resided in the heart.[113]

The lungs and respiration, as I shall later discuss in greater detail, were continuing topics of debate. Bathurst (1654) defended the ideas that respiration was a simple voluntary motion, and that suffocation was due to a lack of a nitrous food drawn from the air.[114] Stubbe (1657) explained how the circulation took place through the lungs, and his fellow disputer of the same year, Elisha Coysh, asserted that the proper function of respiration was the depuration of the blood.[115] Jeamson (1664) defended the seemingly more traditional concept that the aim of respiration was to ventilate the heart.[116]

Unfortunately, the texts of only a few of these *disputationes* have survived, so we know little of their contents beyond what may be read into their tantalizing titles. What is perhaps most interesting is the rapidity with which anatomical innovations, both within and without Oxford, were taken up and examined in these supposedly archaic academic forms. Pecquet's discovery had, as we have seen, an immediate effect on discussions of the liver. The publication of Harvey's *De generatione* in 1651 had an equally rapid impact. In the same year Bathurst asserted the Harveian proposition that the fetus was not nourished by the maternal blood.[117] In 1653 Page defended the Harveian ideas that there was no spirit distinct from the blood, that all animals originated from an egg, and that the embryo sucked *in utero*.[118] In 1657 Quartermaine denied that a nutritive *succus* was distributed by the nerves and proposed that the liquid observable in the newly discovered lymphaducts was a residue from nourishment; both were ideas discussed in Glisson's *Anatomia hepatis* of three years before.[119] In 1661 John Lamphire affirmed that the primary seat of arthritis was in the brain, a concept whose mechanism his fellow disputant, Thomas Willis, had explicated in his Sedleian lectures of that same year.[120] In 1669 and 1672 Jeamson and John Masters debated the characteristics and locus of the ''accension'' of the blood, a concept discussed by Willis and by Richard Lower in their books of 1669 and 1670.[121]

Many of these disputants—Nourse, Arris, Acland, Page, Fielding, Edward Stubbe, Sheldon, Gerard, Hele, Jones, Alvey—are deservedly obscure. There is little or no evidence of their subsequent scientific activity. Others, such as Bathurst, Clarke, Lydall, Quartermaine, Coysh, Lamphire, Castle, Jeamson, and Masters, played parts, more or less important, in the development of the series of physiological ideas that characterized the Oxford scene during the Commonwealth and Restoration periods. Yet the latter group is not understandable without the former, for the larger audience of the Oxford medical community at once created and reflected the new definitions of problems and the new attempts at their solution. Taken individually, these disputations were neither greatly important nor influential. They were even at best not experimental reports, but ''review articles,'' summing up the evidence and suggesting possible lines of explanation. But taken *in toto*, along with the evidence afforded by Petty's lectures and Stephens's anatomy classes, they were symptomatic of the profound and continuing effect that the new discoveries in anatomy, especially Harvey's, had upon the Oxford scientific community.

RALPH BATHURST AND THE PROBLEM OF RESPIRATION

It was no accident that in his disputations *in Vesperiis* on 8 July 1654, Ralph Bathurst chose to discuss the nature of respiration. These shorter exercises were but a precis of the ideas he had argued at much greater length just a few weeks before in his ''Praelectiones tres de respiratione'' and which were to have a continuing influence in Oxford physiological thought for the next two decades.

Bathurst, unlike Highmore, chose to stay on at Trinity after Oxford's surrender. To get a ''tolerable livelihood,'' as he later explained, ''I knew no way better than to turn my studyes to physick.'' For a time he sequestered himself from the university, ''to follow my study of physick, either at London, or in the country.''[122] He eventually returned to his fellowship, submitted to the Visitors in June 1648, and began chemical work with Willis and John Lydall. Soon he, with Willis and Lydall, gravitated into Petty's meetings at Buckley Hall, and thence into the Wilkins circle at Wadham.[123] All the memorialists of the Oxford club— Wallis, Sprat, Pope, Evelyn, Wood and Aubrey—remembered Bathurst as one of the earliest members of the group. He participated in the Bodleian cataloging project and rapidly became good friends with many of the club's members, especially Matthew Wren, Seth Ward, Rooke, Sprat, and later Boyle, Millington, and Samuel Parker.

His medical career grew apace. In its early days he and Willis attended Abingdon market every Monday in search of patients.[124] In December 1650 he assisted Petty, Willis, and Clerke in the famous "resurrection" of Anne Greene.[125] In April and May of 1653 he went through the procedures to get his D. M.,[126] but was delayed by serving, from June 1653 through winter of 1654, as assistant to Daniel Whistler in caring for the naval sick and wounded at Ipswich, Harwich, and possibly London.[127] For his public services he was commended both by his College and by the Visitors,[128] and allowed to proceed D.M. on 21 June 1654. It is testimony indeed to the latitude granted men of science that a confirmed minister of the Church of England, one who continued to act as Bishop Skinner's archdeacon in Episcopal confirmations at nearby Bicester, should be given, by a supposedly anti-High Church commission, a certificate that permitted him to carry on his medical practice while holding a college fellowship reserved by statute for men in orders. By 1658, Bathurst, along with Willis, was a physician described as "eminently known in the *University*."[129]

Bathurst was from the very beginning an adherent of the new philosophy at Oxford. In 1650 he wrote verses praising Thomas Hobbes's mechanistic book *Humane nature*, and in 1651 wrote to Hobbes urging him to publish his *De corpore*.[130] The same inclinations may be seen in the set of propositions Bathurst disputed at the Act in June 1651.[131] In the first he denied that the fetus was nourished by the maternal blood, drawing his arguments from Harvey's *De generatione*, published just a few months before. Bathurst sketched the early growth of the human fetus, described the development of the placenta, and emphasized, as Harvey did, that there was no continuity between the blood supply of the fetus and that of the mother. Citing Harvey, Fabricius, Riolan, and Vesling among others, Bathurst argued that the fetus was instead nourished by the chorionic fluid.[132] The second proposition, that *omnis sensus sit tactus*, argued that nerves were activated only by motion, which in turn excited a "phantasm" in the brain. Just as sound was nothing but a sequential percussive vibration of the air, and light nothing but a similar motion of a subtle body that communicated its action to the optic nerve, so were the traditional "qualities" and "species" of Aristotelian philosophy nothing but metaphysical labels for mechanical phenomena.[133] In the third disputation, Bathurst turned chemist to reject the traditional notion that heat produced digestion in the stomach and to explain it rather as the action of a "fermentum acidum," or saliva, secreted from the walls of the stomach. Acids were, he said, notable in their ability to digest various substances, including bone and stones. Vomit was clearly of an acid nature. Only an acid substance had the power to leach out the sulfurous substances and volatile salt that were the essential part of food. In all of these arguments, Bathurst drew not only upon his own experience in chemistry, but also upon the writings of Helmont, Gassendi, Harvey, and van der Wale.[134]

The same combination of chemical processes and mechanistic matter-theory, applied to physiological questions, appears in even greater clarity in Bathurst's lectures on respiration of June 1654. According to the Laudian statutes, Bathurst could choose to give, as part of the exercises for D.M., a course of either three longer or six shorter lectures on a "Galenic" subject. Bathurst chose the former. Manuscripts of his "Praelectiones tres de respiratione" survived long enough into the eighteenth century to be printed among his other literary remains.[135] They constitute, to my knowledge, the only such seventeenth-century English medical lectures ever to appear in print.

Bathurst's beginning was quite orthodox. He proposed to devote a lecture to each of the traditional divisions of the subject: the mode according to which respiration took place; its immediate cause, the motion of the muscles; and the end or purpose which such respiration served.[136] He began the first lecture by rejecting the traditional explanation that the lungs moved of their own power. They were, he said, rather like air bags, not capable of moving themselves. Neither were they moved by the blood impelled through them. Both of these points could be proved "*in vivorum dissectione*," when the thorax was opened and the lungs observed to sink. The lungs were moved rather by the diaphragm in most ordinary respiration and by both the thorax and the diaphragm in extraordinary breathing.[137] After a lengthy exegesis of this position, buttressed at various points by citations to the work of Harvey, Highmore, and George Ent, Bathurst summarized that a genuine philosophical explanation of the lungs' motions could only proceed mechanically, not according to "*facultates aut qualitates.*"[138] He rejected the notion that the lungs "attracted" air and proposed his own explanation. Air had weight and substance, and it was according to this weight that respiration took place. When the diaphragm moved downward, pushing the viscera, it created an impulse which was communicated through the ambient medium, thus pushing air into the lungs.[139] Once again he buttressed his argument with appeals to open thorax vivisection, the findings of his friend and colleague Highmore, and pathological evidence.[140] Bathurst even mentioned at one juncture a wound he had tended during his recent service as a naval physician.[141] He closed his lecture with an affirmation that respiration was a universal life process, found among insects, plants, and even among fish. He based his explanation of insect respiration on passages in Harvey's *De generatione* and his belief in piscine respiration on the probability that air was dissolved in water.[142]

The initial part of his second day's lecture, dealing with the muscular basis of respiration, was quite prosaic. He accepted the classical theory that muscles contracted because inflated with animal spirits conveyed to them by the nerves.[143] But even in an otherwise derivative discussion, his work with the Oxford group enabled him to adduce an interesting, if bizarre, experiment as proof. If one blew a bladder full of air, he noted, its expansion could be made to support a great weight.[144] He seems to have had in mind a variant of the experiment done by the Petty group in late 1650, described by Petty in his letter to Hartlib and later analyzed statically by Wallis.[145]

With no little dramatic flair, Bathurst reserved the more interesting parts to conclude his lecture. What, he asked, of the vulgar opinions that inspiration served to refrigerate the heart, and that expiration served to blow off fumes, or *fuligines*. The first, he said, was very doubtful. The blood of fish could be seen to have very little heat and would need no cooling. Or, if it did, water could perform the function better than air. Yet, fish could not live without air; if they were put in a container sealed from the outside air, they died. Secondly, he argued, the cold or warm quality of the air seemed to have little relation to the heat of the heart. Thirdly, nature, which always aimed at the optimum, would not have put a heat in the heart so excessive that it constantly required cooling. In fact, Bathurst seems to suggest, respiration encouraged heat rather than suppressed it.[146]

As to the second putative function, Bathurst thought it much more likely. The heart was like a lamp of life and needed to throw off its fumes. Its heat could be regulated, as was clear from the way the chemist regulated the heat of his furnaces, by controlling the amount of fumes and "atoms of heat" that were given off. Bathurst adduced an experiment of Helmont to clarify his point. One could heat charcoal in a closed vessel, yet the fire would pass through the vessel and render the charcoal incandescent. In the same way, the air could pass through the lungs, into the blood, and ventilate it of its dross. In fact, he even implied that the lungs served to separate the dross from the blood and made it easier for the air to carry off the waste products. As an example of how this separation took place, he cited the action of shed blood left to stand or heated; it separated into the same kinds of parts. In all of these arguments, as in his previous ones, he referred at various points to the neoterics for corroboration, especially Helmont, Gassendi, and Ent.[147]

Bathurst saved his finest revelations for the third day. What, he asked, was the primary use of *inspiration*? It served, he asserted boldly, to supply to the animal the *pabulum nitrosum*, the nitrous food by which the blood was tempered and made fit for life. This nitrous food was continually required by the blood and, without it, the blood was not nourished. If this *pabulum nitrosum* were missing in the air, respiration availed for nothing.[148]

Although Bathurst used several different terms to indicate his aerial substance— *halitus nitrosus* (nitrous vapor), *spiritus nitrosus* (nitrous spirit), as well as *pabulum nitrosum*—the entirety of the final lecture was devoted to characterizing its ubiquity, its universal use in the repiration of plants and animals, and how it conduced to life in man. It was to be found, Bathurst asserted, throughout nature, in air, water and earth. He implied obliquely that it existed in the air as a volatile salt, while the mineral *niter* was its fixed-salt form. In fact, this accounted for how one could make niter from the air.[149] He then attempted to show its use to plants, as indicated by the fact that onions would not germinate in a vacuum. He followed this with several pages of examples which, he said, demonstrated the absolute necessity of this *spiritus nitrosus* to plant life.[150] Animals too needed the *pabulum nitrosum*, as demonstrated by the fact that men in closed rooms, or in submarines, or in diving apparatus, needed a constant supply of fresh air. But could not this be due to the accumulation of fumes? No, Bathurst said, as was shown by the difficulties of breathing on high mountains: the *halitus nitrosus* sprang from the ground and was not easily elevated to great heights. Fish too required the *pabulum nitrosum*. The purpose of passing the water over the gills was not to cool the heart, but to extract the *spiritus nitrosus* from the water. This was clearly shown, he noted, by the fact that fish put in a closed vial soon died.[151]

Finally Bathurst came to the most crucial of all the physiological questions: what did this *pabulum nitrosum* do, such that it was absolutely necessary to life? First, he rejected firmly the suggestion that the cooling properties of niter were involved; he had already shown that the purpose of respiration was not to cool the heart. Rather, he said, this *halitus nitrosus* served as the food of the vital and animal spirits. Men died when excluded from fresh air because they were deprived of the source of their spirits, much in the same way a lamp wick went out when deprived of its alcohol. Moreover, he argued, it was not necessary to believe that these spirits were generated in the heart. In the same way as blood made over chyle into blood, so did the spirits in the blood make over the imbibed *pabulum nitrosum* into spirits. It was for this reason that the branches of the veins and arteries were so ramified into the lungs—to allow the blood to imbibe the *spiritus nitrosus* copiously. If a doubter asked how this passage was made, Bathurst would retort that the nitrous spirit in snow was able to penetrate a glass or metal vessel in order to cool summer wine; if it could pass through the pores of such an impervious substance, surely it could pass through the much roomier pores of a pulmonary vessel. And finally, if one did not deny that the *fuligines* from the serous part of the blood were able to exit through the branches of the lungs, surely the entrance of the *pabulum nitrosum* must be conceded.[152]

Although Bathurst obviously culled a great deal of evidence from diverse sources to buttress such a sustained argument, the origins of the *pabulum nitrosum* are of the greatest importance. As Bathurst recorded in his lecture notes, he drew the vague outline of the notion from George Ent's *Apologia pro circulatione sanguinis* (1641).[153] Speaking of the respiration of fish, Ent had stated that air, in the form in which we know it, was found in water only in the smallest degree. But the nitrous virtue for which we breathed the air was also contained in water, and fish lived upon it. For air, Ent thought, was nothing other than an effluvium of water, impregnated with the spirits of the earth that were necessary to plants and insects and which animals breathed for their vitality.[154] In a later passage Ent explained how blood absorbed air in its passage through the lungs of animals and carried it to the left ventricle, where, as in ordinary combustion, it maintained a fire in the heart. If this fire were deprived of air, the vital heat was extinguished like any other fire when the air was cut off. This took place because the fire was cherished by the niter of the air—*ab aeris nitro*. The more niter there was in the air, the more strongly the flame burned.[155]

The concept of an "aerial niter" has a long history whose chemical content and physiological significance will be analyzed in the following chapter. But it is worth inquiring here how Bathurst happened to seize upon a few obscure passages in a book published more than a decade before and to elaborate them into a coherent physiological theory of respiration. The answer seems to lie in Bathurst's London sojourns c. 1646–1648 and 1653–1654. In the much revised second edition of his *Apologia* published in 1685, Ent numbered Bathurst among his good friends and colleagues.[156] Since Bathurst spent almost all of his life after 1654 either in Oxford or in Wells (where he was Dean of the cathedral chapter 1670–1704), it is highly likely that he became acquainted with Ent either during his "study of physick" in the 1640s or during his term as a naval physician during the first Dutch war. The concept of an "aerial niter," extracted from his older colleague's defense of Harvey, fitted neatly into Bathurst's chemical presuppositions and offered him a way to solve a cluster of physiological problems made pellucidly clear by their restatement in Highmore's codification of Harvey's discoveries.

But if the similarities between Bathurst's and Ent's conceptions seem striking, so were the differences. Ent had written of a niter in the air; Bathurst spoke of a *spiritus nitrosus*, a decidedly less substantive concept. It seems that Bathurst, experienced chemist that he was, was bothered by the unlikelihood that so gross a substance as actual niter could exist in the air. Moreover, while Ent spoke of a virtue, Bathurst phrased his account in terms that implied that his nitrous *pabulum* was divisible into particles, thereby transforming the notion into the conceptual categories of the mechanical and corpuscular philosophy. But the physiological differences were even more striking than those of chemistry or matter-theory. Ent had used the traditional idea of a fire in the left ventricle and made his aerial niter sustain it. Bathurst's locus of concern, as would be quite natural for any close reader of Harvey's *De generatione*, was the blood. The *halitus nitrosus* served a process in the blood, not in the heart. Moreover, it served a completely different function; it replenished the vital spirits of the blood, rather than stoking some cardiac fire. Bathurst had clearly borrowed the outlines of the notion from his good friend, but in doing so he transformed it almost entirely. He elaborated it extensively and redefined both its *substance* and *function*.

Bathurst's redefinition of the *pabulum nitrosum* was tied, in turn, to a reinterpretation of pulmonary and cardiac function. Highmore had rejected the entry of air into the body because he had found no airy spirits generated in the left ventricle; in doing so, he had followed out the logic of Harvey's doubts. But since Bathurst believed that gases had substance and weight and that they could be dissolved in liquids, he could reverse his friend and argue that the aerial substance was absorbed into the bloodstream in the lungs and served a process that was localized there, not in the left ventricle. And, whereas Highmore had not been willing to go beyond Harvey's belief that respiration served to cool the heat in the blood, Bathurst, in his second lecture, had decisively rejected refrigeration as a function of respiration. The process at work here is clear: a piecemeal rearrangement and replacement of the component parts of classical explanations, one man reacting to what he saw as the unfinished work of his friends.

Bathurst also, in the process of assembling the evidence to support his notion of a *pabulum nitrosum*, drew upon his experience in the Oxford group. In arguing its necessity to plants, he cited a botanical experiment of John Lydall.[157] In pleading for its use by animals, he mentioned an experiment Wilkins had performed on frogs.[158] He even drew upon the work of absent friends; in giving his reasons for difficulty of breathing on mountains, he cited the same author, Acosta, whom Highmore had used to argue an opposite point.[159]

Bathurst's Oxford colleagues, in turn, seemed to have served as a critical audience for the notions he had expressed in his lectures. During the two years following the Act of 1654, Bathurst made some corrections and additions to the lectures, seemingly with the aim of publishing them.[160] Robert Boyle read them immediately after his arrival in Oxford in late 1655[161] and seems to have volunteered to see them through the press. However, in late March or early April 1656 Bathurst changed his mind and wrote to Boyle asking him not to publish them. Boyle, at that time in London with Wilkins, wrote back to express his disappointment at the "unwelcome orders" concerning Bathurst's "excellent lectures." Wilkins, Boyle said, had "solemnly dispensed" him from promptly obeying Bathurst's desires, and Boyle did not intend immediately "to have those jewells sent backe to Oxford." Boyle would, however, gladly send greetings to "Dr. Willis, Dr. Ward, and the rest of those excellent acquaintances of yours, that have been pleased to tolerate me in their company."[162] A month later, in late May 1656, Boyle was still in London with Wilkins, and Bathurst's reluctance was still a matter of concern. Samuel Hartlib recorded in his "Ephemerides":

> Mr. Boyle knows one that hath an excellent Ms. De Respiratione which hee will not publish. Mr. Boyle.[163]

It is not known precisely why Bathurst stopped Boyle from publishing the lectures. We do know that Bathurst, like Harvey, was quite diffident about publication; perhaps this is just another example of that unobtrusiveness.

But Bathurst's reticence to have his lectures published did not preclude their extensive circulation in manuscript among his friends. Thomas Willis read them, as did his protégé, Richard Lower. Even into the early 1660s Bathurst continued to allow his friends access. Nathaniel Highmore visited Bathurst at Oxford after the Restoration, read the lectures in manuscript, and, upon returning to Dorsetshire, wrote his old friend a letter that was a combination of bread-and-butter note and textual critique.[164] As might be expected, the greatest number of Highmore's criticisms centered around the *pabulum nitrosum*. The specific arguments do not concern us at this juncture. More important is the hint such correspondence provides us about the degree to which attention to a central scientific notion, the *pabulum nitrosum*, was mediated by the links of friendship which bound the Oxford group together. Over the next two decades, in the hands of successive members of that group, this idea was to prove a powerful tool in attacking the problems raised by the Harveian legacy.

BUILDING A TRADITION

From the early anatomical work of William Petty and Nathaniel Highmore to the physiological preoccupations of inceptors in medicine of the Commonwealth and early Restoration period, there runs a continuous theme of interest in those problems defined by the new anatomical discoveries, especially those of William Harvey. Moreover, it was a theme that placed these problems upon a new basis of natural philosophy, one radically different from that with which Harvey would have been familiar or comfortable. The idea of the corpuscular basis of biological materials, argued so forcefully by Highmore and Boyle, one in public and the other in private, provided a new way of attacking the many Galenic assumptions that had passed, unnoticed, into the Harveian description of the circulation of the blood. By saying, as these men did, that a structure was composed of discrete parts and its function was traceable to the characteristics of those parts, the mechanical philosophy introduced into the Harveian tradition during the 1650s provided new ways both to define and to study old functions. Highmore's ideas of the blood, Boyle's microscopic observations on insects, and Bathurst's notion of a *pabulum nitrosum* in the air—all relied, either explicitly or implicitly, on the division of biological matter into discrete parts.

This new approach to problems defined by anatomical discoveries, according to new presuppositions, found expression in new theories of respiration. Highmore had argued, by reasoning and experiments, that the admission of air for the generation of vital cardiac spirits was not a function of respiration. But, in eliminating one of the four classical explanations, he, like Harvey before him, had been driven to reaffirm the cooling function of respiration. Bathurst then did just the opposite. He dismissed the cooling function as illogical and inconsistent with simple observations and posited instead that the blood needed a *spiritus nitrosus* to become fit for maintaining life. Between them, they had eliminated both of the two primary classical reasons for respiration, and Bathurst had substituted for these the necessity for the blood to have an aerial nitrous substance. The vague idea of an aerial niter was not new, but in arguing it as he did, Bathurst drew together widely scattered suggestions of his predecessors and welded them into a coherent *physiological* system. The scheme had its weaknesses. The chemical characteristics of nitrous substances were left unexplored. The corpuscular properties, both of niter and of the air, were ill-defined. There was no real process in the blood that the aerial niter served. Finally, the entire system lacked strong experimental confirmation. But it was a cohesive set of starting principles.

Perhaps as importantly, the rapid development of Oxford as a center of scientific research made possible the continued exploration and criticism of physiological ideas such as Bathurst's. An informal but coherent local community of natural philosophers had been built up at Oxford, one whose participants ranged from the mere dabbler in Stephens's anatomy classes, through the disputants of physiological theses for medical degrees, up to the more closely-knit groups around Petty, Wilkins, and Boyle. The interaction of informal milieu, institutional structure, and genuinely interesting scientific problems was generating a research tradition whose strength was to endure for some time.

ROBERT BOYLE
ON NITER AND THE
PHYSICAL PROPERTIES
OF THE AIR

5

C onsider a little more closely the nature of the air,'' Harvey had said in expressing one of the few doubts he had about the cooling function of respiration. Natural philosophers, unknowingly heeding Harvey's suggestion, did exactly that in the three decades from the mid-1640s to the mid-1670s. "Air" and "airs" were the focus of scientific concern as they were not to be again until the second half of the eighteenth century.

This was a great turnabout, for as late as 1640, air was considered an object so little worthy of inquiry that the writers of textbooks in natural philosophy, especially of those reprinted for use in English universities, gave it only a few pages.[1] They agreed with Aristotle that it was one of the four simple bodies, or elements, and that it originated from the qualities of moist and warm. Thus air had a natural place on the terrestrial globe between the water below, and the fire of the heavens above. Being thus intermediate, it could be condensed back into water, or inflamed into rapid motion to produce fire. Air did have weight and substance, but only in comparison with the lighter element of fire, and the more dense and incompressible elements of water and earth. No creature could live without it, although they were often adversely affected—man included—by the "exhalations" raised up into the air from water and earth. Similar aerial exhalations caused the phenomena of weather, such as wind, rain, thunder, and lightning.[2] Most of all, early seventeenth-century writers agreed almost to a man with Aristotle in his denial of a vacuum, and hence in his rejection of the air as Democritean particles moving in a void.[3] It was not conceivable, they intoned, that there could be space without matter to fill it.

Such simple—almost naïve—concepts of the air suddenly became moot. In 1644 the Italian mathematicians Viviani and Torricelli completely filled a glass tube three feet long with mercury, inverted it into a bowl of the same, and found that the mercury was supported to a height of only twenty-nine inches. This column of mercury was counterpoised, they believed, by the weight of an equal but very much higher column of air pressing downward on the surface of the mercury reservoir. The remainder of the tube enclosed at the top what seemed to be a true vacuum. The experiment was publicized in France by Mersenne in 1645, and in 1646–1647 was reproduced by groups of Frenchmen at Rouen and Paris, including Pierre Petit and Blaise Pascal. They published the results and touted the Torricellian demonstration as an *experimentum crucis* that disproved the Aristotelian concept of a vacuum.[4]

The interpretation of the experiment rapidly became the center of a controversy between Aristotelians, Cartesians, and atomists, most of which is beyond the scope of this present work. Suffice it to say, not all philosophers conceded that the "Torricellian experiment" produced a true Democritean vacuum. Descartes and some peripatetics proposed that the space above the mercury, although devoid of gross air, was filled with a subtle matter, or aether, that enabled the space to transmit light and magnetism. Other Aristotelians argued that the experiment simply showed the great powers of expansion of air, a small amount of which, remaining behind, rarefied to fill the space left by the falling mercury column. This could be shown, Roberval argued in 1647, by placing a flattened carp's bladder, tightly tied, in the head of the Torricellian tube before it was filled with mercury. When the mercury fell, creating the "vacuum," the carp's bladder expanded to a taut little ball, inflated by the great expansion of residual enclosed air. This and other experiments by the French researchers rapidly made it clear, however one interpreted the Torricellian experiment or its carp-bladder variant, that the air had both weight and great powers of rarefaction, and that these properties were best interpreted by corpuscular notions of matter, whether atomistic or Cartesian.[5]

Knowledge of the Torricellian controversy soon reached England. Wallis recalled that the London club considered "the weight of the Air, the Possibility or Impossibility of Vacuities, Natures abhorrence thereof," and "the Torricellian Experiment in Quicksilver."[6] This was almost certainly in late 1647 or early 1648, for in March 1648 a founding member of the London group, Theodore Haak, wrote to Mersenne to thank him for his description of the Torricellian experiment sent to England in the summer of 1647, and to report that "we have made two or three trials of it, in the company of men of letters and rank, with much pleasure and astonishment." Although the members of the "Company" did not "yet wish to declare that it is a true vacuum in the glass beyond the mercury," they wanted "to see several more trials with all sorts of glasses," in which endeavor Haak wished to encourage them.[7] Three months later Haak reported on further mercury experiments that had been performed by the group, which included John Greaves and a number of nobles.[8] The more Baconian circle of Haak's friend, Hartlib, was also kept informed. Charles Cavendish, an English emigré in Paris, described the carp-bladder experiment in a letter to Petty of April 1648, which was duly copied by Hartlib and a pertinent extract sent on to Boyle at Stalbridge.[9]

By the early 1650s descriptions of such pneumatical experiments were available to Englishmen in several forms. Gassendi had described them in his *Animad-versiones* (1649) and drawn the predictable Epicurean conclusion: they not only demonstrated the existence of a vacuum, but also showed that particles of the air had weight sufficient to support the mercury.[10] Charleton, in his exposition of Gassendian atomism of 1654, recapitulated Gassendi's description of the Torricel-lian experiment and expanded at length on its meaning.[11] In addition to announcing the discovery of the thoracic duct, Jean Pecquet's *Experimenta nova anatomica* (1651) also described a series of mercury experiments—including the carp-bladder variant—contrived to demonstrate, he said, that air had not only weight, but "elater," or elastic properties.[12] This work prompted Henry Power to do a series of similar experiments at Cambridge in 1653.[13] As we have seen in examining the subjects of *disputationes* at Oxford, Pecquet's book was widely read there as well. In 1654 both Wilkins and Seth Ward praised Gassendi and the "Atomicall" hypothesis, and noted that it had its "strenuous Assertours" at Oxford.[14] Two years later Seth Ward again praised the revivers of corpus-cularianism, especially Gassendi and Descartes, and specifically used an atomistic concept of the air and vacuum to refute some of the plenist theories of Hobbes.[15]

In sum, by the early 1650s there were numerous reasons for natural philosophers, both at Oxford and elsewhere, to conclude that air had many heretofore unexplored properties, among them weight and possibly "elater" or spring, and that these characteristics were best explained by conceiving of air as a collection of particles moving in (according to one's philosophical inclinations) either a vacuum or a fine penetrative aether. Bathurst, in speaking of his *pabulum nitrosum* as particles in the air, was merely casting a chemical and physiological concept into the terms of a matter theory familiar to his Oxford colleagues.

EARLY CONCEPTS OF THE AERIAL NITER

Although Boyle's name is popularly, if somewhat misleadingly, associated with a Law concerning the physical properties of gases, it was initially through chemistry that his interest in the air was seriously aroused. This emerged c. 1654–1656 out of his research on the chemical properties and composition of a favorite substance in the chemist's armamentarium—niter, what we now recognize as potassium nitrate (KNO_3), probably including a large admixture of sodium nitrate ($NaNO_3$). In exploring these properties of "niter," and in relating them both to the corpuscular philosophy and to Bathurst's concept of an aerial *pabulum nitrosum*, Boyle came to recast into a new form some of the older Paracelsian concepts of a "vital principle" or "aerial niter."

Recent scholarship, especially the excellent work of Guerlac, Partington, and Debus, has shown that, in contrast to the barrenness of early modern scholastic concepts of air, the chemical tradition arising out of Paracelsus contained diverse and complex speculations about the properties of air.[16] This complexity was due in turn, I believe, to the contradictory properties that seventeenth-century chemists could assign to niter.

Sodium and potassium nitrates, rich in oxygen, are powerful oxidizing agents. Since the late Middle Ages in the Latin West, they had constituted the precious and essential active ingredient in gunpowder. Insofar as the concept of *energia* was just evolving into clarity, niter was recognized as the source of that quality in gunpowder. Niter could also serve the same purpose in the standard operations of laboratory chemistry. In the protocols of seventeenth-century chemists, including those of Petty and Stahl, a *regulus*, or fusion of substances, was often created by mixing in a quantity of saltpeter before heating in order to make the experiment proceed more thoroughly by oxidizing the main reagent. Saltpeter was, in both warfare and in chemistry, the only commonly available source of energy and ''heat.''

But niter also has an exactly opposite property: it cools. Saltpeter has a negative heat of solution, and, when dissolved in water, can lower the temperature rapidly. This property was also well known by the first quarter of the seventeenth century. Francis Bacon referred to it both in his *De augmentis* and in the even more popular *Sylva sylvarum*, and by the middle portion of the seventeenth century, this demonstration of amazing ''frigorifick'' power was part of every practical chemist's repertory of effects.[17]

Niter, by virtue of its fixed nitrogen, is also a fertilizer. Seventeenth-century chemists knew it could be extracted from dung-piles, and even as a pure salt it was greatly beneficial to the growth of plants. Dissolved niters, especially in such rivers as the Nile, were held to account for its marvelously fecundating effects. Bacon, for one, believed that niter mingled with water and honey would even cause severed buds to sprout. Niter was, he said, ''the life of *Vegetables*.''[18]

The nitrates were recognized to have a number of other minor properties. Niter was useful in curing meats, since it very effectively inhibited decomposition. It had certain associations with the color red; it turned meats red, and its distilled spirit, nitric acid, fumed red vapors. Finally, niter had a certain sharp taste to the tongue and, when crystallized out from a supersaturated solution, did so in long pointed needles.

As might be expected, the history of functional explanations using niter had a diversity and equivocality that reflected the different properties of the substance itself. The German chemist and physician Paracelsus (1493–1541) seems to have been the first to speculate on a niter in the air, and in his various works one finds both ''heating'' and ''cooling'' properties emphasized. In the one work, the *Grossen Wundarznei*, Paracelsus asserted that the matter of thunder and lightning was a heavenly saltpeter-sulphur, which arose as an emanation from the stars.[19] However, in another work, the *Liber Azoth* (now thought to be spurious, but accepted by the seventeenth century as genuine), Paracelsus appealed to the cooling properties of niter. The Cagastic soul, a mystical entity which Paracelsus thought dwelt in the water of the heart capsule, was in constant need of cooling. It accomplished this by feeding on the nitric salt in the outer world.[20] The explanation was by no means clear and straightforward, a situation not helped by the fact that for Paracelsus, as for many alchemists, the term "niter" designated not only the familiar chemical reagent, but also a generalized, occult, less substantial principle of activity. But the thrust was clear: Paracelsus believed there was a certain nitrous something in the air, which could have either cooling or incendiary effects, as needed.

Although Paracelsus had many followers in the late sixteenth and early seventeenth centuries, only a few of them seem to have taken up his suggestion of a nitrous substance in the air. Those who did had somewhat muddled ideas of what functions it performed. Joseph Du Chesne (1544–1609) wrote in vague terms of "Salt" as the true "balsam," or nourishment of nature. A certain type of this salt was nitrous in nature and rose and fell from the earth as "dew." This dew, Du Chesne implied, was the cause of growth in plants.[21] Blaise de Vigenère (1523–1596) referred to a vital "salt," and also to the fact that "saltpeter is appropriated to the air,"[22] but his context shows he believed neither that this vital salt was exclusively nitrous nor that the saltpeter was the cause of the thunder and lightning.

Two other Hermetics, Alexander Seton (obit 1604) and Michael Sendivogius (1556–1636 or 1646) seem to have had rather clearer ideas of the aerial "niter." In Seton's *Novum lumen chymicum* (1604), he referred to a secret food of life in the air called dew. In the accompanying tract, *De sulphure*, Sendivogius went on to point out the necessity of air to all animals, plants, and to the continued combustion of fire. In another widely separated passage, Sendivogius spoke of the waters of the dew as the source of the "Salt Peter of Philosophers." This dew, brought to earth, joined with the salt niter of the earth to give fecundity to plants.[23] Although the explanation was cast in less mystical terms than most Paracelsian exegeses, the linkage of concepts was still quite weak. Comments on the necessity of air for life and fire were commonplaces, even among the Aristotelians, and the "dew" as a secret food in the air was a stock item of the Paracelsian cabinet. When the connection of a vital "dew" to something nitrous was made, it was clearly in terms of the alchemists' "Niter," a substance that included so many properties as to be almost meaningless. Nowhere were the two concepts of respiratory "dew" and aerial "niter" linked directly.

The earliest systematic attempt to explain some functions of an aerial nitrous substance was made by Daniel Sennert (1572–1637) in his *Epitome naturalis scientiae* (1618); oddly enough in light of the later physiological and chemical fate of the idea, he was concerned, not with biological phenomena, but with the older Paracelsian problem of explicating atmospheric phenomena. Lightning and thunderbolts, Sennert explained, were both made of a tenuous and fiery nitro-sulphurous spirit, differing only in the arrangement and quantity of the particles of sulphur and niter.[24] A later work, the *Hypomnemata physica* (1636), gave the same explanation in greater detail. Just as the "antipathy" of sulphur and niter accounted for the explosion of gunpowder, so thunder and lightning resulted from the union of sulphurous, nitrous and aqueous vapors dispersed into the "minutest atoms" and raised aloft from the earth to join furious battle on high.[25]

By the 1640s then, there were several traditions of nitrous aerial substances, each serving a set of explanatory purposes and each drawing upon certain of the properties of niter. One was biological: Ent's allusions to a nitrous virtue in the air that served to maintain the heat in the heart. Although there is some evidence that Ent, in turn, derived his ideas from the Paracelsian "dew,"[26] both the difference in conceptual clarity and the exact physiological use to which Ent put his aerial niter distinguish it so clearly from Paracelsian notions as to make it a source independent of the fragmentary allusions from which Ent pieced his concept together. The second tradition was Sennert's meteorological theory, embodied in a decidedly atomistic interpretation.

Both Ent's and Sennert's works were well known at Oxford in the 1650s. In several of John Locke's early notebooks he quoted at length from Sennert's *Opera omnia*.[27] In a notebook kept over the period 1655–1657, Locke not only wrote out extensive citations to Harvey's *De generatione* (1651), but also to Ent's *Apologia* (1641).[28] Most importantly, the two traditions were united in Bathurst's 1654 lectures; in addition to the borrowings from Ent's book analyzed previously, Bathurst also had numerous references to Sennert and to Thomas Browne who had adopted the meteorological niter sans atomism. The exposure to Sennert was, in fact, a natural result of Bathurst's medical training, for Sennert was a popular writer on pathology and therapy. It could also have been the direct outcome of Bathurst's Oxford education; an edition of Sennert's *Epitome naturalis scientiae* was published at Oxford in 1632 for use as an elementary Schools text.[29] It seems clear, then, that whatever intellectual genealogy the idea of an aerial niter had before the speculations of Sennert and Ent, it was to these formulations, and not to the earlier Paracelsian vagaries, that the Oxonian notions referred their origins.

It is clear, also, that the use of aerial niter in physiological explanation was as equivocal as its use in chemistry. On the one hand, if one wished to posit a role for niter in the production of heat—niter as "energy"—one had to specify (1) *how* heat was generated and the role of niter in this process, and (2) *where* this process was localized. On the other hand, if one chose to stand by the Galenic concept of innate heat, the cooling properties of aerial niter could be invoked to moderate the internal fire. The properties of saltpeter could be made to fit the needs of the man who sought to apply them. As can often happen when concepts are imported from the physical sciences into biological explanation, the notion of an aerial niter was a double-edged sword. It did not dictate its own use, but rather depended upon physiological *assumptions* with which it was approached.

The Oxford group firmly rejected the notion of innate heat, but its members encountered difficulty in attempting to account for the locus and process of heat production. In the short run, the known properties of niter only seemed to compound their problem. Ent had specified neither how the heat was generated nor how niter served to maintain this generation; moreover, he placed this fire in its traditional location, the heart. Bathurst was quite clear about the locus—the blood—and rather vague about the process. Willis, as we shall see later, was evolving a candidate, fermentation, for the process, but still tying it firmly to the traditional locus—the heart. What was needed was a clearer conception of the *chemical* nature of niter, one which was amenable to integration with the corpuscular philosophy, with the characteristics of combustion, with fermentation as a biological process, and with the blood as a locus of nitrous action. Robert Boyle supplied the basis for that conception.

ROBERT BOYLE AND THE PROPERTIES OF NITER

Boyle's interest in the properties of niter had three roots, two practical and one theoretical. The first was one he shared with all laboratory chemists: the fact that niter was an indispensable reagent. Boyle's "Philosophical Diaries," as they were kept in a fragmentary way from 1650 on, record how his growing laboratory sophistication demanded use of saltpeter as a reagent. The first saltpeter entries, in 1650−1652, were very medically oriented, recording niter's use in receipts for plasters, poultices, and ague medicaments.[30] But by 1655−1657, when Boyle's attention had turned to procedures of synthesis, extraction, and purification, the use of niter became more frequent. In early January 1655, while in London, he recorded "Mr. Smart's" procedure for making sal antimony; it required niter as a reagent to be burnt off in the crucible.[31] A few days later Boyle recorded the use of niter and spirit of sal ammoniac to extract a tincture of steel filings.[32] In early January 1656, after his arrival in Oxford, Boyle used a mixture of tartar and saltpeter to extract pure lead from lead ore.[33] The examples could be multiplied dozens of times in Boyle's research notes of the succeeding years. Saltpeter, whatever its strange chemical and physical properties might be, was a most useful laboratory reagent.

The second practical reason for the attention paid to saltpeter arose from the needs of state. Saltpeter for gunpowder was always in short supply during the Commonwealth, and much ingenuity was expended by Hartlib's circle in attempting to find ways of "generating" it. In 1646 Boyle's colleague in the "invisible college," Benjamin Worsley, was granted a patent for a new way of extracting saltpeter from dung. The project was carried forward with such vigor that Boyle wrote to Worsley from Stalbridge a year later, complaining good-naturedly that those "undermining two-leged moles, we call saltpetermen," had dug up his pigeon house.[34] Throughout the 1650s Boyle, and especially Hartlib, explored schemes for artificial generation of large amounts of commercial saltpeter. In doing so, they too were led to consider whether there was not an aerial nitrous substance, which might be, as it were, "mined." Hartlib collected packets of manuscripts "About Saltpeter & miscllanys," "Concerning Salt-Peter," and even a tract by Worsley c. 1654 on "Salt the cause of Life & Vegetation."[35] In the spring of 1654 Hartlib recorded that a "Major Saunderson"—one claimant among many—would undertake to make "saltpeeter out of common salt and seawater."[36]

While in London in January 1655, Boyle wrote in his notebook how he and Hartlib's son-in-law, Frederick Clodius, tried George Starkey's method of multiplying saltpeter. In a barrel one interlarded layers of dirt and saltpeter to act as seed. For four months one "wet the whole matter about once a weeke with stale urine passt first through horse-dung," then allowed it to stand. The hogshead was then dismantled, "and then it will appear how much hath beene converted into pure and good saltpeter." Hartlib dutifully wrote out almost exactly the same promising protocol.[37] Boyle recorded a few days later how Clodius was trying to make saltpeter from sea-salt, aqua fortis, and a seed of niter.[38] And some weeks thereafter Boyle noted that Clodius thought he had succeeded.[39] The obscure Mr. Smart even believed he had a way of making saltpeter from the air, as Boyle recorded later that month.

> Mr. Smart tooke quicke lime & potash & put them in a wooden dish in the Aire, (but protected from the raine & sun) & after a while he had saltpeter on the outside of his vessel: & the like befell him in a wooden vessel wherein he had put Antimony & potash.[40]

Thus did practical processes for "generating" niter, real and illusory, merge imperceptibly into the assumption of an aerial nitrous substance. Although much attention has been paid to the possible Paracelsian and mystical roots of English interests in niter during the 1650s and 1660s, it seems much more likely that, if Boyle and the circle around him in London and Oxford are any indication, the concern arose initially as much from practical schemes as from an interest in the mystical and alchemical arts.

The third root of Boyle's interest in niter was more abstract: its chemical properties might well illustrate the scientific usefulness of the corpuscular philosophy. Once Boyle had, to his own satisfaction, demonstrated in the early 1650s that the hypothesis of atomic structure might be shorn of its atheistic implications, he set out to apply it to his old love, chemistry. About January 1654, while tending his affairs in Ireland, he hinted to Hartlib that he was at work on "a short essay concerning chemistry, by way of a *judicium de chemia & chemicis*."[41] This essay survived in a copy made by Oldenburg and was later expanded into the *Sceptical chymist* (1661), in which Boyle proposed that a scheme of chemical corpuscles might usefully be substituted for the outdated systems of both Aristotle's four elements and Paracelsus's three principles.[42]

But perhaps the clearest statement of Boyle's early hopes for the application of the corpuscular philosophy to chemistry is to be found in the series of five tracts on matter-theory published in 1661 as *Certain physiological essays*.[43] These were written c. 1655–1659, that is, largely at Oxford, and outline a scheme by which the properties of saltpeter might be explained using atoms and corpuscles.

Boyle shrewdly began by refusing to avow either the Gassendian or Cartesian versions of the mechanical philosophy. His own chemical experiments would vindicate the details of neither. In fact, Boyle said, although he had consulted the works of Gassendi and Descartes, he had not systematically studied either of them, so that he might keep his judgment "unpossess'd" until he knew enough experiments to assess them properly.[44] He had, however, framed for himself "some general, though but imperfect, Idea of the way of Philosophizing my friends esteem'd."[45] This was not based on any attempt (laudable but premature) to explain all properties and processes in terms of "the more primitive and Catholick Affections of Matter, namely, bulk, shape and motion," such as both Descartes and Gassendi had claimed to do. Boyle entertained only the more modest hope of tracing back phenomena to the intermediate causes on which both atomists and Cartesians agreed: that is, to "little Bodies variously figur'd and mov'd."[46] This compromise, this common ground, Boyle designated "the corpuscular philosophy."

Boyle's first test of his "corpuscular philosophy" was in the small tract that comprised part of the *Physiological essays*, entitled "A Physico-Chymical Essay, Containing an Experiment with some Considerations touching the differing Parts and Redintegration of Salt-Petre."[47] The essay was composed between mid-1654 and mid-1656, after Boyle had returned from Ireland and while he divided his time between London and his new lodgings in Oxford.[48] He circulated it among "Learned Men," with whom it found "so favourable a Reception," that Boyle was "much encourag'd to illustrate some more of the Doctrines of the Corpuscular Philosophy, by some of the Experiments wherewith my Furnaces had suppli'd me." One of these men, Boyle said, had been so pleased that he had translated it "into very elegant Latin."[49] This was Henry Stubbe at Oxford, who recalled later that not only had he been happy to translate it into Latin, but that Willis had used the essay to demonstrate many of the same kinds of conclusions.[50] Just as Boyle had read Bathurst's lectures in manuscript, so had he circulated his own literary efforts among the Oxford group.

Boyle's essay began with a simple, yet novel, experiment. Common niter, after being purified by recrystallization, was put into a crucible and heated. When it had melted, Boyle cast in a small live coal, which was immediately enkindled and consumed. He continued to put in live coals until the saltpeter no longer enkindled them. He then took the hot mass and divided it into two portions. The first half was dissolved in water, and the resulting solution was mixed with spirit of niter (nitric acid) until reaction ceased; it was then exposed to the air. The second half was dissolved directly in spirit of niter and exposed to the air. When the first portion had crystallized, the long, thin shoots of saltpeter crystals could be discerned. The second portion seemed to give mixed results, part niter and part ordinary salt.[51]

We now know that such a deflagration of potassium nitrate gives a mixture of products that are in themselves compounds, and which in turn react with the nitric acid to reconstitute potassium nitrate. The exchange and recombination of ions is nowhere near what Boyle visualized. But he had grasped the essential point that by thinking of a chemical substance as composed of corpuscles, each of which linked two different atoms, one could explain how a reagent could apparently be destroyed, and reconstituted again. It seemed indeed that saltpeter was one such substance.

The remainder of the essay was given over to discussing possible protocol variations, detailed observations on the course of the experiment, and, most importantly, its theoretical significance. Boyle saw the experiment as demonstrating the following propositions:[52] (1) niter, or saltpeter, was composed of two parts, a *volatile salt* that flew off in the course of the reaction with coal, and a *fixed salt* that remained behind; (2) while niter itself was neutral, its constituent parts were exactly opposite—the volatile salt was *acid* and the fixed salt was *alkalizate*; (3) both the destruction of saltpeter and its subsequent synthesis, or in Boyle's word, *redintegration*, argued that a unit of saltpeter was composed of a particle of *acid volatile salt* and another of *alkalizate fixed salt*; (4) the qualities of saltpeter proceeded from the determinate matter and disposition of its constituent parts; (5) air *may* have something to do with the redintegration of saltpeter.

Boyle's purpose in adducing the first three propositions was reasonable; one had a series of anomalous analytical results that could not be explained on the assumption that niter was a unitary substance with inherent properties. So Boyle introduced the alternative assumption, that saltpeter was composed of parts, and lodged the resultant chemical properties (acidity, alkalinity, fixity, volatility) in the component particles. The fourth proposition was gratuitous; Boyle had no way of knowing whether chemical properties, of the whole or of the parts, proceeded from the figure and disposition of the parts. But it was a cardinal point of faith in the corpuscular philosophy, and Boyle acquiesced.

On the fifth point, the role of air in the experiment, Boyle was justifiably cautious. He conjectured that, since the crystals of niter did not form until the solution had been exposed to the air for some time, perhaps the air was impregnated with "promiscuous steams" that contributed to the process. But, he said, he was unwilling to assert this, both because he could not determine whether the air or the quietness contributed more effectively to crystallization, and because he had forgotten to stopper one of the glasses as a control.[53] But he invited his reader to investigate the problem further:

> . . . but whatever the Air hath to do in this experiment, I dare invite you to believe, that it is so enrich'd with variety of steams from Terrestrial (not here to determine whether it receive not some also from Coelestial) bodies, that the inquiring into the further uses of it (for I mean not its known uses in Respiration, Sayling, Pneumatical Engines, &c.) may very well deserve your curiosity.[54]

As if such a recommendation were not enough, Boyle added, apparently in the late 1650s, a tantalizing comment:

> Whether the air have any great interest in the figuration or in the re-production of

nitre, the author hath since examined by particular trials; but in vessels and by ways not to be easily described in a few words, and therefore the further mention of them is reserved for another discourse.[55]

Boyle, astute reader of Bathurst's lectures that he was, had the aerial niter, or *pabulum nitrosum*, lurking in the back of his mind, but he was too careful an experimentalist to admit it to his discourses without strong justification.

Boyle's fascination with saltpeter did not stop with the "Essay." By 1659 he had collected enough further observations on saltpeter to make it the focus of another tract, also published in the *Physiological essays*, "The History of Fluidity and Firmnesse."[56] Melting saltpeter showed that bodies could be fluid and yet not wet other bodies. The very passage of niter crystals from a solid to a liquid state under the influence of heat showed that the particles of a fluid body must be agitated in order to flow past one another.[57] Firmness, on the other hand, such as in fused saltpeter, seemed to proceed not from the hooked or branched nature of corpuscles, as traditional atomists had believed, but from the exclusion of air, which thus pressed on the outside of the body and forced the niter particles into congruent juxtaposition.[58] Each simple experiment on niter provided Boyle with the occasion for excursuses to explain how all the properties of matter, whether solid, liquid or gaseous, resulted from its corpuscular structure. Once begun, Boyle did not even stop with the essay of 1659. He continued to collect "Notes on ye essay concerning Saltpetre"[59] that once again formed the basis of another work, *The origine of forms and qualities* (1666).[60]

It was not long after completing the first essay on saltpeter that Boyle, c. 1657–1658, started to write up his ideas on a set of cognate topics: the "History of Heat and Flame." This, in turn, implied an investigation of the phenomena of cold, a subject on which Boyle did experiments in London and at Chelsea, most of them carried out and written up from mid-1660 to the winter of 1663. Niter, in a different guise, could be involved in cold as well as in heat.

Many philosophers, Boyle said in his *New experiments and observations touching cold* (1665), believed that there was a *primum frigidum*, a body that was of its own nature supremely cold and imparted this quality to all other cold bodies.[61] Although doubting even the existence of such a substance, Boyle duly examined the claims put forward for earth, water, and air as the *primum frigidum*,[62] as well as those of the newest candidate, niter. Gassendi, as well as some "eminently learned" Englishmen, had proposed that cold proceeded from the "admixture of Nitrous exhalations, or Corpuscles," that entered into air, water, and other bodies. As much as Boyle respected these men and their opinions, insofar as "they pitch upon Nitre, as the grand Universal efficient of cold," he confessed he could "not yet fully acquiesce in that Tenent."[63] Removing a calorific agent would, in many cases, Boyle said, cause a body to become cold, "without the introduction of any Nitrous particles into the Body to be refrigerated." Even though the "halitous part of Nitre" was disposed to fly up into the air and then dive down into the sea, how many "Nitrous Atoms" would it take to refrigerate the great expanse of the oceans? Yet, when one evaporated sea water, one obtained not saltpeter, but common salt.[64]

Moreover, Boyle felt that this idea of frigorific nitrous corpuscles was not supported by the experimental evidence. Although Gassendi had asserted that niter, dissolved in water, caused it to freeze, Boyle had performed this experiment many times and in many different ways, without the predicted result.[65] Using a thermometer, he had even measured the drop in temperature when niter was dissolved in water, but it fell far short of freezing the mixture.[66] Spirit of niter, "which is a liquor consisting of the volatile parts of that resolved salt," was not cold to the touch; on the contrary, it was potentially hot.[67] Gassendi thought that the figure of frigorific atoms was either tetrahedral or pyramidal; but the crystals into which niter shot were "Prismatical having their base Sexangular."[68] In sum, Boyle said, he who was "no stranger to Nitrous Experiments," could not produce the effects purporting to prove that niter was "such a wonderfully cold Body," and therefore he would remain sceptical of it as the *primum frigidum*.[69]

Boyle's ideas on niter, especially those embodied in the essay on saltpeter, were both circulated in manuscript among his London and Oxford friends and bruited internationally after publication in 1661. Oldenburg, during his pleasant year at Oxford in 1656–1657, had been tutor to Richard Jones, Boyle's nephew, to whom the *Physiological essays* were written; Oldenburg therefore almost certainly read them in manuscript. When he and Jones went abroad for the grand tour from 1657 to 1660, Oldenburg continued on the lookout for new information about saltpeter and about the chemical properties of air, which he reported back to Boyle and Hartlib.[70] Henry Stubbe, who had translated the essay on saltpeter into Latin, was moved to an intensive study of niter, beginning in the late 1650s.[71] Robert Sharrock, who after 1660 was living at Deep Hall and working at chemistry with Peter Stahl,[72] seems to have translated into Latin other parts of *Certain physiological essays* dealing with niter.[73] Thomas Henshaw, a friend of Hartlib and Evelyn, had been living in Kensington since the late 1640s, cultivating his interest in chemistry;[74] the publication of Boyle's *Physiological essays* in mid-1661 prompted Henshaw to gather together his own notes on the making of saltpeter, which were presented to the Royal Society in August 1661.[75] Henshaw's observations were later published by Sprat and subjected to heavy scientific criticism by Stubbe.[76] In late 1661 and early 1662 Oldenburg publicized the *Physiological essays* among his Continental acquaintances, noting to one that the work maintained "a profound silence concerning the Philosopher's Niter";[77] that is, Boyle discussed saltpeter and its component parts and not the more general active substance often mentioned in Hermetic tracts. Oldenburg also sent a copy to Spinoza, who disagreed with some of Boyle's conclusions, although more on philosophical than chemical grounds.[78]

The less exact notions of the "Philosopher's Niter" were well illustrated in a "Discourse" given at Gresham College by Boyle's older relative, Kenelm Digby, to the newly organized Royal Society in January 1661 and published in August of the same year, just as Boyle's *Physiological essays* were coming out.[79] In a discussion of the vegetative growth of plants, Digby digressed to praise, as Bacon had, the fertilizing effects of saltpeter when dissolved in water.[80] However, it was not, Digby said, the niter itself that nourished the seed. The saltpeter fertilized because, like a magnet, it "attracteth a like Salt that Foecundateth the Aire," and that had prompted Seton to say that "there is in the Aire a hidden food of life." Airs that were most impregnated with this "balsamick Salt" were healthful, while those abounding in earthy exhalations or marshy vapors were unsound. This salt was "the food of the Lungs, and the nourishment of the Spirits." Cornelius Drebbel had, Digby believed, succeeded in concentrating this balsam, which he had used to refresh the air in the submarine he had built. The Pope and the Duke of Bavaria had used this same principle of drawing the niter from the air to replenish their saltpeter mines.[81] This salt was everywhere—in the air and in the earth. It was, in fact, a "Universal Spirit," the "Spirit of Life" in plants and animals.[82] "Gold is of the same Nature as this aethereall Spirit; or rather, it is nothing but it, first corporifyed in a pure place, and then baked to a perfect Fixation."[83] Such a salt could be extracted from a boiled decoction of crayfish, which, when recombined with the distillate and incubated in a moist place, would give rise to many animals.[84]

One can understand why Boyle's silence on such a hermetic niter was so profound. Any body found both in the air and in the earth, that fertilized plants and nourished animals, that could be transmuted into gold or reconstituted into small animals, was more in the nature of a quasi-vitalistic seed than a delimited substance definable in corpuscular terms and manipulatable by chemical experiments. Boyle was not so enthusiastic for chemistry as to see an Aristotelian system of vague concepts and jargon replaced by one equally fanciful.

Indeed, the entire import of Boyle's work on saltpeter in the years after 1655 was, on the one hand, to examine its properties by a continuing program of focused experimentation, and on the other to understand those properties by conceiving of them in corpuscular terms. Such experimentation cast doubt, for example, on the Gassendian idea of nitrous particles as a "frigorific," thereby leading Boyle to concentrate even more on niter's heat-generating properties. Moreover, Boyle had clearly taken into his work on niter the assumption, derived in general terms from the Paracelsians, and more specifically in physiological terms from Bathurst's lectures, that a nitrous substance existed in the air. His conclusions on the bipartite nature of saltpeter had seemed not only to support that notion, but to define it more precisely in corpuscular terms. By the late 1650s, Boyle had come to suspect that the aerial niter might be the volatile, acid particle of saltpeter that was driven off in deflagration, and that was replaced by mixing the resulting fixed alkali with nitric acid.

Boyle might well have gone on to develop these *chemical* concepts of the air, had he not been diverted into a line of experimentation that focused his attention instead upon its *physical* properties. It is not without justification that Boyle's name is known to every schoolboy for his findings on the spring and weight of the air.

BOYLE, HOOKE, AND THE VACUUM PUMP: THE PHYSICAL PROPERTIES OF THE AIR

Boyle had had a long-standing, if not greatly active, interest in pneumatics. In early 1647 he wrote to Hartlib from Stalbridge that he had been reading Mersenne's account of a pneumatical engine, or wind-gun, which would use compressed air to shoot a bullet. Boyle thought it might be possible

> . . . by the help of this instrument to discover the weight of the air; which, for all the prattling of our book-philosophers, we must believe to be both heavy and ponderable, if we will not refuse belief to our senses.[85]

In the spring of 1648, Hartlib wrote to Boyle in Stalbridge and passed on an extract of Cavendish's letter to Petty in which Cavendish described the creation of a Torricellian vacuum and how a flat bladder could be made to expand in it.[86] By the early 1650s Boyle would have read of the Torricellian experiment, both in Gassendi's *Animadversiones* (1649) and Charleton's *Physiologia* (1654), as well as of the carp-bladder variant and the concomitant concept of air's elasticity in Pecquet's *Nova experimenta anatomica* (1651). By this time he had also done the Torricellian experiment himself and drawn the expected conclusion that it disproved the Aristotelian idea of a *horror vacui*.[87] At Oxford in November and December of 1657, he set up the Torricellian experiment and kept it in the frame for five weeks, noting that the mercury level rose and fell a full two inches. The "eminently Learned Men" to whom he explained the experiment could not account for this variation, although "that excellent Mathematician Mr. Wren" suggested that such a barometer might be used to test the Cartesian hypothesis that the tides were due to the pressure exerted by the moon on the air.[88] At about the same time Boyle even had a special glass tube blown with a hole and lid at its upper end, which could be sealed with putty or plaster. He hoped thereby to remedy the major difficulty of experimenting with the Torricellian apparatus—the fact that very few objects could be introduced into the vacuum created by the falling mercury column.[89]

Boyle was helped out of his difficulty by Otto Guericke. In 1647 the Magdeburg mayor had built an air suction pump using a cylinder and piston with two flap valves. At Regensburg in 1654–1655, he had used the pump to evacuate copper spheres, a feat reported by Guericke's friend, Gaspar Schott, in his *Mechanica hydraulico-pneumatica* in 1657.[90] By early 1658 Boyle had heard about Guericke's accomplishment, for Hartlib wrote to Oxford, "You still speak of the *German* vacuum as of no ordinary beauty."[91] But he must have experienced difficulty getting details. In later 1658, when his young nephew, Charles Boyle, Viscount

Dungarvan, then a gentleman-commoner at Christ Church, left Oxford for the grand tour, Boyle had still not read Schott's book, even though he told Dungarvan that he was "much delighted with this experiment," since it rendered the force of the external air so very obvious.[92]

But when Boyle did get a copy of Schott's book he was disappointed. Guericke's apparatus had two major deficiencies that made it impossible to extend the mayor's experiments. First of all, the pumping system was so inefficient that to create a reasonable vacuum it required "the continual labour of two strong men for divers hours." Second, and more significantly, the receiver in Guericke's apparatus was a solid globe of glass or metal, so that nothing could be put inside it.[93] The partial vacuum could be controlled, but its effects could not be manipulated. Since Boyle was interested in the air pump not as an end in itself, but as a tool for investigating the physical and physiological properties of the air, this was a serious difficulty. To solve the twin problem of Guericke's apparatus, Boyle set Ralph Greatorex, a London instrument-maker, and Robert Hooke, his Oxford assistant, to work on the question.[94] It was the latter who came up with the solution.

Hooke was a young man admirably suited, by temperament and background, for the task. When he came up to Christ Church from Westminster in 1653,[95] he had been one of Busby's favored pupils. Encouraged by the Headmaster to study mathematics, Hooke had mastered the first six books of Euclid in a week, learned to play the organ, and invented "thirty severall wayes of flying."[96] About 1655 he joined the meetings at Wadham. He showed Wilkins his designs for aviation, made models that actually flew, and even tried to design artificial muscles to supplement the insufficient power of man. He learned astronomy from Seth Ward, participated in the Oxford group's barometric experiments, and designed spring watches.[97] At about the same time he also became a laboratory assistant to Thomas Willis, the older physician very naturally choosing a bright young man from his own college. According to Hooke's good friend Aubrey, Willis then recommended Hooke to Boyle to assist him in his chemical laboratory, once the latter had established himself at Deep Hall.[98] By 1656 Hooke had done a fair amount of reading in the new natural philosophers of the Continent, especially in Descartes,[99] although it is unlikely that, as Aubrey suggests, he had to teach Cartesian philosophy in turn to Boyle. There may be, however, a modicum of truth in Aubrey's comment that Hooke taught Euclid to Boyle, as Hooke had shown early talent in geometry, while Boyle, although interested, was no mathematician.[100] The two complemented each other perfectly. Hooke had a quick, but not particularly patient or painstaking mind, coupled with a marvelous mechanical skill; Boyle was thorough and cautious almost to a fault and, although he was skilled in chemical operations, seems to have had no interest or talent for mechanical devices.

The construction and application of the air pump in 1659 mirrored not only the complementary talents of Hooke and Boyle but also the way in which Boyle's work was encouraged by his Oxford and London friends, and the results disseminated out along the same network. Greatorex first produced for Boyle an air pump that did not have to be kept under water to develop a seal, but Hooke felt it was "too gross to perform any great matter"[101] and perfected a pump of his own.[102] On a wooden frame he mounted a brass cylinder with milled valves; within the cylinder a piston was moved up and down by means of a geared ratchet. Some of the parts for the pump were not available in Oxford, so Boyle sent Hooke to London to have them made up.[103] More important for later success was the vacuum receiver that Boyle mounted on top of the pump. It was large—about thirty quarts— and easily accessible through a four-inch opening at the top. Although this necessitated cementing the lid every time the receiver was to be evacuated, it allowed Hooke and Boyle to put a large variety of objects into the receiver and then observe the effect of the vacuum on them. This convenience was enhanced by a smaller hole in the middle of the lid, into which a twist stopper could be inserted. By attaching strings or wires to the inside of this stopper, manipulations could be performed within the receiver without breaking the vacuum (see p. 131).

The apparatus was probably completed, at least in rough form, at London during the early months of 1659, since Boyle mentions a preliminary experiment made at that latitude "in Winter, in Weather neither Frosty nor Rainy."[104] Boyle then took it with him when he returned to Oxford in March. By the late summer of 1659 good results were forthcoming. Boyle sent reports of these to Paris,[105] where Walter Pope was sojourning[106] and where Boyle's nephews, Dungarvan and Jones, accompanied by Oldenburg, were attending meetings of Montmor's academy. Letters back to Oxford from Paris[107] informed Boyle that Montmor's virtuosi were at that moment "very intent" upon examining the role of air in the Torricellian vacuum, and were therefore highly appreciative of Boyle's new experiments.[108] By November 1659 Boyle could write to Hartlib from Oxford that he was "prosecuting some things" with his engine whose results would be not "unacceptable to our new philosophers," although "we have not yet brought our engine to perform what it should."[109] A few days later Hartlib passed on to Boyle the message from Oldenburg "that more experiments de vacuo are greatly longed for by your *French* philosophers."[110]

Boyle acceded to the growing volume of requests for information by dictating out, in the form of a letter to Dungarvan, his *New experiments physico-mechanical, touching the spring of the air, and its effects (made, for the most part, in a new pneumatical engine)*. He composed the book at Oxford in the late autumn, and finished it at an inn in Beconsfield, 20 December 1659, on his way from Oxford to London to visit his brother (and Dungarvan's father) the Earl of Cork.[111] Sharrock saw the work through the press at Oxford during the winter and spring of 1660, while simultaneously supervising its translation into Latin.[112]

Boyle, in the meantime, had taken his pump to London. There he worked on the *Sceptical chymist,* prepared the *Certain physiological essays* for publication, and showed pneumatic experiments to his friends. Robert Wood had the "satisfaction" to see some of the experiments there in early 1660, and later, upon

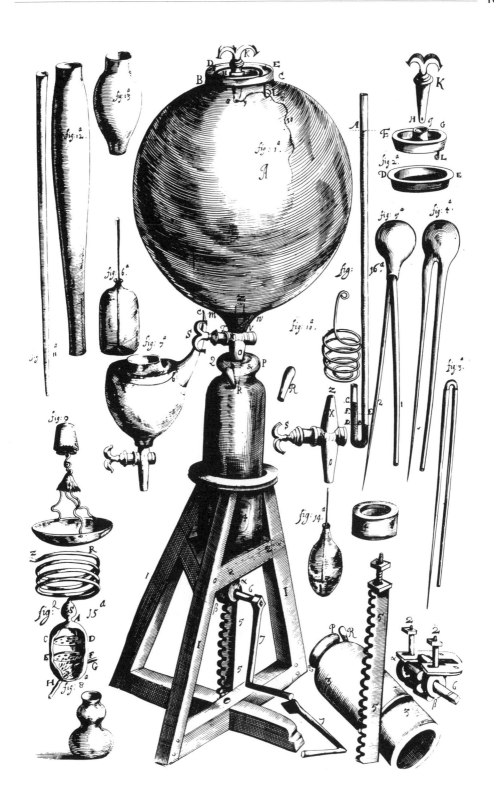

reading *New experiments,* noted approvingly that Boyle did not "affirme all the truth he might have done."[113] Oldenburg wrote from Paris in March that the virtuosi "long much" to see Boyle's book of experiments made with his "pneumatic Engine."[114] Evelyn too visited Boyle in Chelsea, where he saw Boyle's "*pneumatic* Engine performe divers Experiments."[115] The printing of the book was completed by June, and the work was available in the booksellers by August.[116]

Although a few of the preliminary experiments in this, Boyle's greatest single work, were done in London, most were performed at his lodgings in Oxford, where he had the benefit not only of Hooke's technical assistance, but the "presence of Persons, diverse of them eminent for their Writings, and all for their Learning."[117] For one crucial experiment Boyle specifically named Wallis, Wren, and Seth Ward as his "judicious and illustrious Witnesses."[118] When his receiver once showed a luminous cloud-chamber effect upon sudden evacuation, Boyle immediately sent for Wallis, who was "not then above a Bow shoot off," and who "made haste to satisfy his Curiosity." Repetition of the experiment at first failed to give the effect, and then unexpectedly the engine gave a flash and then subsequent ones, thereby convincing Wallis.[119] Although some of the experiments might seem superfluous, they were not, Boyle assured his readers. They were done to answer objections brought up in the course of research by ingenious men whom, out of modesty, Boyle would leave unnamed.[120] In all, it seems clear that Boyle's research agenda was carried out with the collaboration and conceptual guidance of those who gathered at Deep Hall.

Boyle's theme was aptly expressed in the title of his book—the spring of the air. Almost a quarter of the book was taken up directly with experiments that showed, in many ways, air's capacity both to exert pressure and to expand.[121] The very resistance encountered in pulling down the piston of the pump was due to the imbalance of pressures, or "spring," between the outside atmosphere and the interior of the receiver.[122] In a variant of Roberval's carp-bladder experiment, Boyle showed that a partially filled bladder expanded even to bursting when placed in the evacuated receiver.[123] By several different techniques he found that air could expand up to 150 times its original atmospheric volume.[124] The power of this expanding air was so great that an exploding globe cracked one of Boyle's receivers.[125] The atmospheric pressure upward was so great that it required a force of 122 pounds to pull the piston down from a highly evacuated receiver.[126]

All of these results could be explained by hypothesizing, Boyle said, that "there is a Spring, or Elastical power in the Air we live in":

> Air either consists of, or at least abounds with, parts of such a nature, that in case they be bent or compress'd by the weight of the incumbent part of the Atmosphere, or by any other Body, they do endeavour, as much as in them lies, to free themselves from that pressure, by bearing against the contiguous Bodies that keep them bent.[127]

When the pressure was removed or reduced, these particles unbent or stretched out to expand the volume of the air. He compared the air to a fleece of wool, made up of many flexible and slender hairs. These "Aerial Corpuscles" yielded easily to external pressure, but remained capable of expanding again. Another way of explicating this spring was that of the "most ingenious Gentleman, Monsieur Des Cartes." The Frenchman believed, Boyle said, that air was a jumble of flexible particles, of different sizes and shapes, raised from the earth by the heat of the sun, and whirling in the aether. On this model, Boyle said, spring would result from every corpuscle endeavoring to beat off every other that came within its sphere of rotation. Boyle put his finger on the most significant difference between the two models: one was structural and the other, kinetic. The first required air to have the structure of springs, the second only that its corpuscles be kept in "vehement agitation" by the fluid aether. Although Boyle was not willing "to declare pre-emptorily for either of them against the other," and he felt neither could explain all difficulties, he would use the first because it seemed "somewhat more easy."[128]

Boyle's inclinations definitely lay with the more Gassendian version of the corpuscular philosophy and had for some years. In an essay written about 1657, he had cautiously commended the idea that the resiliency of a blown bladder resulted from the "spring of those Aerial Particles wherewith the Bladder is fill'd."[129] In his tract on fluidity and firmness, written in 1659, probably while the pneumatic experiments were under way, he again alluded to "diverse Experiments"—with bladders, with the Torricellian apparatus, and with the adhesion of two smooth surfaces—that proved air consisted of innumerable "little Bowes or Springs," whose motion of restitution exerted pressure in all directions.[130]

According to either model, Boyle said, the normal atmosphere showed spring because of air's other property—weight.[131] He reported experiments that estimated the weight of air as 1/1000 that of water. Therefore, to counterbalance the mercury column of a barometer, one would need a column of air of uniform density at least 35,000 feet high. But since one knew that air was compressible and less dense at the tops of mountains than at their bases, the atmosphere may rise to some 100 miles.[132] Thus the results of Pascal's famous Puy de Dôme experiment in 1648, in which the French philosopher had observed a barometer to fall when taken to a height of almost a mile. This had been repeated, Boyle said, by a friend and great virtuoso (probably John Ball) "in the lower and upper parts of a Mountain in the West of England."[133]

Most of Boyle's other experiments—save those physiological ones to be discussed in the next chapter—examined the behavior of objects in his vacuum and related the results to the physical properties of the air, especially its spring. He managed to seal a Torricellian barometer into his receiver, and he, Wallis, Ward, and Wren observed the gratifying result that, as the container was evacuated, the level of the mercury fell.[134] Boyle did the same with a water barometer.[135] He found that a pendulum swung normally in a partial vacuum and that magnetism passed easily through it, but that the propagation of sounds was dependent upon air.[136] He found that water would freeze under reduced pressure and that exsuction did not produce any heat; the receiver may even have gotten colder.[137] Smokes acted like fluids and flowed to the bottom of an evacuated receiver.[138]

Two findings were to be particularly important, one on fire and the other on water. Boyle put wax and tallow candles, burning charcoal, and military slow-matches into receivers, evacuated them, and observed that the combustion was rapidly extinguished. When he put the items into closed receivers and did not evacuate them, the flames lasted longer.[139] Boyle was puzzled. He concluded:

> Whereby it seem'd to appear that the drawing away of the ambient Air made the Fire go out sooner than otherwise it would have done; though that part of the Air that we drew out left the more room for the stifling steams of the Coals to be received into.[140]

A red-hot iron, on the other hand, was unaffected by a vacuum, and a primed flintlock could, with some mechanical difficulties, be made to go off in a vacuum.[141] Liquids showed puzzles of another sort. As a receiver was evacuated, water placed in vials inside began to bubble, sometimes even so vigorously as to simulate a boil. Other liquids, such as salad oil, turpentine, and spirits of vinegar (acetic acid), urine, and wine (alcohol) did the same.[142] Boyle concluded that "in Water, there may lurk undiscernible parcels of Air."[143] He clearly preferred this explanation to the Aristotelian conclusion that water had been turned into air. Besides, Boyle said, the elastic power of air proceeded from its texture. He couldn't imagine how one could change particles of water to give them permanently the structure requisite to a spring, although it did seem that water vapors could sometimes emulate the spring of true air.[144] As a chemist he would, however, allow that "in general air may be generated anew." Volatile salts, "copiously" ascending into the air, seemed, for example, to be the cause of iron rusting. Boyle had himself generated a true air evolved from the action of dilute oil of vitriol (sulfuric acid) on iron nails. He collected this by water displacement, allowed it to stand for four days, and could demonstrate that it must be air: its spring resisted the rise of the liquid, and it dilated like air when warmed by hands applied to the container.[145] Boyle had generated a very new "air" indeed, hydrogen, but because it conformed to the expected *physical* properties of common air, it did not even occur to Boyle that it might have different *chemical* ones.

BOYLE AND HOOKE DEFENDING THE PHYSICAL AND CHEMICAL PROPERTIES OF THE AIR, 1660–1665

The years immediately after the publication of *New experiments,* from early 1660 to mid-1664, were ones in which Boyle and Hooke, working in London, elaborated, each in his own way, the physical and chemical properties of the air along lines roughed out at Oxford in 1655–1659. Boyle was active in the Royal Society from its very inception on 30 November 1660. He attended meetings faithfully and was a regular participant in experiments and discussions. He brought out five major scientific works and two smaller tracts, and rose rapidly to a position of international eminence. Hooke's emergence was slower. Until his appointment as Curator of Experiments at the Royal Society in November 1662[146] he continued to be in London what he had been at Oxford—Boyle's paid assistant, talented junior collaborator, and general factotum. After joining the Royal Society in a similar capacity, he rapidly took charge of its experimental agenda. When Walter Pope, Hooke's fellow Westminster and Oxonian, left London in June 1663 to shepherd Evelyn's nephew on the grand tour,[147] Hooke took over his lecturing duties as Gresham Professor of Astronomy and moved into the College permanently in August 1664.[148] In May 1664 he stood for the Gresham Professorship of Geometry, and after a contested election, he was finally confirmed in that position in March 1665.[149]

Before they dissolved their highly fruitful partnership, Boyle and Hooke amplified their ideas of the physical constitution of the air in three short books. In 1661 Hooke published a tract in which, expanding on the results of one of Boyle's pneumatical experiments, he suggested that the rise of water in capillary tubes, as well as the differing shapes of menisci in mercury and water, could be deduced from the pressure of the air and especially from "the Shape of the Springy Particles of the Air."[150] Unlike Boyle, Hooke strongly preferred the Cartesian version of the corpuscular philosophy, in which a "circum-ambient fluid aether" pervaded, moved and shaped the grosser particles of matter.[151] The other two books, by Boyle, were the result of controversy.[152] He had hardly finished getting out his *Sceptical chymist* and *Physiological essays* in mid-1661, when his pneumatical experiments were attacked by Franciscus Linus and Thomas Hobbes. Linus denied the existence of a vacuum and proposed instead that the suspension of mercury in the Torricellian experiment and the force drawing up the piston on Boyle's machine were due to a "funiculus" in the enclosed space, pulling up on the mercury and the piston.[153] Hobbes's attack was more vitriolic and diffuse, his main objection being to Boyle's implied assertion of the existence of a vacuum.[154]

Although Boyle liked to think of himself as an experimentalist, unpossessed of elaborate theories, he seems to have been drawn unwillingly into controversy, not to defend the pneumatical experiments themselves, whose accuracy was conceded by Hobbes and Linus, but to vindicate his notions of the spring of the air. He painstakingly answered the specific criticisms of each opponent[155] and showed in detail how Linus's hypothesis of a "funiculus" was "partly *precarious*, partly *unintelligible*, and partly *insufficient*, and besides *needless*."[156] He answered Hobbes's sarcastic comments about the company at Gresham College by pointing out that both the experiments and the expressed opinions were his own, published some months before the Society was founded.[157] John Wallis, in a rambling diatribe against Hobbes's book, written in the form of a letter to Boyle, likewise testified that Boyle's experiments had been done at Oxford before the inception of the Gresham Society.[158]

In answering these critics, Boyle reaffirmed and elaborated his concept of the air. Its ponderability was proved by experiments weighing bladders, aeolipiles, glass globes, and by Guericke's Magdeburg experiment.[159] Hobbes's notion of the air as a homogeneous penetrative fluid or aether moving circularly was not, Boyle said, consistent with experiment; Wallis scored the same point against his old adversary.[160] Although Boyle admitted to having rejected plenist ideas, he had not claimed to have proved or disproved a vacuum.[161] Nor had Boyle misunderstood Descartes. In addition to the "grosser and more solid Corpuscles" with which the atmosphere abounds, there might indeed be an aetherial matter; Boyle was not willing to declare either for or against the atomic or Cartesian hypotheses of the air.[162] But it was nonetheless clear that air had spring, a point on which Wallis agreed vehemently.[163] Boyle then examined Hobbes's and Linus's objections to the concept of spring and refuted them.[164] He even suggested a more detailed structural mechanism. Let us suppose that particles of air consist of long, thin, slender, and flexible *laminae*, coiled up like the spring of a watch. Assume also that these coils had an "innate circular motion" or spin. Particles so shaped and moved would necessarily resist compression and tend to expand when the force incumbent upon them was lessened. The rarefaction of air by heat was also explicable by the same model. As "Atoms of fire" flowed through the air, they would accelerate the rotary motion of the coils, causing them to "flye wider open," and thus occupy more space. Or, if his critics disliked this model, Boyle proposed the alternative Cartesian one: air was composed of "long, slender, flexible particles," agitated by the subtle aether, whose motions outward create pressure.[165] Although Boyle clearly preferred his whirling coils, either model would account for the obvious spring of the air.

It was into this context that Boyle introduced what was later to be known as his "Law." To prove that the force of the air's spring is reciprocally related to its volume, Boyle brought forward some quantitative results of experiments on its compression and dilation. He took a J-shaped glass tube, closed on the short end, and poured mercury into it in such a way that he could measure the compression of the trapped air in the shorter arm of the tube, and compare it to the weight of the column of mercury in the longer arm. The results agreed almost exactly with the hypothesis that the volume varied reciprocally with the pressure. Experiments done with a modified Torricellian apparatus proved that the relationship held for dilations as well.[166] Although Hooke seems to have made some preliminary experiments in the summer of 1660 on the relation of volume and degree of compression,[167] recent scholarship, especially that of Webster, has made it clear that the impetus for Boyle's experiments came from Henry Power.[168] He and his patron, Richard Towneley, both then living in Halifax, read Boyle's *New experiments* in 1660 and were moved to repeat and extend the pneumatical experiments Power had first made at Cambridge in 1653. By April 1661 they had obtained results on the dilation of air that suggested the reciprocal relationship. In July Power sent these findings to William Croone at Gresham College, who passed them on to Boyle.[169] Hooke immediately set up the J-tube he had used in his earlier trials and confirmed the Towneley-Power hypothesis for compression.[170] Boyle reported these results to the company at Gresham in September 1661,[171] devised his own experiments proving the same relationship for dilation, and published his and Hooke's results in May of 1662 as part of his *Defence* against Linus.

In all, it seems clear that during the two and a half years following the completion of *New experiments*, Boyle came more and more to focus on the physical properties of the air, especially its spring. These intensifying theoretical concerns were, at least for those years, not balanced by a complementary diversification of experimental research utilizing his pneumatic engine. When he used the pump at all, it was largely to repeat past experiments for visitors, in one case even for the king himself.[172] Nor was experimental work continued at Gresham. Vacuum trials were discussed there, and in May 1661 Boyle even presented the company with a copy of his own air pump; but, before the summer of 1662, it was little used except to repeat experiments already published by Boyle, or to entertain such visiting dignitaries as the Danish and Genoese ambassadors.[173] In all, success and controversy had caused Boyle to "see" the air in a new way.

It was Hooke rather than Boyle who recalled attention to the chemical properties of the air and to the hints thrown out by Boyle in his essay on saltpeter. The occasion was his famous book of microscopical observations, *Micrographia*, begun in 1661 at Wren's urging,[174] and completed by June of 1664.[175] Air, Hooke wrote, was "a kind of *tincture* or *solution* of terrestrial and aqueous particles," most especially saline, dissolved in an all-pervading Cartesian aether that kept them in perpetual agitation. The chemical analogy explained air's great powers of rarefaction and condensation. Just as a few grains of salt could disperse into water, yet still be found by precipitation, so could the air, dissolved and agitated by the fluid and agile aether, disperse and expand itself into a vast space. Hooke even speculated, based on this analogy, that one might filter out or precipitate air from the aether.[176]

What were these saline particles in the air, and what was their function? Hooke digressed from his microscopic observations of charcoal to answer this question with an hypothesis which, Hooke said with customary frankness, had not "been publish'd or hinted, nay, not so much as thought of, by any." The air in which we lived, moved, and breathed, Hooke said, was the "*menstruum*, or universal dissolvent of all Sulphureous bodies." These bodies could only be dissolved when heated, which resulted in fire, light, and in turn more heat. Hooke asserted further that

> *the dissolution* of sulphureous bodies is made by a substance inherent, and mixt with the Air, that is like, if not the very same, with that which is fixt in *Salt-Peter*, which by multitudes of Experiments that may be made with *Saltpeter*, will, I think, most evidently be demonstrated.[177]

These "dissolving parts of the Air" must be, Hooke surmised, relatively few, because a small quantity of melted saltpeter, which abounded with these

"Dissolvent particles," could consume a large sulphureous body very quickly and violently.[178]

Hooke's hypothesis of combustion caused by nitrous saline particles is most interesting.[179] First of all, it was clearly chemical and physical, rather than physiological in nature; he promised that he would elsewhere "shew the use of the Air in respiration,"[180] but he made no such attempt in the *Micrographia*. It was a corpuscular concept, but more Cartesian and kinetic than Gassendian and structural. The notion of a *menstruum*, or solvent, and fire as dissolution was clearly modelled on practical chemistry. The nitrous substance in the air, although necessary for the reaction, was seemingly passive. There was no real explanation of what combustion *was* beyond saying that it was a dissolving of sulphureous particles into the air, and that a niterlike substance was required for this process. Hooke was, in his discussion, clearly more concerned with showing how his hypothesis accounted for the structure of flame, the salty and sooty parts of vapors the ash left behind, and the rate of combustion,[181] than in explicating the role of the niterlike substance. No doubt when he spoke of the "multitudes of Experiments" which would verify the role of niter, he had in mind Boyle's famous saltpeter experiments. When he spoke of the niter-substance as being the same as "that which is fixt in *Salt-Peter*," he seems moreover to have had in mind Boyle's distinction of saltpeter into volatile and fixed parts. In fact, in many ways Hooke's hypothesis reads like a gloss on the Boylean experiment, seen from a different perspective. Boyle and Hooke were interested in arguing different points, and so approached the same data in different ways.

Hooke lost no opportunities to publicize his theories. Oldenburg praised the *Micrographia* in his newly founded *Philosophical Transactions*,[182] and, although he did not mention niter specifically, said that Hooke had offered "some considerable *Hypotheses*," among which was one "for making the *Air*, a dissolvent of all *Combustible Bodies*."[183] Hooke pushed the more specific point about niter in his capacity as Curator of Experiments for the Royal Society. At the first meeting of January 1665, while the *Micrographia* was still in press, Hooke performed the experiment of placing a live coal under a glass vessel. It soon seemed to go out. But when the glass vessel was removed and the coal exposed to the free air, it recovered its burning. This experiment tended to show, Hooke said, "that air is the universal dissolvent of all sulphureous bodies, and that this dissolution is fire; adding, that this is done by a nitrous substance inherent and mixt with the air."[184] One member objected that combustion might need agitation of the surrounding air—the coal went out because the air around it could no longer move. Disproving this suggestion experimentally took Hooke the next few weeks of meetings. He heated substances red-hot without exposing them to air and noted no combustion. He blew coals with bellows within closed containers; although the air within the box was agitated vigorously, the coals went out.[185] Boyle and Hooke then tested Hooke's conjecture that since rarefied air made burning bodies go out, so compressed air would keep them burning longer. Hooke's speculations were borne out at the next meeting; a lamp burned in compressed air for 15 minutes, while in uncompressed air "not above 3 minutes."[186] Both results were consistent with Hooke's unstated assumption that time of combustion depended on the concentration of aerial nitrous particles. Hooke went on in February and March 1665, pursuing a line of experiments to prove that niter could supply the want of air in combustion. He showed that niter could revive a flagging fire, that gunpowder could burn without air, that niter and sulphur would burn both inside and out of the vacuum pump, and that even tin filings cast upon molten niter would flame.[187] By June of that year, Hooke's experiments supporting his theory of air as a "dissolvent of combustible bodies" had become set pieces, used to entertain two visiting Frenchmen.[188]

Viewed from one perspective, the work of Boyle and Hooke on niter and on the physical properties of the air over the decade 1655–1665 displayed a tension that was not easily resolved. Boyle was a skilled chemist and a dedicated corpuscularian. His experiments on niter in the late 1650s strongly suggested to him that it conformed to the Gassendian model of two separable atoms—one alkali and fixed, the other acid and volatile—united to form a single molecule. Boyle and Hooke were equally aware that saltpeter, and most probably its volatile acid component, played an important role in generating the heat of combustion. Niter's cooling characteristics, they had determined, were clearly secondary. Yet the calorific properties of these volatile acid corpuscles were, in their aerial form, curiously refractory to experimental manipulation. One could prove that fire needed a supply of something in the air; one could deduce that this substance, like all air, was corpuscular; and one could demonstrate that the molecular form, saltpeter, could supply the lack of air in combustion of solids. But, in the chemical and experimental realm at least, one could for the moment go no further.

Conversely, Boyle and Hooke had in their vacuum pump apparatus a powerful new tool for examining many aspects of the air. But their very skill and dedication in using it pushed them, Boyle especially, inexorably toward ever more physical concepts of the air—those very properties of gases that pneumatic instruments in general are most capable of analyzing. In this they were spectacularly successful. We look upon "Boyle's Law" as a major scientific achievement. Yet in a sense it was merely a by-product of his program of pneumatic experimentation, his ever more sophisticated elaboration of the concept of the spring of the air, his search for correlations between chemical and physical properties, and, most generally, his firm commitment to corpuscular explanations. In the following chapter we shall see how these new investigative tools and physicochemical concepts were turned on one of the most ancient of biological problems—respiration.

NEW EXPERIMENTS ON RESPIRATION 1659–1665 6

When, at Oxford in late 1659, Boyle came to apply his air pump to the problem of respiration, he did so with growing scientific maturity and anatomical sophistication. In the five or more years since his preceptorship with Highmore and Petty, Boyle had continued to learn anatomy and physiology, both in London and at Oxford. He attended dissections and postmortems, which only convinced him of the variability of the supposedly uniform human body.[1] His subjects ranged from lusty young thieves to, as John Ward recorded, deceased college servants. In one he discovered blood flowing from a cut nerve; in another, he saw adhesions of the lungs to the ribs and took in the disagreement among the participating physicians as to how common this was. He came to believe ever more strongly that the anatomy and physiology of man could best be illuminated by animal dissections and vivisections.[2] Had not, he said, the discoveries of Aselli about the lacteals, of Pecquet on the thoracic duct, and of Bartholin on the lymphatics, been "first made in Brute Bodies, though afterwards found to hold in humane ones."[3] In the quest for understanding of the human body's parts and their functions, it was important, he wrote, to dissect and perform experiments on many different "Creatures." An organ might, in different animals, be absent or present, differ in size, situation, figure, or connection with other parts, such as to render its function "more conspicuous, or at least more discernible" to the inquisitive naturalist.[4]

Accordingly, Boyle carried out during the mid- and late 1650s a wide range of dissections and experiments, all of them informed by current debates in human physiology. He made embryological observations on the chick which he discussed with Highmore on visits to Stalbridge.[5] He confirmed both in hens' eggs and in fish that the heart and vessels were filled with red blood long before the liver had any appreciable color. He observed the regeneration of tails in lizards, and claws in lobsters.[6] He investigated at length the possible ways of preserving anatomical specimens, of which the best, his experiments showed, was by immersion in alcohol.[7] He did dissections and experiments to confirm the recent speculations that the digestive powers of gastric juice were due to its acidity.[8] To reconcile the "Controversie" between "Dr. *Harvey's* Opinion" and "That of the Cartesians" on the cause and manner of the heart's motion he excised the hearts of snakes and fishes to observe their motion. He took out the heart of a flounder, cut it transversely into two parts, carefully wiped off all the blood with a linen cloth, and observed that the severed and bloodless parts continued to contract "for a considerable space of time."[9] Clearly, Boyle did not accept the Cartesian idea that the heart's motion was caused by the expansion of the contained blood. His experiments also disproved that the heart was the "principal seat of Life and Sense." He could cut out the hearts of frogs, stitch up the wound, and observe them "to leap more frequently and vigorously" than before. Conversely, when he severed the head and brain from the body of a completely fledged chick embryo, the motion of all the parts ceased, save that of the heart. He did similar trials on snakes and insects.[10]

Experiments led him to speculate on the manner and use of respiration. When vivisecting vipers, he noted that their heart and blood, "even whil'st it circulates," was "actually cold." This observation gave him "just occasion to inquire a little more warily whether the great use of Respiration be to cool the Heart." Another puzzle of respiration was raised by an experiment with frogs. Although they had lungs, Boyle said, and breathed as well as other terrestrial animals, they could be held under water "for very many hours (sometimes amounting to some days) without suffocation." He used vivisection to illuminate the mechanical nature of respiration. For example, when one made a large wound in the side of a dog, though the lungs remained untouched, the one on the injured side would collapse and remain so during normal respiration.[11]

In this program of self-education in physiology, Boyle continued to read and to maintain his contacts with fellow anatomists. Essays written in the mid- and late 1650s cite not only a large array of writers on chemistry and materia medica, but also many in learned medicine (Sanctorius, Sennert, De Riviere, Helmont) and in anatomy proper (Colombo, Fabricius, Bartholin, Tulp, Borel, and Ent).[12] Boyle referred to the practice of "Oxford physicians" and praised the skills of Timothy Clarke.[13] He recounted stories about Harvey and must have been saddened when Hartlib wrote to him at Oxford in June 1657 to tell of the death of England's "Aesculapius."[14] Boyle and "that dexterous Dissector" Joyliffe even got together in London about April or May 1656 to perform the "noble" experiment of "taking out the Spleen of a Dog without killing him." To verify that this supposedly impossible experiment had been properly done, Boyle did part of it himself, holding the spleen in his hand while Joyliffe cut asunder the vessels "that I might be sure there was not the least part of the Spleen left unextirpated." The dog recovered within a fortnight and was often seen by Richard Jones as it romped around the Ranelagh house, whence, because of the poor dog's fame, "he was stol'n."[15]

Boyle clearly had both the anatomical knowledge and the experimental sophistication to see that the vacuum pump could do much to extend knowledge of respiration.

BOYLE AND VACUUM PUMP EXPERIMENTS ON RESPIRATION—OXFORD, 1659

Experiment 41 of Boyle's *New experiments* was done, he said, "to satisfie our selves in some measure, about the account upon which Respiration is so necessary to the Animals, that Nature hath furnish'd with Lungs." He and Hooke put a lark into the receiver of their apparatus, closed it quickly and plied the pump. For a while the bird appeared "lively"; but, upon a "greater Exsuction of the Air, she began manifestly to droop and appear sick," went into convulsions, and died, all within ten minutes, a good part of which had been spent in cementing down the lid of the receiver. Since the dead lark had been injured during capture, Boyle thought it best to repeat the experiment with a hen-sparrow caught with bird-lime and hence, unharmed. The receiver was evacuated until, within seven minutes, the bird seemed dead. Fresh air was let in, and the bird revived within fifteen minutes. But, upon beginning exsuction again, the sparrow was killed within five minutes. In a similar experiment a mouse was dead within about eight minutes, even though Boyle had expected that "an Animal used to live in narrow holes with very little fresh Air, would endure the want of it" better than the birds. Boyle even had the experimental animals opened; their lungs seemed very red, but this was normal, especially for birds.[16]

Boyle was puzzled both by his results and by their physiological implications. Although the air inside the receivers had been rarefied, there still remained, even at the end of the experiments, "no inconsiderable quantity." More importantly, it would seem that

> by the exsuction of the Air and interspersed Vapors, there was left in the Receiver a space some hundreds of times exceeding the Animal, to receive the fuliginous Steams, from which, expiration discharges the Lungs; and, which in the other cases hitherto known, may be suspected, for want of room to stifle those Animals that are closely pent up in too narrow Receptacles.

If animals in closed containers were killed by accumulation of wastes, why didn't they live longer in evacuated containers? Instead, it seemed that the death of the animals "proceeded rather from the want of Air, then that the Air was over-clogg'd by the steams of their Bodies."[17] As a control he had another mouse shut up in a tightly closed but unevacuated receiver. It lived above three-quarters of an hour and would have lived longer had Boyle not been visited by a ''Virtuoso of quality,'' whom Boyle obliged by showing how the animal could be killed within ten minutes by exsuction. Boyle later repeated the control by shutting a mouse in an unevacuated receiver overnight. This must have been done sometime in November or December 1659, for Boyle had to keep the engine by the fireside all night to protect the mouse from ''the immoderate cold of that Frosty Night.'' The next morning, over twelve hours later, the mouse was not only alive and well, ''but had devour'd a good part of the Cheese that had been put in with him.''[18]

With these results in hand, Boyle was ready to bring a degree of order into the theories of respiration mooted at Oxford, and in England generally, over the previous decade. In a long ''Digression containing some Doubts touching Respiration,'' he examined contemporary notions of its manner and function and brought his results to bear on them.[19] The importance Boyle attached to the subject is shown by the very length of the ''Digression''—almost forty pages, more than was devoted to any other subject in the book save its central concept of the spring of the air.

Boyle, like Bathurst, dealt first with the mechanism of respiration. The lungs moved not of themselves, he said, but only because they followed the motion of the diaphragm and thorax. As evidence he could cite not only the anatomical observation that the lungs were destitute of muscles and fibers, but also Highmore's vivisectional observation—which Boyle had himself confirmed—that a puncture in the thorax of a dog resulted in the collapse of the lung on that side of the mediastinum. The Galenists believed that the lungs attracted air, and the Cartesians that the impulse of the expanding thorax, communicated through the plenum of the air, forced it into the lungs. ''Our Engine,'' Boyle said, furnished an easy solution to this controversy. Both were wrong. When the diaphragm and thorax dilated the lungs, the pressure or ''spring'' became less on the inside than on the outside. The contiguous external air therefore pressed in at the open windpipe. Thus was solved a long-standing controversy: the chest was like a pair of bellows, filling because it was dilated, rather than like a bladder, dilating because it was filled. Having introduced his concept of the ''spring of the air'' to explain the mechanics of respiration, Boyle had to deal with an objection making use of the same concept. Could not the death of the animals in the receiver be due, not to a lack of air, but to the springy expansion of the air already contained within the thorax, which thereby forcibly compressed the lungs, impeding respiration and obstructing the pulmonary circulation? An observation and some reading provided Boyle with an answer. First of all, the animals *did* continue to move their lungs, even as they died; so the motion of respiration had not been blocked. Second, most of the animals killed in the air-pump were birds, Boyle said, and he remembered that Harvey in his *De generatione* had affirmed that in the lungs of birds there were small perforations in the pleura that allowed a free communication between the external air and that in the abdomen, and thus such a compression could not take place.[20]

Boyle then moved, rather more warily, on to his main subject: the use of respiration. As he wisely noted before he began, it was a topic on which "both Naturalists and Physicians do so disagree, that it will be very difficult either to reconcile their opinions, or determine their controversies." He began by reviewing the three major traditional explanations. The first, that respiration cooled and tempered the heat of the heart and blood, was held not only by most of the schoolmen, but was also well received among some of the "new philosophers," especially Cartesians. While acknowledging that the absence of lungs in certain cold-blooded animals seemed to confirm this explanation, he pointed out that the comparative argument could also provide counterexamples. Such animals as frogs had an obvious need of respiration, yet were destitute of any sensible heat. Moreover, temperate air, not cold, was best fitted for respiration. Most important, nature would not have been so profligate as to install an immoderate heat that needed perpetually to be cooled. Here he repeated exactly those arguments that had led Bathurst to the same conclusion. The precise similarity of the phrasing in the two works strongly suggests that Boyle, in arguing against cooling as a function of respiration, simply abstracted his own notes of Bathurst's "Praelectiones."[21]

The second traditional explanation was that the "very substance of the Air" was brought down by the vessels of the lungs into the left ventricle, not only to temper its heat, but to provide for the generation of vital spirits. Boyle was doubtful. In all the dissections he had seen, it was difficult to discern how the air could be conveyed into the left ventricle, especially since the respective motions of the heart and lungs were so far from being synchronous. Besides, the spirits of the blood were but its "most subtle and unctuous Particles," of a very different nature from the "lean and incombustible Corpuscles of Air." As for other objections against this opinion, he referred his reader to those that had been "propos'd, and prest by that excellent Anatomist, and my Industrious Friend, Dr. Highmore."[22]

The third use of respiration was to ventilate, not the heart, but the blood, to disburden it of those "Excrementitious Steams" arising from blood's "super-fluous Serosities." This was a quite common opinion, he noted, held among the moderns by Moebius, "and is said to have been that of that excellent Philosopher Gassendus." On this opinion Boyle could bring his air-pump experiments decisively to bear.

There were, said Boyle, two ways of elaborating this hypothesis. According to the first, air was a purely passive body. Just as a flame went out in a confined space because the ambient air could absorb no more fuliginous steams, so did an animal die in the same circumstances because it lacked a yielding ambient body to receive the sanguinary excrements, to "depurate the Mass of Blood, and make it fit both to circulate, and to maintain the vital heat residing in the Heart." According to a second possible interpretation, the air was more active. It not only passively received the wastes, but by reaching to the very ends of the bronchial tree, the air associated itself with the exhalations of the circulating blood and carried them away.

Boyle was quick to point out that his air-pump experiments disproved the first interpretation; in exsuction, increased space was left to absorb the waste products, yet the animals died more quickly. Boyle's results were, however, congruent to the second notion, that "there is a certain consistence of Air requisite to Respiration." If the air were too thick and already overcharged with vapors, it would be unfit to unite with the vapors of the blood and carry them off. If the air were too thin or rarefied, "the number or size of the Aërial Particles" was "too small to be able to assume and carry off the halituous Excrements of the Blood, in such plenty as is requisite." Confirmation for this could be found in the fact that men had difficulty breathing, both in the deep mines of Devonshire where the air was too thick, and in the high Andes, where it was too thin.[23]

This very physical interpretation of the action of respiration, dependent on the *pneumatic* as opposed to the *chemical* properties of the air, obviously held a great attraction for Boyle. It fitted very nicely with his own notions of the *spring* and *weight* of a homogeneous air, as well as with his *structural* interpretation of the source of these phenomena. One can see Boyle imagining that his aerial springs had to be in a certain state of tension in order to carry off the particles of waste from the blood. But he was also much too good an experimentalist to allow himself to be carried away by his own model. Although his results inclined him to believe that "the Ventilation and Depuration of the Blood" was "one of the principal and constant uses of Respiration," he suspected that air did something more. He simply couldn't imagine that the momentary inability to throw off "superfluous Serums" should be able to kill perfectly sound and lively animals so quickly. In a small receiver, birds could be killed within one and a half minutes, while others shut up in the same small receiver without exsuction lived between forty and forty-five minutes. Boyle had done such an experiment several times, which gained it the advantage of having "Persons of differing Qualities, Professions and Sexes, (as not onely Ladies and Lords, but Doctors and Mathematicians) to witness it."[24]

Boyle could not help but conclude "that there is some use of the Air, which we do not yet so well understand, that makes it so continually needful to the Life of Animals."[25] Boyle was naturally very drawn to a physical interpretation of the role of the air, one which limited its function to carrying off waste products. Yet his familiarity with Bathurst's speculations on an aerial niter and his own experimental results would not allow him to take the easy way out. Both speculation and experiment pointed unerringly to the fact that air not only took something *from* the blood, it also contributed something *to* it, a something which was not comprehended under the old discredited idea of vital spirits generated in the left ventricle.

To suggest some answers to this problem, Boyle examined briefly some theories that had previously been put forward. He cited the Paracelsian idea that only a vital quintessence in the air was absorbed in the lungs, and the rest was rejected. But, as Boyle pointed out, such an assertion not only failed to explain *how* this vital quintessence was necessary, but also ran into the same objections he had previously brought forward against the generation of vital spirits from the air. More suggestive were the ideas of Cornelius Drebbel. In the time of King James, Drebbel had built a submarine to carry twelve rowers and sundry passengers. Boyle had asked Drebbel's son-in-law, a medical man, how the adept had planned to supply air for the rowers. The physician replied that Drebbel conceived that it was not the whole body of the air, but only a certain "Quintessence," that made it fit for respiration. The remaining grosser part, or "carcase," of the air was unable to cherish the vital flame residing in the heart. Drebbel had prepared, the physician had said, a chemical liquor which, when unstopped, would replenish the vital parts of the air and make it respirable again, but he had never disclosed what this liquid was.[26]

It was tales of inventions like these, Boyle admitted, that sometimes inclined him to think favorably of the chemists' idea that the air was necessary "to ventilate, and cherish the vitall flame, which they do suppose to be continually burning in the heart." Moreover, Boyle said, his experiments with the air-pump tended to confirm this notion: "For we see, that in our Engine the flame of a Lamp will last almost as little after the Exsuction of the Air, as the life of an Animall." Yet he was still not convinced. Although his apparatus had shown "a new kind of resemblance betwixt fire and life," the theory of a vital flame had some difficulties, even contradictions. In the hearts of many animals, the blood might be warm or even hot. But it was, for Boyle, not easy to conceive either how the substance of the air could get to the heart and its blood or how, once there, it would be able to increase the heat. Air blown on hot liquids cooled them; it did not heat them. Although some "eminent Naturalists" had proposed that the flame in the heart was a temperate and almost insensible one, like that created by a cloth dipped in alcohol, this was based upon a false analogy; alcohol flames were very hot indeed.[27]

Boyle was obviously relieved to turn from this vexing conundrum to two other areas of respiration upon which he could bring new experimental evidence to bear: respiration of the fetus *in utero*, and of fish. The starting point in the first problem was a bit of Boyle's reading in *De generatione*:

> Our English *Democritus*, Dr. *Harvey*, proposeth this difficult and noble Problem to Anatomists, *Why a faetus, even out of the Womb, if involvèd in the secundines, may live a good while without Respiration; but in case after having once begun to breathe, its respiration be stoppèd, it will presently die?*

To test this proposition, Boyle and Hooke took a bitch about to whelp, hanged it, and then took one of the four puppies out of the womb. After they freed it from the membranes, placenta, and umbilical cord, it began to breathe. He and Hooke then opened the abdomen and thorax, cut the diaphragm, and observed that it still attempted to breathe. Turning to the other three, who had not breathed at all, they found the pups quite dead with no motions to their hearts. In contrast, the heart of the puppy that had respired continued to beat for some time, the auricle for about eight hours.[28] Although Boyle rightly refused to draw any conclusions from this vivisection, it did lead him to consider the larger problem of fetal respiration in the uterus. Since nature had, Boyle said, contrived certain peculiar and temporary vessels in the fetus such that the circulation of its blood bypassed the lungs, it seemed unlikely that fetuses respired in the ordinary sense. But his own earlier experiments had proved, he pointed out, that since all kinds of liquids "abound in interspers'd Corpuscles of Air," it was not altogether absurd to say that the fetus may actually exercise "some obscure Respiration." Boyle said he had even considered testing these ideas by putting a pup, with enclosing amnion and chorionic membrane, into the receiver, but he had not been able to obtain a bitch at the proper stage.[29]

Boyle had thus taken the initial tentative steps in the use of his discovery of dissolved gases to explain fetal respiration; he was even more successful in applying the phenomenon of dissolved gases to the respiration of fishes, where he could exercise experimental controls. Citing the experiments of "divers authors"— probably including Bathurst—he noted that fish died in ponds that froze over or in glasses that were stopped up. His experiments on dissolved air in water, Boyle concluded, made it seem "not impossible that Fishes may make some use" of that air, either "by separating it, when they strain the Water thorow their Gills, or by some other way." An experiment seemed to partially confirm this speculation. Boyle put an eel into a water-filled jar and placed the container into the receiver, which was then exhausted. The eel soon turned over on its belly and seemed dead. But upon taking it out, Boyle found that it was still alive. He was more than customarily cautious in interpreting this experiment as negative, especially since it had yielded a disappointing result; he concluded only that it was impossible to decide whether eels need only very little air or no air at all.[30]

Boyle's experiments with animals in evacuated receivers, and the "Digression" to which these trials gave rise, were the beginning of experimental respiratory physiology. Harvey had hinted, Highmore had argued, and Bathurst had speculated, but Boyle had experimented. As he was all too aware, his apparatus could not solve all the problems relating to respiration. In deciding against cooling and the generation of vital spirits as functions of respiration, Boyle perforce had relied upon the qualitative and comparative arguments of Bathurst and Highmore. But it was his "pneumatic engine" that enabled him to disprove conclusively the ancient theory that suffocation resulted from an accumulation of noxious wastes. He could substitute for it the hypothesis that ventilation of the blood in its pulmonary transit was dependent upon a requisite spring—not too little, not too much—in the air. He brought forward good evidence for believing that air dissolved in liquids could explain such diverse phenomena as fetal and piscine respiration. Perhaps even more importantly, the dramatically quick and convulsive deaths of animals in his evacuated receivers led him to suspect "that the Air does something else in Respiration, which has not yet been sufficiently explain'd."[31] Boyle's experiments further suggested that this "something" was related to combustion, although he, as an experienced dissector, was clearly uncomfortable with the vulgar notions of a cardiac fire fed directly by air. Harvey, in following the logic of the circulation, had been led to deny admission of the air into the body. Boyle, following the implications of his own experiments, had reintroduced it.

FURTHER EXPERIMENTS IN RESPIRATION—BOYLE, HOOKE, AND THE ROYAL SOCIETY

In the two and a half years following the completion of *New experiments*, from December 1659 to August 1662, Boyle and Hooke—and hence the Royal Society—did almost no further experiments on respiration. Boyle's time was taken up with preparing the *Usefulnesse* for the press, in bringing out the *Physiological essays*, the *Sceptical chymist*, and in defending his pneumatic experiments against Hobbes and Linus. Hobbes, citing the reputed abilities of divers to remain underwater for an hour, had scoffed at Boyle's contention that animals were killed by lack of air. In riposte, Boyle reported some new experiments in which he had drowned mice and birds in less than a minute.[32] He reiterated the same point in response to Linus, recounting an experiment in which moles were drowned and were killed by exsuction, in approximately the same length of time.[33] Other than these experiments, and one of a frog in an evacuated small receiver done in June 1660, Boyle and Hooke left off their respiratory inquiries. The Royal Society did the same. In April 1661 John Evelyn saw "divers Experiments in Mr. *Boyls Pneumatic* Engine" in which a chick was killed outright by evacuation, but a snake merely sickened.[34] A few months later Evelyn likewise viewed some trials made of a diving engine at Deptford, but the device entailed no fundamental physiological problems.[35] In April and May 1662 experiments of placing pigeons, chickens, mice, and vipers into the air-pump were put on the Society's agenda but never carried out.[36] The topic was in limbo.

The impetus for a renewed attack on the problem of respiration came from Boyle. In August 1662, his literary efforts completed, he began a new series of evacuation trials that continued, with an intermission in February and March, until May 1663. Although these "New Pneumatical Experiments about Respiration" were not published until 1670, the dated experiments within them, as well as Boyle's own testimony, place them firmly within this ten-month period.[37] Unlike the *New experiments*, in which Boyle drew far-ranging conclusions from a select but limited number of trials, these articles on respiration were uncompromisingly experimental. He reported over sixty experiments and drew no hypotheses whatsoever.

Boyle's major objective was to examine the differential response to evacuation of many different kinds of animals. His experimental subjects ran down the scale of both size and perfection: kittens, mice, ducks, linnets, larks, greenfinches, sparrows, vipers, snakes, frogs, tadpoles, gudgeon, crawfish, oysters, caterpillars and butterflies, grasshoppers, beetles, snails, leeches, flesh and ordinary flies, gnats, ants, and mites. Within this hierarchy he reported some consistent differences. Warm-blooded animals could be killed by evacuation within a matter of minutes.[38] He could wedge a full-grown duck into his receiver, and it would be dead within three minutes of beginning evacuation. A mouse put into a receiver so very small that it could be evacuated in one suck would die in less than half a minute. By way of contrast, cold-blooded, air-breathing vertebrates, such as snakes and frogs, could survive at greatly reduced pressure for hours, in some cases, even days.[39] Terrestrial invertebrates, such as snails and leeches, could similarly survive for days.[40] Insects were often stunned initially, but could live many hours.[41] Even the very smallest creatures, such as ants, gnats, and mites, required air.[42]

Boyle's results led to some generalizations about suffocation. They confirmed, as he had reported to Hobbes, that the times needed to kill by evacuation and by drowning were of the same order of magnitude.[43] A duck subjected to each procedure died within about six minutes. A snake lasted almost twenty-four hours under evacuation and could live for more than four hours under water. He found that warm-blooded vertebrates, such as linnets, larks, and greenfinches, could survive pressures of one-half normal, but not much beyond. This agreed, Boyle said, with the observations of travelers whom he had queried about the difficulty of breathing on high mountains.[44] His experiments also seemed to demonstrate that animals such as mice might be made accustomed to breathing air made unfit for respiration by rarefaction.[45] He reported that his experiments also supported Harvey's assertion that newborns—in Boyle's case, kittens—who were recently accustomed to live without respiration, could better stand the want of air.[46] Boyle even attempted to carry this acclimatization, or "assuefaction," one step further by growing animals within a vacuum. Unfortunately, tadpoles put into water under reduced pressure died before they could metamorphose into frogs. Water insects, however, would still turn into gnats under evacuation, although the gnats did not fly.[47]

But, even if rarefaction caused death, this did not necessarily imply that *any* air of normal pressure would be respirable. In an important set of experiments, Boyle showed that air could be rendered unfit for respiration, yet retain its "wonted pressure." He sealed a mouse and a mercury air-pressure gauge into an unevacuated flask. At the end of two and a half hours, the mouse seemed dead to spectators, yet the level of the mercury had not changed. He repeated the experiment with a bird; again the animal died without changing the reading of the manometer. To prove that the bird did not die from lack of cooling, he sealed yet another into a small flask, and, when it began to be distressed, he immersed the flask in water cooled by sal ammoniac, to no avail.[48]

Boyle confirmed further that liquids had interspersed air and that this was probably important for respiration. He put a bowl of swimming gudgeons into the receiver, evacuated it, and observed bubbles to rise from the water around the fish. Although within a few hours the fish were visibly distressed, they were not easily killed.[49] Fresh lamb's blood put under reduced pressure also bubbled and boiled, although at first not as manifestly ''as the Spiritousness of the Liquor made some expect.'' Cow's milk and gall did the same. This led Boyle to consider the possibility that death of the animals in the vacuum resulted, not from the removal of air and ''what the Airs presence contributes to life'' but rather from the multitude of bubbles that would block the smaller vessels and hinder the circulation of the blood.[50]

An allied point was also brought under experiment; just as in the 1650s Boyle had satisfied himself that the heart may beat without blood, so could he prove that it did not need air. He put the excised heart of an eel into a small receiver, plied the pump, and observed that, although the heart bubbled and grew tumid, it continued to maintain a regular beat. In repeating the experiment he found that when the heart's beat flagged after an hour, it could be revived by warming; only after the third hour did its motion cease beyond recall.[51]

Boyle's only overt and extended discussion of the function of respiration came in recounting an experiment that illustrated what he had ''not found denied by any, though considered by very few; namely, the office of the Air to carry off in Expiration the fuliginous steams of the Lungs.'' He distilled human blood and obtained a blood-red liquid that consisted, he said, of ''Saline and Spirituous particles.'' When unstoppered, a bottle of this liquid (probably ammonium hydroxide of some form) emitted copious white smoke. If an open vial were placed in a vacuum, it would cease to emit these visible fumes and begin again only when fresh air was readmitted into the receiver. The very contact of air, Boyle concluded, gave the corpuscles of moist bodies a peculiar volatility or facility to emerge in steams. These steams ceased rising in a stoppered vial, just as they did in a vacuum; in one case the air was clogged with steams, in the other there were, he implied, not enough air corpuscles to carry off the fumes. Although to avoid controversy, he did not speculate further, Boyle clearly saw this as strong evidence for one explanation of respiration: air became unrespirable either when clogged with steams or when too rarefied to volatilize these same steams from the blood.[52]

Such an extensive program of research naturally aroused interest in the meetings of the Royal Society at Gresham College, especially among Boyle's one-time Oxford confrères. On 20 August 1662, within a few days after recommencing respiration experiments with the air pump, Boyle offered to bring in his account of a duck in the air pump.[53] This initiated a flurry of respiration trials, the momentum of which was maintained by Hooke's appointment in November as the Royal Society's curator of experiments, and which only faded in July 1663, when

Boyle's private investigations had similarly run their course. The instrument used in a number of the experiments was the air pump, a duplicate of Boyle's own, with whose vagaries only Hooke among the Greshamites was prepared to cope. Many of the experimental themes were the same, while others supplemented Boyle's private and more strictly pneumatical experiments. In all, the two curves of inquiry match perfectly.

The most important line of research at the Royal Society, albeit not a very spectacular one, explored the possible relationship between combustion and respiration. In early October 1662 Wilkins proposed the experiment "of sinking a lamp in a glass-vessel under water, to see how long it would burn there."[54] At the next meeting Jonathan Goddard, who a few weeks earlier had devised an experiment to measure the maximum volume of the human lungs, was asked to help Wilkins with his inquiry.[55] On 29 October they reported their results. A lamp burning under water in a vessel of four gallons would last, they said, eleven minutes. The connection to respiration was made clear at the end of the report; Wilkins and Goddard were desired to repeat the experiment several times, "as also to try it with some live creatures."[56] Although the two Oxonians were pressed several times during the following month to continue their experiments, nothing happened until January 1663, when Boyle appeared again at Gresham after a prolonged absence.[57] On 21 January 1663 Hooke performed an experiment of trying how long the same air could be used for respiration "without the supply of fresh air." Breathing into a bladder, he found that it served for five inspirations, "though with difficulty." A lively discussion on respiration ensued. Christopher Merrett mentioned instances of divers who could stay long under water without respiration and told of a man reputed to know "an art of keeping new-born infants alive, without respiration, for a good while." Boyle reported his observations on a viper kept alive under water for five hours and urged "that the lungs of amphibious animals, and particularly of frogs and tortoises, might be carefully dissected, and the structure thereof well observed." Then, at the end of the meeting, "Mr. Hooke proposed an experiment against the next meeting of shutting up an animal and a candle together in a vessell, to see whether they would die at the same time or not."[58] The conjunction is important. Previous experiments had tested only the separate effects of enclosure on combustion and respiration. What could be more logical than to combine the two?

If Hooke and Boyle had hoped this experiment would vindicate the similarity of combustion and respiration, they must have been disappointed. First performed at the meeting of 28 January 1663, it was a resounding failure:

> Mr. HOOKE made the experiment of shutting up in an oblong glass a burning lamp and a chick; and the lamp went out within two minutes, the chick remaining alive, and lively enough.[59]

A repetition of the experiment a fortnight later, using a chick and live coals, yielded the same negative results: "the chick, after it had been kept in it [the glass] about 2¾ minutes, was taken out alive and well; but the coals were extinguished about a minute before.''[60] Boyle's speculation, expressed in *New experiments*, that the air-pump had pointed out some similarities between combustion and respiration, seemed clearly false. Air could be unfit for combustion, yet still support life. Little wonder that Boyle left this concept out of the respiration experiments then in progress, and emphasized all the more the importance of pressure.

But even such negative results did not fail to spark discussion. Attention turned to fish. Charleton thought they died because the motion of their gills was impeded. Merrett believed gills served the ''office'' of lungs, and that the pattern of blood's circulation in fishes supported this concept. Matthew Wren told of carp kept out of water for a week, needing only to have their gills moistened once or twice a day.[61] Thus began a long series of experiments, carried out from February through June 1663, on the respiration of fishes, in which the main participant was Hooke. Goddard reported that a tench shut up in a glass with water had lived some fourteen or fifteen hours.[62] Another tench in water was subjected to the air pump. It was immediately ''put into much disorder,'' and died within an hour and a half; when opened, its swim bladder was discovered to have burst.[63] Hooke later repeated the experiment with a ''middle-sized carp'' and was helped in the post-mortem examination by Timothy Clarke.[64] A like experiment on two young eels seemed, however, to show no effect.[65] In the meantime, Hooke was engaged in using the air-pump to repeat and extend Boyle's earlier findings that water did indeed contain air, of which it could be purged.[66] Thus, even though the work at the Royal Society suggested that Boyle may have been wrong about combustion and respiration, his speculations about fish ''breathing'' the parcels of air in water were better founded, even if the evidence was a bit muddled by the artifact of bursting swim-bladders.

In the midst of these trials Hooke came up with another mechanical marvel. Late in 1662 he had proposed some experiments testing the effects of compressed air on insects,[67] and by April 1663 he had designed a ''new engine for the condensation of the air.''[68] Among the inquiries he proposed for this ''compressing engine'' were ''What animals will live in it, and what die?'' as well as ''Whether fishes will live in water under a pressure?''[69] The initial experiment of ''killing a mouse in the condensing engine'' failed ''because the vessel was not staunch.''[70] but it was obviously expected to bear out Boyle's theory that air too thick was as deadly as air too thin. Certainly other recent evidence did. A few months before, Henry Power had sent in from Halifax a long account of the suffocating nature of mine damps, which were thought to thicken the air; Power's conclusions were confirmed soon thereafter by Robert Moray.[71] Unfortunately, after the initial experiments with Hooke's new machine for ''thickening'' the air, Boyle left London to visit his sister at Leese, Essex, for over a month; the trials with the compressing engine stopped and were not revived again until March 1664.

The results achieved at the end of 1663, as well as the questions yet unanswered, exemplify how Boyle's machines, in revolutionizing research on respiration, had also tended to push it in a very distinct direction. The parameter of pressure was the one most easily manipulated, so that when experiments were designed and the results conceptualized, it was most often in congruence with Boyle's favorite notion of the "spring" of the air. There was also a deeper source of conceptual momentum: the corpuscular philosophy.

Boyle was both a skilled chemist and a well-read corpuscular philosopher. He believed firmly that many phenomena could be explained by blending the two—by conceiving of matter as organized into particles, and as possessed of definite chemical properties. Each aspect of matter was simply one side of the same coin. When presented with a solid or a liquid, something with which one could develop an easy familiarity in the laboratory, Boyle tended to see its particles primarily in their chemical properties. But by the mid-1660s, when presented with a gas, Boyle's natural reaction, reinforced by the very success of his pneumatic experiments, was to think first of the physical properties of the aerial particles, and only secondarily of possible chemical characteristics.

This is perhaps best exemplified in seeing Boyle through other eyes. In May and June 1663, just as Boyle was winding up his respiration experiments, the junketing French virtuoso Balthasar de Monconys passed through London. On his first visit to "l'Académie de Gressin" he was greatly impressed with Boyle's engine, and recorded in detail how it showed that particles of air lurked in liquids. Monconys saw Boyle at these and later meetings, bought and read the *New experiments,* and a few days later visited Boyle at his lodgings in Chelsea. Boyle's conversation there was largely on the pressure of the air, on air particles in water, and on the role of the physical properties of the air in respiration—subjects all very congruent to the lines of work he was then carrying out privately, and which he and Hooke had also carried into the Royal Society. When Boyle returned Monconys's visit a few days later, they again talked about the weight of the air and about Boyle's instruments for measuring its variations.[72]

Yet, despite the dominance of this physical theme, the possible chemical properties of the air were not forgotten in Boyle's circle. Monconys recorded Oldenburg's affirmation that Drebbel had extracted "un esprit subtil de l'air" which, when added to the grosser parts of the air, rendered it suitable for respiration. A few days later Oldenburg and Monconys traveled three miles east from London to Stratford-le-Bow to call on Drebbel's son-in-law, Kuffler, who described the famous submarine and praised the *"Quintessence de l'air"* that Drebbel had isolated to make the boat's atmosphere respirable.[73] Another friend of Boyle and Hooke, Christopher Wren, had attended Royal Society meetings in the spring of 1663, and, when asked that summer by Brouncker for suggestions for experiments for the king's visit, wrote from Oxford with a more specifically Oxonian version of the same concept:

> It would be no unpleasing spectacle to see a man live without new Aire, as long as you please. A description of ye Vessell for cooling & percolating the Aire at once I

formerly showed the Society, & left wth Mr. Boyle. I suppose it worth putting in practise. You will at least learne thus much from it, that something else in Ayre is requisite for life, then that it should be coole only, & free from the fuliginous vapours and moysture, it was infected with in Expiration, for all theise will in probability be separated in the circulation of the breath in the Engine. If nitrous fumes be found requisite (as I suppose) ways may perhaps be found to supply that too, by placing some benigne chemical spirits that by fuming may impregnate the Aire within the Vessell.[74]

At much the same time Hooke was hard at work on his *Micrographia,* with its own calefactory version of the aerial nitrous substance. Clearly, the chemical approach to the air had by no means been killed in the vacuum of the *machina Boyliana.*

THE MECHANICAL THEME AND HOOKE'S EXPERIMENT OF 1664

Boyle's ''Digression'' on respiration, although it examined well the received explanations, especially as seen from the viewpoint of the Oxford group, was incomplete. He failed to take into account one explanation for respiration that had been suggested in passing by Harvey, and which had grown increasingly popular among European physiologists in the 1650s and 1660s: that the prime function of the lungs was, by their constant motion, to prepare, mix, and otherwise alter in a *mechanical* way, the blood passing through them. To some degree this popularity resulted simply from the exclusionist logic of scientific explanation. As the major traditional functions of cooling and generation of vital spirits fell into disrepute, uses formerly considered as minor, such as waste disposal and mechanical trans- mission, were reexamined for possible unperceived explanatory utility. But, to an even greater extent, the popularity of a mechanical explanation resulted from the interaction of traditional presuppositions with recent anatomical discoveries.

One such innovation was Pecquet's discovery of the *ductus thoracicus.* His announcement in 1651, that nutritive chyle from the lacteals bypassed the liver and flowed directly into the venous system near the vena cava, created a chaos in physiological explanation second only to Harvey's discovery of the circulation. Physicians across Europe, including Glisson's students at Cambridge and Willis's colleagues at Oxford, recognized that the liver had been dethroned from its privileged position as the *officina,* the factory, of the blood. Where then was the blood made? The question was made even more pressing by the deep-seated traditional assumption that every structure had its function and, more importantly, every function had its own structure. Since the function of sanguification could no longer be assigned to the liver, it had to be transferred elsewhere. This dilemma called forth two kinds of reaction. Some thinkers, breaking with the Galenic symmetry of structure and function, proposed that the blood itself made over the chyle into new blood: *sanguis sanguificat.* Others, unable or unwilling to change the patterns of their thinking processes, transferred sanguification to the heart and lungs. Had not Harvey proved that blood passed incessantly through the heart and lungs? What better organs could there be for mixing, kneading, and heating the food components and thus changing them into blood? A new justification could thus be given for the well-known necessity of respiration.

The implication of Pecquet's thoracic duct was reinforced by the Italian Marcello Malpighi's discovery in 1660 that the blood, in passing through the lungs, always remained within minute capillary vessels. Although it has often been assumed by historians that this finding "completed" the evidence for Harvey's circulation, the capillaries were certainly "seen" many times before Malpighi demonstrated their existence in the lungs. No one remarked upon them because they were not perceived as important in any physiological way. Harvey himself believed that the connection between arterioles and venules was made by blood flowing through small, open cavities, or sinuses, in the flesh, but he would as well allow the transit to be made by minute versions of the Galenic "anastomoses" believed to exist between arteries and veins.[75] John Locke observed such capillaries, but attached no particular significance to them; he wrote in his commonplace book during the late 1650s:

Circulatio   Take a frog [and] strip it you may
Sanguinis   see ye circulation of bloud if you
                 hold him up agt ye sun.[76]

Conversely, Malpighi diligently searched out the capillaries of the lungs because he believed they had a function.

In his *De pulmonibus observationes anatomicae* (1661) Malpighi argued that respiration did not serve to temper the imagined excessive heat of the heart, but rather to mix the mass of the blood. Blood consisted, he said, of an infinite number of serous and red particles that were not naturally endowed with fluidity, tending instead to unite and coagulate. The function of the lungs was to prevent solidification by mixing thoroughly the two kinds of particles—an explanation confirmed by the discovery of the thoracic duct and of the lymphatics, both of which dumped their contents into the blood. After receiving a rude mixing in the right ventricle of the heart, this composite was carried to the lungs for a more perfect amalgamation. There, by creeping through the smallest vessels, the blood was broken up, mingled, and "kneaded" by the compression of air in the surrounding vesicles, or alveoli. Gills and the placenta served the same purpose of mixing and mingling for, respectively, fish and the fetus. The entire scheme, Malpighi announced almost offhandedly, could be verified by looking at the dried lungs of a frog with a microscope: blood was always found within minute capillary vessels.[77]

Malpighi's mechanical theory of respiration, and the anatomical discovery that was its outcome, did not come quickly into England. Although the *De pulmonibus* was published as a small pamphlet in 1661, no copy of this exceedingly rare tract has been traced outside of Italy.[78] More likely it was introduced into England in late 1663 as an appendix to a work by Bartholin.[79] In the meantime, Englishmen had come to similar conclusions independently.

In late spring of 1663, Nathaniel Highmore was invited, most likely by John Wilkins, to communicate with the Royal Society. Knowing Highmore's interests, Wilkins included in the invitation a summary of the Society's recent work on piscine respiration. Highmore responded by offering what help he could and by praising the "noble experiment" that showed "the necessity of aire" for fishes. He analyzed the results at length, lauded Boyle's pneumatic experiments, and then went on to give his own views on the use of respiration. The key was the fetal state. If nature had contrived such wondrous vessels as the *foramen ovale* and the *ductus arteriosus* to bypass the motionless fetal lungs, it seemed logical that in the adult the lungs served some purpose with reference to circulation. Thus, Highmore argued, in inspiration the lungs admitted the blood from the right ventricle, and in expiration they emptied their enclosed vessels into the left ventricle. Lowering the ambient air pressure pumped up the blood and blocked the circulation across the lungs. A similar process occurred in fishes under reduced pressure. In both cases, Highmore said, the circulation "would be at a stande," and death would ensue.[80]

Highmore reiterated many of the same points, at about the same time, in critiquing Bathurst's concept of an aerial *pabulum nitrosum* taken up by the blood in respiration. Highmore visited Oxford in the early 1660s, where he called upon Willis and Bathurst, and where he read the manuscript of Bathurst's "Praelectiones" of 1654. Upon returning to Sherborne, Highmore wrote out a long letter to Bathurst noting that the lectures had given "much content in the reading them," praising his fellow Trinity man's treatment "de modo et instrumentis respirationis," but demurring as to its "finall causes." That a nitrous food was either in the air, or communicated to the blood, and thus constituted "the chiefest ende of respiration I do not see." Highmore iterated no fewer than seven reasons why such a salt could not exist in the air, and why it could not have the properties Bathurst ascribed to it, much less that it should somehow "repair" the spirits of the blood. Rather, respiration served to relieve the blood of heat, and even more importantly to allow the blood to circulate freely across the lungs. In arguing this mechanical theory Highmore cited not only the fetal evidence he was later to mention to Wilkins but also adduced the failure of pulmonary circulation as the reason why animals were killed so quickly "in Mr. Boyles engine."[81]

A related mechanical theory, similarly based on the pneumatical experiments of Boyle, was put forward a year later by Nathaniel Henshaw in his *Aero-chalinos* (1664).[82] Nathaniel, the younger brother of Thomas, had been educated in medicine and anatomy at Padua and Leyden. He was on the periphery of London science in the late 1650s, knew Evelyn and Boyle, and was active in the Gresham society during 1660–1662. His main contention in *Aero-chalinos* was that the chyle, after entering the venous system from the thoracic duct, was carried to the lungs, "where, by the reciprocations of that part," it was "yet more perfectly mixed with the blood."[83] The lungs were the true *officina sanguinis*.[84] Their churning motion aided a "fermentation" that perfected particles of food into particles of blood.[85] The function of the air in respiration was, by its spring, to mediate the reciprocal motion of the lungs, and by compressing all the pulmonary vessels equally, to mix the blood and to squeeze it on its way to the left ventricle.[86] Henshaw rejected entirely the putative role of the lungs in cooling and generating vital spirits. To the notion of respiration discharging wastes, he had no objection, noting only that this was not the primary purpose.[87] The crucial respiratory property of the air was, however, its "elastic power," a fact clearly demonstrated in the recently invented pneumatic engines.[88] Henshaw therefore closed his book with a description of a proposed pneumatic "chamber," in which a patient would be treated by increasing or decreasing the ambient air pressure.[89] In Henshaw's fancy, Hooke's machines were to be transformed from death-dealing to life-giving.

Several lines of inquiry at the Royal Society, carried out largely in the spring and summer of 1664, seemed congruent to this mechanical theory of respiration. One set of experiments utilized Hooke's "compressing-engine." In the first and most dramatic of these trials, Hooke put a bird in each of three containers, one containing air at normal pressure, one at a reduced, and one at an increased pressure. The bird in the "common air" remained well, that in the rarefied was panting, while that in the compressed was dead.[90] Subsequent experiments subjecting birds to twice, thrice or four times normal pressure had mixed results; in some cases the animals sickened and then recovered, in other cases they died.[91] Such results prompted William Croone to try a related experiment in June. He choked a chicken until it showed no signs of life and then revived it by blowing air into its lungs through a slender glass pipe inserted down the animal's throat.[92]

Thus, when Boyle left London in late July 1664 to return to Oxford via Stalbridge, he may well have puzzled over the problem of respiration and the difficulties of approaching it by experimental means. He thought to visit the mines at Mendip, Somerset—most likely to study their damps at first hand—but had to forego the trip to tend agricultural affairs at Stalbridge. He was, however, visited there in early August by his neighbor, "the ingenious Dr. *Highmore*." As befitted friends who were both virtuosi and country gentlemen, they discussed anatomy and gardening. Boyle, in reporting the visit to Oldenburg, mentioned that Highmore had told him of "some odd anatomical observations"—tantalizingly left unspecified.[93]

When Boyle arrived at Oxford in the middle of August, he found anatomy even more the subject of investigation. Willis, Lower, Needham, and Millington were all involved in experiments and dissections, and Boyle was quickly drawn in. He found one recent Oxford experiment particularly interesting. Needham had discovered that in a vivisected dog, even if the lungs had collapsed and the heart stopped, its beat could be restarted. One blew air into the *receptaculum chyli,* whence it travelled through the thoracic duct into the vena cava and then into the heart. More significantly, the heartbeat could be renewed without the concurrence of the lungs.[94] Here was an experiment that suggested, contrary to the accumulating evidence, that air did have some role in the body, especially relating to the heartbeat.

The finding must have seemed all the more crucial in contrast to a recent reminder of the mechanical theory that posited just the opposite. On 7 September 1664 Hooke sent a copy of Henshaw's book to Boyle at Oxford and noted in his letter the next day that Henshaw had referred to Boyle, "and altered many things in his book."[95] Clearly, Boyle and Hooke had been privy to its earlier drafts.

Many—and sometimes contradictory—lines of inquiry were converging: the vacuum and compression pump experiments of Boyle and Hooke; the mechanical theories of Malpighi, Highmore, and Henshaw; the inflation experiments of Needham and Croone; and the aerial niter theories of Hooke's recently printed *Micrographia*. Did air, or some portion of the air, enter the body through the lungs? Or was the mere motion of the lungs, perhaps in conjunction with the requisite atmospheric pressure, the important aspect of respiration?

The opportunity for experimental adjudication came quickly. Six months earlier, Boyle, Hooke, Wilkins, and "All the physicians of the society" had been constituted as an anatomical committee to guide the Society's work in this area.[96] On 5 October 1664 it was ordered to meet at George Ent's house "to consider of the particulars to be observed in the next anatomical administration."[97] The next day Oldenburg wrote to Boyle at Oxford to inform him of the meeting and to solicit suggestions for "some things, yt may be worthy of further investigation."[98] Boyle's response has not survived. We know only that on 12 October, when Charleton reported the committee's deliberations, the first inquiry for the next dissection was: "1. Whether there be any visible passage of the air into the heart."[99] Given what we know of his concerns and his immediate environment at Oxford, it may well represent the absent Boyle's "suggestion." Unfortunately, when the dissection took place at Gresham on 22 October, Charleton's misidentification of some structures involved him in a disagreement with Scarburgh and Ent, and the question of the passage of the air to the heart was forgotten.[100]

Hooke, not to be deterred, reintroduced the subject at the meeting of 2 November:

> Mr. HOOKE proposed an experiment to be made upon a dog by displaying his whole thorax, to see how long, by blowing into his lungs, life might be preserved, and whether anything could be discovered concerning the mixture of the air with the blood in the lungs. It was ordered, that the experiment be made between that and the next meeting.[101]

What were the proximate origins of this proposal? We know that over the previous few weeks, Hooke had been dissecting vipers,[102] the very animals whose lungs, Boyle had suggested in the 1650s, might shed some light on the nature of respiration. We know further that Hooke, now living in Gresham, and Boyle, at Crosse's in Oxford, were corresponding regularly; unfortunately, only Hooke's letters to Boyle have survived. But it seems likely that Boyle, deeply immersed in anatomical work at Oxford, impressed with the implications of Needham's experiment,[103] and having only recently discussed anatomy with Highmore at Stalbridge, suggested to Hooke in London that he use Highmore's open-thorax experiment both to explore whether air went to the heart and to test the mechanical theory of respiration.

Hooke performed the experiment at Gresham College on Monday, 7 November 1664, with Goddard and Oldenburg as spectators. It was done, Hooke said, in "prosecution of some inquiries into the nature of respiration in several animals." Hooke attached a piece of cane to a pair of bellows and inserted it into the windpipe of a dog. By pumping the bellows, which alternately inflated and deflated the lungs, Hooke could cut away the thorax and diaphragm and observe the regularly beating heart. He kept the heart thus exposed for "above a hour's time," and could have continued it "as long, almost, as there was any blood left within the vessels of the dog." Hooke found that when he removed the bellows, the lungs grew flaccid and the heart began to have convulsive motions. Upon renewing the motion of the bellows, the heart's motion became normal. Although the motion of the heart and lungs were thus related, they were not synchronous; Hooke did not "in the least perceive the heart to swell upon the extension of the lungs: nor did the lungs seem to swell upon the contraction of the heart." As to the purpose of the experiment, Hooke reported that he "could not perceive any thing distinctly, whether the air did unite and mix with the blood."[104]

Oldenburg was pleased with the results. His letter to Boyle on 10 November included a copy of Hooke's report on the vivisection and urged Boyle to write his reflections on the experiment.[105] In another letter a week later, Oldenburg thanked Boyle for his anatomical comments (not specified) and noted that Millington and Needham had visited the Royal Society the day before, where Needham had recounted "yt at Oxford they had by blowing into the receptaculum Chyli continued ye pulse of ye heart, wthout (if I mistook not) ye exercise of ye Lungs." Oldenburg rejoiced "to find Anatomicall Experiments and Observations so well poursued, both here and at Oxford." He was persuaded, he said, that "we shall at length find out more for ye use of respiration, and ye account upon wch it is so absolutely necessary, yn ever was done."

> Will it not be made out at last, yt Life is a kind of subtil and fine Flame? to wch ye Aire must be applied, to keep both it in motion, and ye blood, wherein it resideth, wch thence looks florid and sprightly, when the Aire, having been mingled with it, as it were, per minima, passeth along wth it into ye left ventricle of ye heart, and thence into ye Arteries; after wch, when the bloud returns into ye veins, it there begins to change its countenance, and looks dull and torpid, till it come again to ye place wch can revive it.[106]

Oldenburg saw, as he could be certain Boyle would, the implications of the experiment for physiological explanations.

The dissector himself, Hooke, was rather less pleased—not with the results, but with the unpleasant nature of the experiment. On the same day Oldenburg exuded enthusiasm, Hooke wrote to Boyle an account of the vivisection, "which I shall hardly, I confess, make again, because it was cruel." Hooke reiterated how, when the reciprocal motion of the lungs ceased, the heart and all the other parts of the body were seized with convulsive motions; "but upon the renewing the reciprocal motions of the lungs, the heart would beat again as regularly as before, and the convulsive motions of the limbs would cease."

> But though I made some considerable discovery of the necessity of fresh air, and the motion of the lungs for the continuance of the animals life, yet I could not make the least discovery of this of what I longed for, which was, to see, if I could by any means discover a passage of the air out of the lungs into either the vessels or the heart.

Although there were several Fellows of the Society that were "much awakened by this experiment," and intended "to prosecute it much further," Hooke himself could "hardly be induced to make any further trials of this kind, because of the torture of the creature." The inquiry was very noble, he said, but he doubted there was any opiate that would stupefy a dog under those conditions.[107] In the next few weeks Hooke therefore turned his attention back to the more insensible snakes. He reported to Boyle a fortnight later not only about Needham's recent visit, but that in dissecting a viper he had found an "infinite company of veins and arteries" in the lungs and even some small vessels that seemed filled with "several chains of small bubbles." But he was not yet able experimentally to determine whether this air "be conjoined with the blood in the vessels, or with the open air."[108]

If Hooke hoped his colleagues at Gresham would pursue the vivisectional approach to respiration, he was mistaken. Although the physicians of the Society were exhorted to consider, "whether and how the experiment might be farther improved,"[109] none did so. Ent and some of the other physicians helped Hooke dissect the viper's lungs,[110] but otherwise there was only talk. Goddard was assigned to review Henshaw's book and gave the fair summary that the author considered only the density and rarity of the air to affect men's health. Ent noted "that the author of this book made respiration chiefly serve for the speedier circulation of the blood from the one ventricle to the other," and added cryptically "that all phaenomena did not agree with this hypothesis."[111]

Hooke himself immediately took up, in January and February 1665, another aspect of the air—proving his theory, now published in *Micrographia,* that "air is the universal dissolvent of all sulphureous bodies," that "this dissolution is fire," and that "this was done by a nitrous substance inherent and mixt with the air."[112] In this he had the help and encouragement of Boyle, who had returned to London in mid-December 1664, where he stayed, regularly attending meetings of the Society, until driven back to Oxford in early June 1665 by the virulence of the plague.

The success of this line of work on niter, air, and combustion, discussed in the foregoing chapter, once again prompted trials about the similarity of respiration and combustion. Ent suggested in late January 1665—possibly in ignorance of the failure of such experiments in 1663—"that some animal and a burning candle might be included together in a close glass-vessel, to see, whether they would live one as long as the other.'' The results this time were more equivocal. A bird and a chafing dish of coals were put together into a receiver. At about the same time the coals had been extinguished, ''the bird also began to die, but being let out into the open air, recovered.''[113]

If common air acted this way, what of the ''air'' generated by chemical reactions? In early March, immediately after Hooke's experiment showing that niter could render even tin combustible, Wilkins brought forward Wren's suggestion that the air generated by effervescing liquids should be collected in a bladder.[114] This was done the following week, 15 March 1665, with Wren now present, when the air (carbon dioxide) generated by the action of aquafortis (nitric acid) on ''pounded oister-shells'' (calcium carbonate) was collected in a bladder. As the gas was being collected, Wren explained how this experiment illustrated by analogy the motion of muscles by explosion. Boyle then suggested putting this air to some use. He proposed that an animal be put into the receiver of the air-pump and by evacuation be made sickly. Then some ''new air,'' generated inside the receiver by the action of distilled vinegar (acetic acid) on coral (calcium carbonate), would be released ''to see, whether by this means the animal could be revived.'' In a similar vein, Wilkins moved that the aquafortis/oyster-shell ''air'' be blown into a dog's or cat's mouth, to see what the effect would be.[115]

Two important experiments were thus carried out the following week at the meeting of 22 March 1665. The first was Boyle's. A bird and the vinegar/coral generator were placed in the receiver of the ''rarefying engine.'' The air was pumped out until the bird became sick, and then the carbon dioxide was released. The bird ''recovered not.'' The second experiment was done in the same manner, using a kitten instead of a bird, and aquafortis instead of vinegar. In this case, however, the dying kitten seemed to recover when the ''nitrous exhalation'' (in reality also carbon dioxide) was released. The experiments prompted suggestions and discussion. How, it was asked, could one be certain that these gases were ''true air''? Brouncker answered ''that a body rarefied by heat, and condensed by cold, was air.'' A bladder of the substance was put to the fire, and it obligingly expanded; when taken away, it grew flaccid. Thus, one can see the Greshamites agreeing, it must be true air. A vial of the ''steams'' produced by vinegar acting on oyster shells was then passed around among the members to smell. It was ''found by most of them incommodious, as it was undiluted.''[116]

The experiments, especially that with the ''kitling,'' held out great promise. Even a new and very green member, Samuel Pepys, saw the significant point: that the ''ayre'' used to revive the kitten was ''made by putting together a Liquor and some body that firments—the steam of that doth do the work.''[117]

Unfortunately, this promising line of work died almost as quickly as it arose. Carbon dioxide was generated at several subsequent meetings, and Hooke once suggested that its effect might be tested on dogs, but the respiration trials were never carried out.[118] Goddard, Wilkins, and Boyle's assistant, Daniel Cox, digressed onto inquiries on the relation of air to vegetation.[119] Boyle himself suggested that putting an animal with a thoracic puncture wound into the vacuum pump might illuminate "whether an animal in an exhausted receiver dies for want of air, or because of the compression of the lungs."[120] But these experiments, carried out in April and June, proved either inconclusive or merely confirmed the well-known fact that a thoracic puncture renders the lung on that side incapable of dilatation.[121] The plague forced the cessation of meetings altogether at the end of June, and many members left London. The final echo of "nitrous exhalations" that spring was in a letter written 20 April 1665 by Wren to Hooke, after the Savilian Professor had returned to Oxford. Wren, whose ideas on nitrous gases had started this line of inquiry in March, expressed his doubts about the possibility of "generating Ayre under water" by "corrosion of Nitrous or animall saltes." The "Ayre" liberated by these "agitated Saltes" would, if collected over water, "returne again to Nitre & water." The gas let loose would be absorbed again "by degrees into the little particles that imprison & condense the Ayre as before," and the purpose of the experiment would be defeated. Hooke may well have taken these objections to heart, as he did no more "new air" experiments after its receipt.[122]

Six years of intense and resourceful experimentation had yielded a picture of respiration at once clearer and more complex. The speculations of the previous decade had been put to the test. Cooling and generation of vital spirits were out, excluded not so much by trial as by consensus. Boyle had refined the idea of waste disposal by seeming to demonstrate that it was an active rather than a passive process, one in which excrements were plucked out of the blood by air of the requisite spring. The mechanical function of respiration seemed all the more confirmed, most especially by Hooke's pulmonary inflation experiment of 1664. The experiment itself was soon made known beyond the Royal Society and its Oxford correspondents by inclusion in Thomas Sprat's *History of the Royal Society* (1667),[123] and its subsequent translations into French and Latin. Perhaps more importantly, the mechanical theory, in relating respiration to the circulation of the blood, also explained the prominent and very mystifying finding of Boyle's vacuum-pump experiments: the extreme rapidity with which animals died in rarefied air.

Yet the question was by no means settled. Scattered and less precise pieces of evidence suggested that air was related to animal heat, and far from diminishing, increased it. Warm-blooded animals were much more highly dependent on respiration than their cold-blooded brethren. Yet the similarity between combustion and respiration was far from a demonstrable identity—witness the fire and bird experiments of 1663 and 1665. And still there was the problematic nature of niter, both solid and aerial. Hooke's experiments of 1665 had supported his contention that fire used an aerial component and that niter could supply the want of air, but they could not complete the circle by proving that a nitrous substance existed *in* the air. The same difficulties applied *a fortiori* to respiration. One could suggest, but not prove, that breathing depended on extracting nitrous particles from the air. Even when air could be generated anew, it could not always be proved to have life-supporting properties. Moreover, if some particles were removed from the air in respiration, why did it retain its normal pressure when made unfit? The gaps of knowledge that separated the physical, chemical, and physiological properties of the air were, even after intensive investigation, still great enough to preclude a unitary theory of its nature.

Just as the experiments of 1659−1665 were rooted in the more diffuse speculations of the preceding decade, so also were the patterns of communication and collaboration in this period based upon the network of friendships built up in the "clubb" at Commonwealth Oxford. All of the contributors were, with the passing exceptions of Henshaw and Croone, sometime participants in the Oxford meetings at Wadham, Crosse's, or Beam Hall: Boyle, Hooke, Oldenburg, Wren, Wilkins, Goddard, Highmore, and Needham. The pace, nature, and direction of investigations mirrored their enthusiasms and prejudices, their comings and goings. This is not to say that the group was "isolated" from other scientific ideas, but merely to reaffirm a sociological commonplace: especially in avocational matters, in questions almost of intellectual "fashion" in an age before professional science, a man's concerns, preferences, techniques, and to some extent his very ways of thinking reflect the ideas and accomplishments of those whom he respects, and with whom he is in active association and communication.

OXONIANS ON ANIMAL HEAT AND THE NATURE OF THE BLOOD 1656–1666

7

Boyle's pumps—both vacuum and compression—had been so successful in clarifying the nature of respiration precisely because they came to this ancient physiological problem, not from the body and its anatomy outward but from the air and its properties inward. But the problem remained in essence biological. Any truly successful treatment had not only to suggest reasons why the air was so necessary, but to articulate those explanations with new physiological conceptions of animal heat, the constitution of the blood, the reasons for its colors, the action of the heart, and even with such subsidiary problems as fetal respiration. The new system of explanation had to be as comprehensive as the old.

The pivotal problem, at least initially, was heat. If, as Bathurst and Boyle had suggested, the primary function of respiration was not to cool an innate heat (whether in the heart or in the blood), the way was open to consider the opinion of the "chemists" that organic heat was generated, and that air served a similar function in feeding both ordinary fire and the *flammula vitalis*. But it was a way blocked by many obstacles. The foremost of these was the palpable difference between animal heat and combustion. Animal heat was gentle, controlled, and constructive; combustion was vigorous, rampant, and most of all, self-destructive. Men skilled in dissection, even two such different ones as Harvey and Boyle, had to admit, each for his own reasons, that organismic heat was different from fire. What was needed was a conceptual system within which the generation of heat, even in such palpably different manifestations as animal heat and fire, could be seen as resulting from a common chemical action. In modern terms, that root process is oxidation. Both metabolism and combustion are actions that consume oxygen from the air and produce heat. Although the rates differ by several orders of magnitude, the process is the same. Oxford physiologists had in nitrous aerial particles a rough conceptual equivalent to oxygen. What they needed was a chemical concept of the root process it subserved. Thomas Willis filled that conceptual void.

WILLIS AND FERMENTATION

Willis's interest in chemistry antedated his membership in Petty's club at Buckley Hall and may even have gone back to the years of the Oxford siege, when he first entered medicine.[1] Before moving out of Christ Church about 1648, Willis "studied chymistry in Peckewater Inne chamber,"[2] Aubrey recalled. In the following year, John Lydall reported to Aubrey the attempts of Bathurst and "Mr Willis our Chymist" to make *aurum fulminans,* one of the few explosives not compounded with niter.[3] By 1652 Willis was one of those who, Seth Ward said, "have joyned together for the furnishing an elaboratory" at Wadham, for which apparatus Willis neatly recorded his expenditures in a notebook.[4]

Through his contacts with Wilkins and with such recurrent visitors as Boyle and Aubrey, Willis's reputation as a chemist spread. Hartlib noted in November 1654:

> Dr. Wellis of Dr. Wilkins acquaintance a very experimenting ingenious gentleman comunicating every weeke some experiment or other to Mr. Boyles chymical servant, who is a kind of cozen to him. Hee is a great Verulamian philosopher, from him all may bee had by meanes of Mr. Boyle.[5]

By June 1656 Willis had completed his first chemical work, a tract "De fermentatione." In the usual fashion, he circulated it in manuscript among his friends; Robert Wood later recalled "having formerly seen some sheets of his upon ye subject, when we met at a Club, at Oxford."[6] In November 1656 Aubrey similarly reported to Hartlib that Willis was "a leading and prime man in the Philosophical Club at Oxford," who "hath written a treatise De Fermentatione" that Aubrey "much commended."[7] This treatise, published two years later as the first of *Diatribae duae medico-philosophicae,*[8] was the linchpin of Willis's chemical philosophy and the basis of his later teaching at Oxford.[9]

Willis's doctrine of fermentation was founded upon a compromise between chemistry and atomism. He saw matter as divisible into ultimate particles, but he ascribed to these particles separate identities based upon their *chemical composition,* and not upon their shape and size, as did the purely mechanical atomists. These particles could be grouped according to their activity into five classes, or "principles": spirit, sulphur, salt, water, and earth. Thus there were, he believed, many differing particles of spirit, but they all shared the common property of being highly subtle and active, always endeavoring to fly away. Sulphurous particles were as a group heavier than those of spirit, but still endeavored to fly away. What we perceived as heat was due to the motion of sulphurous particles in a body. With relatively little motion, they produced maturation and sweetness; with more motion, sensible heat; and with very vehement motion of the sulphurous particles, the body dissolved into flames. Willis's "sulphur" thus encompassed, in modern terms, the large groups of substances capable of being oxidized, most of them containing carbon. At the other extreme of activity, the principles of salt and earth encompassed particles of a more fixed and solid nature.

Change arose, Willis believed, when particles of one principle joined with particles of another. Thus saline particles could become more fixed when joined to those of an earthy nature, and more volatile when joined to sulphurous or spirituous particles.[10] For example, the sulphurous particles in the blood joined with those of salt and spirit to produce a more volatilized salt that accounted for blood's florid color.[11] Willis's conceptual scheme deftly blended the chemist's emphasis on properties having a real existence in matter with the atomist's insistence that matter was organized into natural minimum units and that change resulted from the combination and rearrangement of these particles.

Under this scheme, fermentation was the name given to a process of corpuscular rearrangement that resulted in gross and perceivable change:

> Fermentation is an intestine motion of particles, or of the principles of every body, tending either to the perfection of the same body, or because of it, to a change into another body.[12]

Willis's aim in the remainder of the treatise was to show how this chemical/corpuscular conception of fermentation could account for all changes, especially organic ones. It explained, he believed, the growth of plants,[13] animal digestion,[14] alcoholic fermentation,[15] putrefaction of bodies,[16] and even precipitations and coagulations.[17]

Willis did not originate the general idea of animal "fermentations." During the 1640s a number of works by Angelus Sala (1576–1637), by his son-in-law Anthor Gunther Billich (1585–1639), and most especially by Jan Baptiste van Helmont (1579–1644), had argued that various "ferments" were the origin of organic change.[18] The difference between Willis and the Continentals lay not in the name or in the chemical terminology, but in the underlying conceptual structure and the presupposed physiological scheme. Helmont's digestive fermentations, for example, were orchestrated by a series of *archei,* indwelling directive forces of a kind that were anathema to an Oxford generation that had rejected such quasi-Aristotelian refuges of ignorance. Helmont had ignored the circulation, which all of the Oxonians accepted implicitly. Willis in turn ignored Helmont's vitalistic and obscure metaphysic and recast the Helmontian idea of fermentation into intellectually respectable corpuscular terms.

In doing so, Willis produced the first hints of a linkage between blood, heat, niter, and the air. Blood, like any liquid composed of heterogeneous particles of the five principles, was always in fermentation. To maintain this, nature placed a ferment in the heart. As blood passed through, the bonds of its corpuscles were loosened and the particles of spirit, sulphur, and salt endeavored to break out. This produced an "ascension" or enkindling, even an effervescence, whose heat was diffused through the vessels to the rest of the body.[19] Heat of any kind, and most especially fire, was, Willis believed, nothing but a vehement motion and eruption of sulphurous particles. These particles excited a similar motion in other sulphurous particles, and so the combustion was spread.[20] Niter was a special example of this scheme. It consisted chiefly of salt, with a little sulphur, arranged in such a way that the sulphur particles were confined by the saline ones. Melting niter loosened this conglomerate, but not enough to release the enclosed sulphur particles. But if one added more sulphurous particles, such as were found in charcoal or the common chemical reagent sulphur, the unstable mixture would explode violently.[21] The explosion of *aurum fulminans,* such as Bathurst and Willis had made in 1649, could be explained structurally in the same way. Particles of salt (in this case, sal ammoniac and tartar) were bound by the gold; when they escaped violently, an explosion resulted. The gold/salt complex was unstable in the same way as the salt/sulphur complex that made up niter.[22] Moreover, Willis accepted, *en passant,* the opinion of the "most famous Gassendus" expressed in his *Animadversiones,* that many atmospheric phenomena were caused by interactions of particles of salts and sulphurs, among them, nitrous salts, exhaled from the earth.[23]

The physiological ideas of "De fermentatione" were expressed only *inter alia* in a tract that was largely nonmedical. Willis's real development of fermentation as a physiological concept came in the companion treatise on fevers. Willis continued to work on "De febribus" as late as the autumn of 1658, for it includes accounts of Oxford epidemics written up in September 1657 and in June and September 1658.[24] He then combined "De fermentatione" and "De febribus" with a tract on urines written as a letter to Bathurst, and published them all in late November 1658.[25] Robert Boyle must have been among the first to get a copy, for in mid-December he sent one to Hartlib, who in turn passed on news of the book's publication to Robert Wood in Ireland.[26]

The "De febribus" argued at length a point adumbrated only briefly in "De fermentatione": that all diseases were perversions of natural fermentations in which the sulphurous and spirituous particles in the sanguinous mass were set in too great a motion, and the blood therefore became overheated. The physician, like a vintner, must maintain equable fermentation in his patient.[27] Progress, Willis believed, necessitated this new concept of fever. Harvey's discovery of the circulation of the blood had changed ancient ideas of its motion; therefore, we must also reform our ideas about blood's chief affliction—fever. Only from a proper chemical view of the blood as a fermentative liquid, he argued, could one expect to reap the clinical benefits of Harvey's discovery. Willis intended to do exactly that.[28]

He began with a chemical "anatomy" of the blood. Like any other fermentable liquid, it was composed of heterogeneous particles, of diverse forms and energies, that could be comprehended under his five principles. Blood contained much water and spirit, a modicum of salt and sulphur, and a little earth. The particles of spirit and sulphur were, he thought, especially important. The spirituous particles were the volatile part of the blood, always ready to move rapidly, to expand, and to bring heat to the whole. He concluded that there were many sulphurs in the blood, since men were fed with fatty and sulphurous foods, which were the nutrients of the blood. The red color of the blood was most likely due to its sulphurous particles. This heterogeneity of blood's parts could be seen, Willis said, by allowing shed blood to stand in a bowl. When the heat and vital spirits, which conserved the blood as a mixture, had escaped, it separated into serous and fibrous parts. The pure and sulphurous portion floated to the top and was bright red, while the purple-red clotted, and impure parts held fast to each other at the bottom.[29] Willis had done the same simple experiment as Harvey and had used it to argue, like Highmore, not the vital unity of blood, but its patent divisibility into fractions of different corpuscular character.

How was this unique mixture created from the raw materials of food? Where, in other words, did sanguification take place? Willis, like many of the post-Pecquet generation, transferred this function to the heart. Within its ventricles was established the chief ferment by which the cruder particles of the chyme were enkindled and acquired the necessary volatility. There the spirituous and sulphurous particles were set into greater motion, the blood was expanded and heated, and the acquired volatility altered the color of the blood.[30]

The process by which this incalescence took place was, Willis admitted, difficult *mechanice explicare,* but some "not improbable" explanations had been suggested by those learned men, Descartes, Ent, and Hooghelande. From their opinions Willis concluded that there was a nitro-sulphurous ferment in the ventricles of the heart which, when brought in contact with the blood, loosened the frame of the blood and allowed the spirituous and sulphurous particles to break forth into greater motion, a kind of fermentation, or ascension, that created heat which was imparted to the entire body.[31] In proposing this scheme, Willis had performed a neat sleight of hand, modestly hidden by his seeming dependence upon his predecessors. He had changed each of their ideas significantly. Descartes had indeed believed that the blood grew warm in the heart, but had attributed it to a heat innate in the *heart's substance,* which was communicated passively to the blood, thereby causing it to rarefy and expand greatly. Ent had indeed argued that an aerial niter from the air maintained heat in the heart, but he believed that this non-particulate substance fed a true cardiac flame. Hooghelande had indeed believed that there was a ventricular ferment that caused the blood to effervesce and grow hot, but he had not called it nitrosulphurous. Willis had taken elements of each explanation, expressed them in particulate terms, and used the concept of corpuscular agitation, or ascension, to obviate the common-sense objection that the living heart contained neither a palpable flame nor a visible effervescence.

In a later passage Willis suggested the role of the air in these physiological processes. He reiterated that the atmosphere, which we necessarily draw in for the continuance of life, consisted of a collection of fumes and vapors, sulphurous and saline atoms, which were continually exhaled from the earth. The internal—"intestine"—motions of the structural particles of all animals depended upon the motion and temper of the particles of the air. These atoms, for example, carried away the body's soot and fumes. These also—and here Willis reverted to the older terminology—excited the *flammam vitalem* with their nitrosity and supplied it with its nitrosulphurous food. Unfortunately, Willis went no further. One can only speculate that perhaps Willis saw the *pabulum nitrosulphureum* in the air as related to the *fermentum nitrosulphureum* in the left ventricle, the agent that initiated the internal and incalescent motion of the blood.[32]

Willis's scheme had both its strengths and its weaknesses. By arguing that all exothermic processes were the result of increased agitation of sulphurous particles, and that exactly such a process took place in the fermentation of the blood, Willis had provided the necessary link between animal heat and combustion. Fermentation was established as an oxidation surrogate. But he had failed to develop beyond the merest intimations the role of aerial nitrous particles in this process. Perhaps more unfortunately, he tied the initiation of fermentation, the role of nitrous agents, and by implication, the change in color of the blood, firmly to the heart, and especially to its left ventricle. He thus unwittingly perpetuated the very kind of Galenic preoccupation with the heart that Harvey had striven so long to overthrow. Willis's conservatism in this respect had several roots. He was a busy Oxford clinician and was strongly inclined to treat physiological explanations largely as prolegomena to medical theory. His statements on the fermentation of the blood were intended only as a preface to his study of fevers. In chapter after tedious chapter, he explained how his ideas of the five principles in blood and their interaction in the corpuscular process of fermentation accounted for tertian, quotidian, and quartan intermitting fevers, as well as for continual, ephemeral, putrid, malignant, and epidemical fevers.[33] Perhaps the strongest influence shaping Willis's ideas, at least up until the early 1660s, was Willis's laboratory expertise: it had been in chemistry, not in anatomy or experimental physiology. He may have thought of himself as a follower of Harvey, but the vivisectional methods of his master found no place in "De fermentatione" or "De febribus." It was among Willis's younger friends and protégés—Boyle, Lower, Needham, Millington, and Wren—that a more experimental approach to the heart and blood was developed. One such new technique was devised at Oxford just as Willis was putting the finishing touches on his fermentation tract.

MANIPULATING THE BLOOD: INJECTION AND TRANSFUSION, 1656–1666

Since antiquity, blood has been revered as a vital fluid of great importance and of many powers. It is scarcely surprising that almost from the beginning of medicine, men speculated about ways to change the nature of blood or to replace its loss. Ovid relates the legend of the aged Aeson, whom Medea rejuvenated by draining his vessels, and filling the ancient veins with a brew that renewed his blood. In the sixteenth and early seventeenth centuries a number of writers conjured with the possible rejuvenating effects of putting new blood into old veins. One such, Andreas Libavius, even described in 1615 a procedure using silver tubes to bleed a young man into an old. That this was never actually done may be inferred from the protocol; he directed the physician to bleed from the artery of one into the artery of the other.[34]

The acceptance of the circulation prompted yet more men, independently, to consider the possibility of transfusion. Wallis recalled that the idea of transfusing blood "from one live animal to another" was discussed at Oxford "as far back as the year 1651 or 1652, or at any rate about that time," when he, Wilkins, Ward, Wren, and others "were in the habit of exchanging talk on philosophical matters."[35] One man who took the idea beyond talk was Francis Potter (1594–1678), a friend of John Aubrey and fellow of Trinity until he was ejected by the Visitors in 1648. About 1639, while reflecting on Ovid's legend, Potter came up with the idea of transfusion. He later became acquainted with Harvey, most likely during the Oxford siege, and corresponded with him in the 1650s. In late 1652 and early 1653, Potter carried out, at his rectory in Wiltshire, a series of experiments in which he tried to collect blood in a small bladder and transfer it into another animal by means of various ivory tubes and quills. He detailed these experiments, even though unsuccessful, in letters to Aubrey, who may well have told Oxford friends about them in his frequent visits there.[36] Although these first English trials and speculations came to naught, they were clearly based upon Harvey's discovery of the circulation. It is therefore fitting that when a sustained line of work on injection and transfusion developed at Oxford, it was based, not just upon the circulation generally, but upon a specific Harveian point in particular.

In the sixteenth chapter of *De motu cordis,* after he had stated the direct experimental evidence for the circulation, Harvey noted certain other facts that supported it, as it were, *a posteriori.* The most important was the actions of poisons and medicaments. How else but by the circulation could the effects of a septic wound or the bite of a mad dog or venomous serpent come so quickly to incapacitate the entire organism?[37] Such questions about poisons held a special fascination for Boyle from his youth. He wrote a tract about turning poisons into medicines and, in the early 1650s, did experiments on poisonous vipers,[38] one aspect of which puzzled him. The circulation explained why a dog, when bitten by a viper, died so quickly. Yet when Boyle fed a dog the head, tail, and gall of a viper (the parts popularly supposed to contain the poison), the animal was unaffected. Why should this be, if blood was made from food? Boyle conjectured that perhaps it was the anger of the viper that rendered his teeth poisonous and not some intrinsic poison. He proposed to test this hypothesis by pricking a dog's veins with needles dipped in poison to see whether poisons "carried by the circulated Blood to the Heart and Head" differed in the rapidity or manner of their action from those taken in at the mouth.[39]

About March 1656, Boyle, Wilkins, and Christopher Wren were discussing such questions in Boyle's lodgings at Oxford, when Wren ventured that he could "easily contrive a way to convey any liquid Poison into the Mass of Blood." Boyle obligingly provided a large dog and called upon "the assistance of some eminent Physitians, and other learned Men"—almost certainly Willis, Bathurst, and probably Dickenson—to assist in the experiment. Wren and his helpers freed a large vessel on the hind leg of the dog (probably the crural vein) and ligatured it. They slit open the vein proximal to the ligature, and the "dexterous" Wren, having "surmounted the difficulties which the tortur'd Dogs violent struglings interpos'd," inserted the slender pipe of a syringe into the vein and infused "a warm solution of *Opium* in Sack." Once into the mass of blood, the tincture "was quickly, by the circular motion of That, carry'd to the Brain, and other parts of the Body." The dog was quite stupefied for a time, but soon recovered and even grew fat. Its exploits, Boyle remarked laconically, "having made him famous, he was soon after stoln away from me." Subsequent experiments on other dogs improved the technique. The group soon found that a bladder attached to a slender quill served for injection better than a syringe, and that "unless the Dog were pretty big, and lean, that the Vessels might be large enough and easily accessible, the Experiment would not well succeed."[40]

Wren was quick to publicize his accomplishment. When Robert Wood left Oxford in late June 1656 to seek his fortune in Dublin, he carried with him a letter from Wren to Petty reporting on the recent activities of the club. Wren described their work in optics and on microscopes, their astronomical observations, and most especially their anatomical investigations. They had dissected fish and fowl. They had observed kidneys, brains, and nerves with microscopes and had traced how the lymph ducts from all parts of the body emptied into the *receptaculum chyli*. But the "most considerable" experiment, Wren said, was that of injecting a dog with "Wine and Ale into the Mass of Blood by a Veine, in good Quantities, till I have made him extremely drunk, but soon after he Pisseth it out." Wren had also injected two ounces of an infusion of *crocus metallorum,* an emetic, which had the predicted effect: the dog "immediately fell a Vomitting, & so vomited till he died."

> It will be too long to tell you the Effects of Opium, Scammony & other things that I have tried this way: I am now in further pursuit of the Experiment, which I take to be of great concernment, and what will give great light both to the Theory and Practice of Physick.[41]

Wren seems to have done exactly that. He related some of the results to his Wadham friend and contemporary, William Neile, who later recalled that Wren "did propose the giving a man nourishment in at a kind of Tappe into the guts," and even attempted transfusing blood, although "without much successe." Neile also recollected that Wilkins knew about these trials.[42]

In August 1657 Wren was appointed Gresham Professor of Astronomy and moved to London. With his temporary departure from All Souls, the history of injection and transfusion divides into two stories, one centered in London, the other in Oxford. Once in the capital, Wren communicated his technique to a fellow Oxonian, Timothy Clarke, who had left the University about three years earlier.[43] Clarke expressed great interest because he had himself been, according to his own account, "busily investigating the nature of blood."[44] This preoccupation must have been long-standing, because in 1653 he had argued a thesis at Oxford that denied sanguification took place in the liver, spleen or veins.[45] Moreover, like other Oxonians, Clarke approached the problem with special reference to "Chymistry," which he considered "one of the best keys to the secret recesses of nature."[46] Clarke immediately applied himself to the prosecution of injection experiments, resolving to inject many different kinds of waters, beers, milk, whey, broths, wines, alcohol, and even blood itself.[47] Clarke and Wren even tried injection on a man. In the autumn of 1657, at the home of the Duc de Bordeaux, the French Ambassador to the Commonwealth, they attempted to inject an infusion of *crocus metallorum* into "an inferior Domestick of his that deserv'd to have been hang'd." But the patient fainted as soon as the injection began, so Wren and Clarke desisted in prosecuting "so hazardous an Experiment."[48]

This failure prompted a return to animal trials. In late 1657 Wren, Clarke, and others did a further series of injections, some of which they prosecuted in the home of Henry Pierrepont, first Marquis of Dorchester, Harvey's old friend and patient, and something of an amateur physician. They found that a moderate dose of the emetic seemed to have no effect upon a large dog.[49] Wren passed the news on to Petty in Ireland, who in turn suggested to Boyle in early 1658 that such experiments would be appropriate for the rejuvenated club that now met at Deep Hall.[50] Boyle heeded the advice and induced an unnamed "ingenious Anatomist and Physitian" at Oxford to try injections of diuretics, with good success.[51] These experiments immediately became well known. Stubbe recalled that injection was "a thing much practised by Dr. *Wren* and others in Oxford," and saw at Boyle's lodgings a "Dog into whose *veins* there was *injected* a Solution of *Opium*."[52] Some even saw the possibility of miraculous cures. Francis Vernon, a Christ Church friend of Hooke, alluding in poetry to the classical legend of Aeson rejuvenated by injections of blood, speculated in 1658 that the physicians of Oxford were so knowledgeable in "finding out the *blood's Maeandring dances*" that they too might be able to effect rejuvenation.[53] Stubbe, in another poem that year, also used the Aeson myth.[54] Hartlib was more prosaic; he noted in early March 1658: "A new anatomy experimented at Oxford to open veines and to spout medecins unto it."[55]

Wren returned to Oxford in late 1658 or early 1659, leaving Clarke as the only practitioner of injection in London. There Clarke continued his own private trials, but was reluctant to publish, both because of his natural caution and because of his growing doubts "of the usefulness of this kind of experiment."[56] The company at Gresham tried to draw him out. On two occasions in 1661 Clarke was requested to turn in "the narrative of his injection of liquors into the veins."[57] He gave a brief report in 1662,[58] and was still carrying out experiments in 1664, when Pepys witnessed privately Clarke's largely unsuccessful attempt "of killing a dog by letting opium into his hind leg."[59]

On only one occasion, in September 1663, was Clarke finally cajoled into reading some account of his experiments. The contents were never registered, and hence do not survive, but the communication sparked a lively discussion. Some discussants objected that no liquid untransformed by digestion could possibly be compatible with the blood, while others believed that chemical distillations of digested products, such as "spirit of blood" might be usefully tried. To minimize this incompatibility, it was suggested—by whom is not recorded, although Wilkins and Hooke were both present—"to let the blood of a lusty young dog into the veins of an old one," using two silver pipes connected by a leather bladder, by which means the blood could be moved from one animal's veins to another.[60] Nothing came of the suggestion until the spring of 1665, when Oldenburg discovered that a German physician, Johann Daniel Major, had published a book advocating injection therapy.[61] The Society once again urged Clarke to make his work public,[62] and in the ensuing flurry of interest in injection, Wilkins suggested "that the experiment of injecting the blood of one dog into the vein of another might be made."[63] On 7 June 1665 Wilkins reported that he, Hooke, Petty, Daniel Cox, and Francis Willoughby had succeeded. They had opened a dog, let five or six ounces of blood from the vena cava into a bladder and then, using a brass nozzle, injected about two ounces of this blood into the crural vein, but without, they reported, "any sensible alteration in the bitch."[64]

With this experiment, the London work on injection came temporarily to a halt, interrupted by the outbreak of plague in the capital and the consequent prorogation of the Society's meetings. But the pattern is discernible. Those who pursued the technique and saw it as a useful device for examining the properties of the blood were those within the Oxford club's nexus of communication. Wilkins, Hooke, and quite possibly Cox (Boyle's operator) had been in Oxford when the experiment was first successfully executed. Oldenburg had arrived soon thereafter. Wren had reported the technique to Petty in Dublin and taught it to Clarke in London. Boyle, in whose lodgings the first experiments had been performed, was present at the Royal Society on almost all the occasions when the subject came up for discussion. The physiological technique of injection, like its pneumatic counterpart, the vacuum pump, remained almost exclusively a tool of the cognoscenti.

In London, Clarke was injection's sole persistent votary; in Oxford, its use spread during 1658−1665 until it was a standard anatomical and physiological tool among members of the club. Wren carried on with injections of medicaments, as

John Ward noted in his diary in April 1659.[65] Boyle did the same whenever he was in Oxford, a fact obliquely reflected in Ward's jotting of February 1661 that "Mr. Boghil" had killed a dog with opium, "as I heard Mr Wren did with Infus. Crocj."[66] But its main devotees were Willis and Lower; their increasingly sophisticated use of the technique may be seen in Willis's anatomical masterpiece *Cerebri anatome* (1664), and in Lower's letters to Boyle.

On the elementary level, injections of ink and other colored fluids proved highly useful in tracing out the course and speed of blood flow.[67] Lower wrote to Boyle in early 1662 of Willis's intention to inject "some kinds of liquors, tinctured with saffron, or other colours," into the carotid arteries of a freshly killed animal, "to try how the blood moves, and how the tincture may be separated in the brain."[68] In 1663 Lower and Willis syringed both milk and ink into certain vessels to test Franciscus Sylvius's recently published conjecture that liquids fell from the brain, via the infundibulum, into the throat and palate. The Dutchman was proved wrong; not the least drop of the tracers came out into the mouth.[69] In 1664 they used injected milk to prove that the hepatic artery went to all parts of the liver.[70] Most of all, Willis and Lower used injections to trace blood flow in the brain. They injected ink into the vessels of the dura and pia mater. They did the same to trace the intricate vascular patterns of the *rete mirabile,* and to show the passage of blood from the carotid artery to the interior of the pituitary. They even injected embryonic preparations to trace the development of the vascular system.[71] But their most important use of injection was to confirm the function of the structure whose eponym was to make Willis's name known to hundreds of thousands of medical students—the "circle" of anastomosed arteries at the base of the brain by which, if any carotid or vertebral artery were blocked, the remaining ones could maintain full blood flow to all parts of the brain.[72] Lower wrote to Boyle in June 1663 that he had tied both carotid arteries of a dog very tightly and observed that circulation to the brain was unimpaired.

> But this I might have told you in a shorter time; for if one artery be syringed with any tincted liquor, all the parts of the brain will equally be filled with it at the same time, as several times we have tried.[73]

As Lower later noted, without syringes, "anatomy is as much deficient, as physic would be without laudanum."[74]

More complicated procedures used injections to replace body fluids. Lower, duplicating in 1662 an experiment done previously by Boyle, injected two quarts of Tunbridge water into a dog. The animal soon "discharged himself," and the experiment set Lower to speculating whether one might keep a dog alive "without meat, by syringing into a vein a due quantity of good broth, made pretty sharp with nitre, as usually the chyle tastes." Such intravenous feeding might well utilize a permanently implanted pipe, which Lower sketched in the margin of the letter, and through which various liquids could be syringed in at will, without the need to make a fresh incision each time.[75] Two years later Lower tried the procedure. He injected a quart of warm milk into a dog, but found that the animal died within an hour. Upon autopsy, Lower found the vessels occluded by "blood mixed with milk, as if both had curdled together."[76] Obviously the difference between the two fluids was too great.

This failure naturally led Lower to reflect, in his letter to Boyle of 8 June 1664, that the problem of heterogeneity might be obviated by using the most compatible substance, blood itself. He planned, "as soon as I can get two dogs of equal bigness," to bleed from an artery of one into a vein of the other, "for an hour's time, till they have whole changed their blood." As a variant, he hoped to bleed one dog "until he be quite faint and cannot stand, and then let the other dog's blood run into him," to see whether the recipient would recover his strength. One could try a similar experiment on a sheep and a mastiff. It might even turn out, Lower suggested, that sheep's blood may agree with human and could be used to save patients who had suffered great losses of blood.[77] Thus, Lower at Oxford, like Clarke and Wilkins in London, was led from injection to the possibility of transfusion.

In the summer of 1665, these two lines of experimentation and speculation—Oxford and London—coalesced. In mid-June, Boyle came down to Oxford and was followed in rapid order by many of the Fellows of the Society, driven, like the Royal Court, out of London by the ever-increasing virulence of the plague. By late summer William Petty, Joseph Williamson, Robert Moray, and Seth Ward, among others, had joined Boyle, Wallis, Willis, Lower, Millington, and Dickenson at Oxford. Regular Wednesday meetings were soon arranged and continued through the summer and fall, mostly at Boyle's lodgings at Crosse's, but sometimes at Wallis's house in Cat Street, a few hundred feet away.

Boyle's immediate preoccupations were anatomical. Lower told him about a recent observation of blood let in phlebotomy, which seemed to separate out into its chylous portions.[78] In early July, Lower and Henry Clerke had told Boyle about dissections of cattle they had made, in which they found the windpipes considerably stuffed with grass.[79] By mid-July, Boyle and Lower had repeated for the group an experiment Boyle had previously related at Gresham, "yt if ye Thorax were sufficiently layd open ye Lungs though unhurt would not play."[80] In early August the group was doing experiments of poisoning dogs with injections of "oyle of tobacco" similar to those performed earlier that spring at the Royal Society, and which had occasioned Wilkins's suggestion of injecting blood.[81] Boyle duly sent reports of their findings to Lower's Westminster schoolmate, Hooke, in Surrey, and to Oldenburg in London, who published them in his *Philosophical Transactions*.

But there was another topic under scrutiny as well. Boyle had been at Gresham on 17 May 1665, when Wilkins had suggested injection of blood. Once he arrived at Oxford and fell in with Lower doing experiments, it seems that they decided jointly to attempt what they both had previously considered, but never tried: transfusion. Therefore in August 1665, as Lower later recalled, they tried "conveying the blood from one dog's jugular vein to the other by pipes," but "found it altogether impossible," because the blood was apt to congeal in the pipes.[82] Boyle found the experiments promising, but blamed the difficulties on the "unsuitability of the apparatus."[83] Unfortunately, the trials did not long continue. Lower had taken his D.M. only a few weeks before, and during July had been negotiating, with the help of his drinking chum, Anthony Wood, for the hand of a widow in nearby Garsington. On 8 August the mysterious "Mrs. H." denied Lower's suit, and he seems to have resolved to seek a wife in his native Cornwall.[84] Boyle was disappointed. He wrote to Oldenburg that Lower's departure "will put a stopp to our Anatomicall Proceedings wherein this weeke we made some Tryalls whose Event I have not now time to sett downe." Besides, Boyle noted, past experience had given reason for reticence:

> Some here are a litle Jealous yt if our Expts be known elsewhere without being before hand registrd by you together wth ye Time of their having been made or proposd, they may beget such claimes & disputes as yt wch wee formerly made here of iniecting into ye Veines of live Creatures.[85]

Two days later Wood and his cronies toasted Lower's farewell at the Castle Tavern, and on 31 August the new Doctor departed for Cornwall.[86]

Although in the following months Boyle did no more experiments on injection and transfusion, the subject could not have been far from his mind. On 25 September the king and court arrived, with Timothy Clarke in the entourage.[87] In early October, Williamson gave Boyle a copy of Johann Sigismund Elsholtz's *Clysmatica nova*, in which the German described his own injection experiments, dating back to the early 1660s.[88] Correspondence with Oldenburg on this pamphlet and on Major's of the year before resulted in Oldenburg writing an article for the *Philosophical Transactions*, in which he defended the priority of the Englishmen.[89] The account was for the most part derived from an earlier narrative by Boyle, with a few additions at Oxford by Timothy Clarke and by John Wallis, the latter of whom saw the issue through Richard Davis's press there.[90]

When Lower returned from Cornwall in mid-February 1666,[91] several of his potential collaborators were gone. Clarke had returned to London with the court in late January, missing Lower completely.[92] Boyle, finding the visits of courtiers intolerably distracting, had retired in mid-November 1665 to Stanton St. John, a village three miles northeast of Oxford, to write and do barometric experiments. Lower thus went to work by himself. Vein-to-vein transfusion had failed, so he modified the procedure to transfer blood from artery to vein, making use of the significantly higher arterial blood pressure to avoid clots during transfusion. First, he bled a dog heavily at the jugular vein, to which he then attached a tube running from the cervical artery of another, larger dog. He continued the process of alternatively bleeding and transfusing until two donor mastiffs were dead. The recipient dog, on the other hand, once its jugular vein was sewn up and its restraints removed, "promptly jumped down from the table, and apparently oblivious of its hurts, soon began to fondle its master, and to roll on the grass to clean itself of blood.'' It recovered well, with few signs of ''discomfort or of displeasure.'' All of this must have been done within about a week, for by the end of February 1666 Lower was able to call in John Wallis, Thomas Millington, and other Oxford physicians, as witnesses, for whom he repeated the procedure.[93]

Boyle, in Stanton St. John, must have heard of the success almost immediately. Unable to maintain total secrecy, he wrote in early March about transfusion to John Beale in the West Country and to Robert Hooke in London. Each old friend responded according to his personality. Hooke, the technician, responded with suggestions about how to arrange brass tubes for proper injection or transfusion.[94] Beale, the visionary, praised the possible therapeutic value of transfusing the ''superabounding blood of healthful young people into the veins of the aged and decayed''—the Aeson legend once again.[95] Thus, when the Royal Society reconvened for the first time on 14 March 1666, transfusion was a natural topic of discussion. Clarke was urged once again to finish and publish his experiments on injection. Moray, who had been at Oxford in August to see Boyle and Lower's first attempts, but had left before Lower returned in February, praised transfusion as a ''considerable experiment, if it could be practised.'' Clarke affirmed that he had tried transfusion ''above two years before,'' but ''found it so difficult, that he gave it over.'' Moray then intimated, ''that Mr. BOYLE was of opinion, that the difficulties of this experiment might be mastered.'' Interestingly enough, John Wallis, who had witnessed Lower's demonstration just a few weeks before, was also present at Gresham. But the mathematician understood well the nature of scientific priority; he kept his silence.[96]

Lower, in the meantime, continued his transfusion trials at Oxford, where their fame spread, at least locally. His Christ Church crony, John Ward, visited Oxford that spring and heard the news at first hand.

> Dr. Lower let one doggs blood into ye bodie of another by opening ye veins of one and ye arteries of another and putting in a quill into each and so letting one blood on ye other side: ye one died, ye other lived: another dog yt had 2 Vena Cavas as they say:—Dr. Lower. They had siringed in beer into ye Crurall arteries and likewise infus. Cros. metallorum.[97]

It was hardly a sophisticated description, but it got across the essentials.

News of Lower's success was spread more slowly outside of Oxford, partly because of a common desire to protect his priority, and partly because of movements to and fro. When Boyle attended at Gresham on 18 April 1666, his first meeting in almost a year, he was immediately pressed for details of his own transfusion experiments. He reported that his siphon had broken while attempting the transfer, but that he still believed the experiment possible and would contact Lower about it.[98] Lower responded from Oxford on 6 June describing the complete procedure in a letter to Boyle.[99] But the letter miscarried, and the Society finally learned about the Oxford transfusion only on 20 June when Wallis, no doubt thinking Lower's priority established, told of his success. Boyle was immediately pressed to obtain a protocol.[100] He wrote on 26 June,[101] but because Lower had already left Oxford, the letter lay at Crosse's until it was forwarded to Cornwall in late August.[102] Lower responded from there on 3 September with a second protocol that missed Boyle in Chelsea and had to be forwarded to him back in Oxford. Boyle sent on the news to Oldenburg in late September.[103] A committee of Cox, King, Hooke, Goddard, Merrett, and Clarke was then appointed to meet at Pope's lodgings in Gresham to replicate the experiment.[104] But it was not until they obtained the Oxford protocol that Edmund King and his helpers were able, on 14 November 1666, to duplicate Lower's results of nine months before.[105] Pepys, always the appreciative onlooker, met Croone at the Popeshead tavern that evening, where they discussed the experiment, which "did give occasion to many pretty wishes, as of the blood of a Quaker to be let into an Archbishop, and such like."[106] Pepys was probably more sober when he met Hooke two days later, who told him that "the Dogg which was filled with another dog's blood at the College the other day" was "very well, and like to be so as ever." Hooke doubted not of transfusion "being found of great use to men."[107]

This further history of transfusion as a *therapeutic* technique, its rapid rise to fashion in London and Paris during 1667 and 1668, and its equally rapid fall into disrepute, has been told too often by historians of medicine to need repetition here.[108] The *physiological* use of injection and transfusion by Boyle, Lower and their friends during the late 1660s, especially as it was used to demonstrate the properties of the blood and the nature of the heartbeat, finds its place in the following chapters. The aim here of tracing at some length the early history of injection techniques and their natural transition into procedures for transfusion has been twofold. First, it delineates clearly how the problem arose out of exigencies created by the discovery of the circulation and, in turn, served to refine and advance that doctrine. Second, such a tortuous historical course shows just how clearly the development of injection and transfusion was inextricably linked to the complex nexus of scientific friendships formed in almost a decade of Oxford history. From the early experiments of Boyle, Wren, Wilkins, and others at Boyle's lodgings in 1656, through the splitting of the history of the technique into London and Oxford segments during the late 1650s and early 1660s, to the unification of these two experimental lines in the work carried out at Oxford in 1665 and 1666, it is clear that changes of technique and objective are understandable only in terms of shifting patterns of collaboration and communication.

"COLLEGIUM ANATOMICUM OXONIENSE" AND THE NATURE OF THE BLOOD

The injection experiments of Boyle, Wren, Willis, and Lower, as well as the transfusion attempts of Boyle and Lower, were only a part of a larger program of anatomical investigations carried out at Oxford during the decade from the mid-1650s to 1668. This research was the common interest of a coterie around Beam Hall that included not only physicians, but many virtuosi whose ultimate careers as churchmen and civil servants took them far from the medical sciences. Such a broad appeal was founded in part upon the enthusiasm of such leaders as Boyle and Willis, and in even greater measure upon mutual interests in chemistry and in the corpuscular philosophy. Physiological problems came to be seen, to an even greater extent than they had been in the 1650s, as soluble by the interpretation of new observational and experimental findings in chemical and corpuscular terms.

Thomas Willis was the leader. His growing local reputation was greatly enhanced by the publication of the *Diatribae duae* in 1659, so much so that a jealous younger medico later accused him of contriving "to block at his Pen, and so by forging of Novelties," to attract patients.[109] His fame as a physician and writer, his staunch Anglicanism, and his brother-in-law John Fell's recent appointment as Dean of Christ Church, made him the natural choice as Sedleian Professor of Natural Philosophy. Joshua Crosse was forced out of the chair by the Restoration Visitors in August 1660[110] and his fellow club member was elected in his stead.[111] Willis took his D. M. two months later and completed the laurels by defending theses on the role of the blood in intermittent fevers at the Act in July 1661.[112] His professional fame and financial position were rapidly consolidated; from a poor boy who had worked his way through Christ Church as a servitor, Willis rose to be one of the richest and most respected men in Oxford. He bought the lease to Beam Hall in 1662 and the manor of Bushelton, in Herefordshire, from Aubrey, in 1663.[113] Later that year he was elected F. R. S. He became an Honorary Fellow of the College of Physicians in 1664. Patients, like Lady Anne Conway, came from afar to consult him.[114] Many did so when passing through Oxford on their way to Bath,[115] so much so that in 1666 he and another physician, Peter Eliot, took out a lease on the Angel in High Street, the better to see fashionable travelers. By 1667 Willis's income was rated at £300 a year, the highest in the city.[116] Scholarly visitors paid him no less heed. Samuel Sorbière and Balthasar de Monconys both sought him out when their grand tours took them to Oxford,[117] although such visits were not always mutually felicitous. Sorbière had difficulty understanding Willis's Latin pronunciation, while Willis, as he later told Sprat, found Sorbière something of a *poseur* who understood little of chemistry and medicine, despite having "the great names of *DesCartes* and *Mersennas* . . . frequently in his Mouth."[118] Reputation brought its share of awkward situations.

Richard Lower, Willis's apprentice, protégé and younger collaborator, also followed the road of the academic physician, albeit one better paved with wealth and honors. His parents, prominent members of the Cornish gentry, had sent him to Westminster, where, under Busby's tutelage, he was a King's Scholar, and was elected to a Studentship (the equivalent of a Fellowship) at Christ Church in 1649. He took his degrees there in due order, rose in seniority, and was Praelector in

Greek 1656–1657 and Censor in Natural Philosophy for the three years from 1657 to 1660.[119] He fared well in the House until 1663, having risen to the top of those not in orders and having, as he said to Boyle, insufficient "favour or friendship" to get one of the two physicians' places, he lost his Studentship. Even then he continued to occupy rooms in college.[120]

Lower's interest in medicine appeared early, for there are receipts and observations by him recorded in John Locke's notebooks dating from the mid-1650s. By 1659 he was an active practitioner whose cases and prescriptions were regularly recorded by John Ward. Among his patients was his drinking chum, Anthony Wood, for whom he prescribed an issue in 1662, and who recorded some of his crony's more bizarre cases.[121] In this practice Lower was often Willis's younger associate, being sent out into the country to attend patients, once as far as Cambridgeshire.[122] For his work Lower usually got a fee equal to half of Willis's; in a case in September 1662 where they both tended one of John Locke's tutees, Willis had £1 and Lower, 10 shillings.[123] Such a junior partnership sometimes had inadvertent advantages. On one trip into Northamptonshire in April 1664 Lower discovered on his return the medicinal character of the waters of Eastrope, Oxfordshire, near King's Sutton, which he and Willis then promoted into a local competitor to Tunbridge.[124]

Lower's interest in anatomy and physiology did not long postdate his emergence as a practitioner. His first known foray dates from about October 1660, when Ward recorded: "Mr. Lower cut a doggs windpipe and let him rune about: hee had a week so: he could note smell but would eat any thing I was told."[125] By the following year, if not before, Lower met Boyle, and began a pattern of collaboration during Boyle's visits, and correspondence when the young aristocrat was in London. Lower continued as an active dissector and vivisector through all his remaining years at Oxford, with such dedication that it is not surprising to find Anthony Wood recording in 1664 that Lower was dissecting away at Christ Church on a calf's head even on a Sunday morning.[126]

The third anatomist, Walter Needham, came to Oxford fully trained, attracted by the fame of Willis and his circle. Needham had also been a King's Scholar at Westminster under Richard Busby, and was elected to Trinity College, Cambridge, in 1650. There he began work on anatomy about 1653,[127] although it got him no fellowship, for which he moved to Queens' in 1655. His early interest in science, as well as the special relation many old Westminsters retained with their headmaster, is shown in a series of letters written to Busby in that year. All science, Needham wrote, proceeded "by joining reason with experience." The peripatetics had done real harm by making a philosophy out of Aristotle's observations. The new ideas founded upon "ye Democriticall & Epicurean systemes," although "they doe depend more upon experiment" and hence have been "very happy in ye explication of divers particular occurrences" previously thought "unsearchable," also tended to veer from the true criterion of experiment.[128] Thus, he believed that "ye soules of brutes" acted from a cause beyond mere mechanism.[129] Harvey's discovery of the circulation, Needham thought, exemplified the dependence of truth upon experience. Had not the Louvain physician Plemp thought Harvey easy to confute "by conclusions drawne from Galen," until he tooke "to consult nature herself, & try axioms she could afford him?"[130] With such inclinations, it is little wonder that in 1658 Croone passed on to Hartlib the opinion of Cambridge's Regius Professor, Francis Glisson, that Needham "is like to prove one of rarest anatomists that hath beene in the world."[131]

Needham met Wilkins at Cambridge in 1659 and left Queens' in the summer of 1660, most likely as the result of political changes.[132] After a time spent apprenticing to a Dr. Smith in his native Shropshire,[133] he came to Oxford, probably early in 1662. His reputation had preceded him there. He had been a long-time friend of another Westminster at Christ Church, Nathaniel Hodges, who was in turn a close associate of Willis.[134] Ward, also at Christ Church, had known Needham by reputation as far back as April 1659, when he recorded Jacob Bobart's recounting of a dissection of a tortoise that Needham had done at Cambridge.[135] Once at Oxford, Needham put himself in touch with his old schoolfellows—Lower, Millington, Wren, and Locke—and with their mentor, Willis. He seems not to have met Boyle until the summer of 1664, when he came back to Oxford after taking his M. D. at Cambridge, and before returning to Shrewsbury to set up practice there.

These three major anatomists were merely the center of a larger coterie active in the experimental medical sciences at Oxford in the 1660s. Henry Clerke, although he had now ceased to be Tomlins Reader, still did anatomical work occasionally—*vide* his observations passed on to Boyle in 1665. Thomas Millington at All Souls was always there to help with dissections and discuss results, as was his fellow collegian, George Castle. Christopher Wren had succeeded Seth Ward as Savilian Professor of Astronomy in 1661 and thus was more concerned with the mathematical sciences and his emerging career as an architect than with anatomy; yet he continued to help with dissections and during the early 1660s seems to have built pasteboard models to illustrate the statics of muscular motion.[136] Wren's friend, Thomas Jeamson, was the Tomlins Reader and a careful anatomist.[137] Samuel Morris worked with Lower on anatomy, a tutelage he gratefully acknowledged in a thesis later presented for the M. D. at Leyden.[138] David Thomas and John Locke, both well-versed in anatomy, were more interested in chemistry. In the practical aspects of this subject, the group had the help of Peter Stahl. The German taught chemistry at Deep Hall from December 1659 to mid-1661, when he moved his laboratory two doors west to Arthur Tillyard's, only to move again in early 1663 a block farther down the High Street. Stahl was especially close to Willis, Robert Sharrock, and Edmund Dickenson, and before his departure from Oxford about 1665, he taught chemistry to scores of local virtuosi, including Wallis, Wren, Millington, Bathurst, Lower, Locke, Williamson, Ward and even the diarist Anthony Wood.[139] Little wonder that by 1664

Sharrock was hopeful that a formal society for experimental philosophy could be endowed at Oxford.

One important focus of activity was Willis's university lectures. From the very beginning of his tenure as Sedleian Professor, he ignored the statutory injunction to expound Aristotle, and instead gave his own opinions on recent anatomical controversies, on fermentation, and on the action of the circulation in various diseases. Thus were all his younger colleagues and students exposed to his ideas, not just through his books, but through his teaching as well. Ward, for example, recorded on 9 February 1661 that "Dr Willis is now reading about ye succus nervosus";[140] the professor doubted that the process of nutrition conformed to Glisson's pet idea. In 1664 Lower sent on to Hooke some observations on the stomachs of ruminants, with which Willis had been so taken "that he will make a lecture of it next term."[141] Lower was likewise responsible for the survival of several of Willis's lectures given at Christ Church c. 1661. He copied them for his own use, and then in November 1662 sent a transcription on to Boyle,[142] among whose papers in the Royal Society they have survived.[143] About 1664 Lower also lent his text to Locke, who abstracted them into one of his notebooks, now in the Bodleian Library.[144] In the Lower/Boyle manuscript Willis explained his opinions on how the function of the cerebrum and cerebellum resulted from the interactions of particulate spirits drawn from the blood. Dysfunctions such as paralysis, insomnia, convulsions, epilepsy, hysteria, vertigo, and lethargy were similarly to be explained as the interaction of spirits with the circulation. True to the intellectual and local portions of his heritage, Willis cited Harvey's *De generatione*, and explained muscular motion using the experiment of raising weights with a bladder,[145] the very one that had been performed at Petty's club in 1650.

Neurological lectures such as these started Willis and his colleagues on a new research program. He was dissatisfied with the excessively speculative nature of his teaching, and resolved to explore the brain by dissections, not ratiocination.[146] By January 1662 Lower could report to Boyle that they had begun to cut up several brains. Willis found "most parts of the brain imperfectly described" and intended "to make a whole new draught thereof."[147] Thus began an intensive period of dissections in which, according to Willis, almost no day passed "without some anatomical administration." Whole "hecatombs" of animals were slain in the anatomical court. They dissected horses, sheep, calves, goats, hogs, dogs, cats, foxes, hares, geese, turkeys, fishes, and even a monkey, in addition to human cadavers. The indefatigable worker was Lower, "the edge of whose knife and wit" Willis gratefully acknowledged. Millington and Wren were frequently present at the dissections to "confer and reason about the uses of the parts." Millington especially was Willis's confidant and critic, to whom he privately proposed his conjectures and observations for scrutiny.[148] Other friends followed the work. Wallis participated in some of the initial dissections.[149] Ward passed through Oxford on his way to Stratford about October 1662 and recorded: "Glandules found in ye Braine by some Dr. in Oxford as I was told by Mr.

Lower.''[150] Lower reported to Boyle their preliminary results on the cerebellum, the medulla oblongata, and the auditory and optic nerves, including the use of anatomical injections. By November 1662 he could write to Boyle that the anatomical part was finished.[151] The following spring Wren drew some ''most excellent schemes of the brain, and the several parts of it, according to the doctor's design,''[152] and by the end of the summer Allestry was printing the *Cerebri anatome*.[153] When it appeared early in 1664, it immediately superseded all previous anatomical work on the brain.

Everywhere one finds evidence of the zest with which anatomical studies were both pursued and applauded at early Restoration Oxford. Wallis later recalled that Willis and Lower gave private courses in anatomy that attracted younger medical students.[154] Practically every one of Lower's letters to Boyle was filled with medical and anatomical observations, not to speak of his letters to Hooke that have not survived.[155] One single epistle of June 1664 passed on to Boyle the results of five recent postmortems.[156] One particularly spectacular postmortem was occasioned when two students of Wadham were killed by lightning in May 1666. Willis, Lower, Millington and Wallis assembled to do the examination, which Wallis later wrote up for the *Philosophical Transactions*.[157] Boyle joined the anatomical work when, as in the latter half of 1664, he spent an extended period of time at Oxford. At that time he participated in a dissection of a body ''having but one kidney,'' learned of Needham's experiment of blowing into the thoracic duct, did experiments of cutting the vagus nerve in dogs (for which his notes still survive), and of stopping a dog's respiration by breaking both phrenic nerves.[158] John Ward was also drawn in whenever he visited Oxford. About February 1664 he noted that ''Dick Lower had been searching ye Testicles of a Boar'' for nerves, to test the theory that the sperm was elaborated from the *succus nervosus*. Ward recorded also that ''They have a dog wch they call spleen because his spleen is taken out.'' When ''Spleen'' finally died in April 1665, Ward was there to dissect him and to see whether the organ ''is any whit growne againe since it was taken out.''[159] Even the divines on the periphery of the Oxford group followed the anatomical questions closely. In 1665 Samuel Parker, a friend and admirer of Bathurst, wrote a long treatise on natural theology in which scores of pages were devoted to proving that recent advances in anatomy and physiology evidenced the deity.[160] A year later he managed to work into such an unlikely topic as the preexistence of souls an account of Willis's findings on the linkage of brain and heart via the intercostal nerves.[161] Robert Sharrock, in a sermon on the resurrection, digressed to discuss Willis's neurological theories, especially the functions of the *corpus striatum* and the *corpus callosum*.[162]

It was therefore quite natural that Lower, Boyle, and their friends should, in the midst of all this work, turn to examine one of the oldest of physiological problems, the nature and especially the color of the blood. In June 1664, just a few weeks after he had written telling of his plans for transfusion, Lower wrote again to Boyle, this time to tell of his intentions to explore ''the reasons of the different colour of the blood of the veins and arteries; the one being florid and purple red, the other dark and blackish."[163] Lower adduced an old experiment. Why, he asked, is blood collected in phlebotomy always capped with a layer that "is always florid and finely red, and that under is always dark and black." Whereas, if one let blood directly from an artery into a porringer, it remained entirely florid even for several days. Lower thought he had an answer. It relied upon distinguishing two modes of transfer between arteries and veins. In the first and more direct, arterial blood passed "into the veins by anastomosis, without any change"; it was kept entirely within blood vessels, and so did not change its color. In the other mode the arterial blood was "circulated through the habit of the muscles, where without it loseth many parts (viz. by lymphatic vessels, &c.) before it is resumed into the veins." In other words, Lower was attempting to distinguish a supposed intracapillary and extracapillary circulation. Thus, he reasoned, the thin florid film on shed venous blood was simply the component that had come through anastomoses, unchanged, while the mass of the blood had, as it were, been filtered through the muscles, and become much darker.

Lower supported his hypothesis with an injection experiment. If one syringed milk into the artery of an animal, "the milk will come forcibly and presently out of the vein belonging to it, mixed with the blood, though not so perfectly mixed but so you may discern it very plainly." A portion of arterial blood came directly through to the veins in the same way, although the "difference of colour cannot be distinguished so soon in the two bloods, because they are of a nearer colour and mixture." But once the venous blood had settled, its arterial part separated, and "swims uppermost," because it was "more spirituous and lighter." Thus, Lower noted, the last venous blood drawn was always the lightest in color, since it had been least despoiled of its parts by the lymphaducts. One might think of it in chemical terms. The "separation of the lympha from the arterial blood, before it is resumed into the veins" was like the distillation of any liquor—that which was left behind was always "more dark and gross; so that the blood in the veins is like the caput mortuum, when the lympha is separated." Whereas Harvey had been forced to explain the blood-color difference as an artifact, Lower, making use of the subsequent discovery of the lymphatics, and especially of the injection techniques that had been developed at Oxford, could argue a much more sophisticated kind of filtration theory. But even then his explanation was clearly incomplete. By positing that certain blood components were lost in the periphery, he could explain why venous blood could be *derived from* arterial. But the explanation, of itself, shed no light on the origins of that difference. Lower, had he chosen to discuss the subject in his letter to Boyle, would almost certainly have opted for Willis's theory of arterial blood created by fermentation in the heart.

Boyle was brought into the problem almost immediately. He arrived in Oxford a few weeks later, in early August 1664, fresh from visiting with Highmore at Stalbridge. In the following months he worked closely with Needham, Millington, and especially with Lower, and may well, as suggested in the previous chapter, have proposed that Hooke use the open-thorax experiment both to judge whether air went to the heart and to test the mechanical theory of respiration argued in Henshaw's book. He also worked on blood. Unfortunately, almost all of the letters Boyle wrote to Oldenburg from Oxford that year have disappeared. We know only what Boyle reported to the Royal Society upon his return to London in mid-December. He told his Gresham colleagues of experiments of "putting volatile and acid salts into warm sheep's blood." The volatile ones, such as spirits of sal ammoniac, of urine, and of hartshorn (all presumably ammonium hydroxide), would "render the blood florid, and keep it uncongealed and sweet, as long as one pleases, heightening also the colour." Alcohol did the same, without enhancing the color. On the contrary, acid spirits, "as that of salt (hydrochloric acid) and of nitre (nitric acid), being in a small quantity put into the blood, and shaken therewith, coagulated it immediately, and changed it into a dirty colour."[164] Boyle's results were clearly equivocal. If the latent assumption was that a volatile, acid salt was responsible for the blood-color difference, then the experiment confirmed only part of the speculation. Here, as in the experiments on respiration, the nitrous substance presumed to be in the air (and the oxygen it contained) could only be manipulated via its salt or its acid.

Such preliminary experiments led Boyle to envision a larger program of research on blood. As was his wont in such circumstances, he sat down and composed a series of "Titles" for investigation. He seems then to have circulated this schema among the "ingenious physicians" of Oxford, both for their comments and as a means of guiding their research.[165] Many such sets of titles survive among the Boyle Papers, and although none may be the exact list that Boyle composed c. 1664–1665, they provide an insight into his intentions.[166]

Three preliminary titles were to inquire into the process of sanguification, the quantity of blood in the body, and the motion of the blood in its circulation. The first grouping of "Titles of ye First Order" dealt with the physical characteristics of the blood, an area keynoted by Boyle's very first item: "Of ye *Differences* between *Arterial B*. & *venal*, at ye first emission & afterwards." Boyle then proposed to consider blood's color, taste, odor, heat or coldness, inflammability or volatility, and specific gravity (both whole and in its serous and fibrous parts). He wished to observe blood's constituent parts: whether there was any basis for the ancient theory of the four humors, whether chyle could be found there or whether it had been converted into serum, and which constituent parts of blood were "discoverable by ye *Microscope*." Just as important, given the recent interest in respiration, was to inquire "Of ye *Aereal Particles* naturally mix'd with H[umane] B[lood] & also found in its distinct parts." Chemical analysis came next. One had to examine the products of distillation, such as the "*Spirit*," "volatile *Salt*," "*Phlegm*," "ye (two) *Oyls*," "*Fixt Salt*," and "*Terra damnata*" of human blood. One should investigate further the coagulation time, as well as the liquors and salts that accelerated or impeded this process. Finally, he proposed utilitarian and comparative studies: the mechanical, chemical, and medicinal uses of blood;

the differences of blood according to age, sex, and race, or those of human blood compared to that of "Quadrupeds, Birds, Fishes, & Sanguinous Insects"; and ultimately, of the comparative natures of blood and of those liquids that were thought to derive from it, such as gall, pancreatic juice, semen, lymph, and milk. One such peculiarly Boylean topic was also not to be overlooked: "What Changes (if any) [are produced in] ye B. of Animals strangled by Suffocation or in the exhausting Engine."

The larger physiological context of Boyle's research program on the blood may be discerned in the notebook jottings of his collaborator, John Locke, to whom Boyle eventually dedicated his *Memoirs for the natural history of humane blood*, when it appeared almost twenty years later. Locke had long maintained an interest in the nature of blood, and as he had matured in the early 1660s, recorded extracts on it from many of the English and Continental physiologists: Harvey, Ent, Glisson, Charleton, Highmore, Willis, Moebius, Velthusius, Deusingius, Wepfer, and even some personal observations of his fellow Westminster and Student of Christ Church, Richard Lower.[167]

Locke then went on to speculate in late 1664 about how blood was related to respiration. "One use of respiration," he wrote, "seems to be for ye carrying away those vaporous excrement of ye bloud wch are usually cald fuliginos." These find a "fit receptacle in ye pores of ye aire" and so are ventilated from the blood in expiration. This explained Henry Power's observation that candles were put out by mine damps. But Locke believed there was another use for respiration as well, which was "to mix some particles of aer with ye bloud & soe volatilize it." A chemical observation drawn in part from Boyle's *Sceptical chymist* supported this notion. Vegetable substances would neither ferment nor yield any volatile salt "without a communication with ye open aer." Similarly, vegetables and even charcoal distilled in closed containers yielded hardly any volatile salt.[168]

By about the spring and summer of 1665, Locke had carried these notebook speculations further.[169] It seemed more certain, from several bits of chemical evidence, that fermentations volatilized substances and in the process created volatile salt.[170] Moreover, he reiterated that "what interest ye aer hath in the businesse of fermentation will be worth enquiry since noe fermentation is without a communcation wth ye aier."[171] Could this be true of the fermentation in the blood? And what was it in the air that performed this volatilization? He noted the travelers' tale that "on pike Tenerif and the Andes in Peru and other great heights," the air was "lesse usefull to respiration." Could it be that air at high altitudes "wants some of those salts which by Mr. Hooke are thought to be Niter which mixes with the blood and helps to its fermentation, and which are found in the aer of lower regions"? Or was it because the pressure of the mountain "aer being lessened by the height of the place, it is scarce sufficient to lift up the lungs and soe respiration is hindred?"[172] Here Locke posed himself the two alternatives: a *chemical* or a *physical* theory of respiration. It is important to note, however, that Hooke in his *Micrographia* had made no mention of the aerial niter in respiration. Locke had either heard the theory privately from his Westminster and Christ Church colleague, or read about Hooke's ideas on the role of aerial niter in combustion, and made the natural elision.

Locke was at the same time privy to Lower's experiments—both physiological experiments generally, and those on the color of the blood in particular. Locke had, for example, recorded the same experiment of breaking the phrenic nerve that Lower had recently done and which had been described by Boyle to the Royal Society in December 1664.[173] Thus he also recorded, probably late in the same year, Lower's continued toying with the problem of blood color:

> Bloud taken out of ye veines and arteries of ye same creature at ye same time very much differs. Yt yt comes out of an opend veine being ye greatst part of it of a darke colour wch they commonly call crassamentum nigrū with a florid red about ye thicknesse of half a crowne on ye top. yt wch comes out of an opend arterie is all of yt florid colour without any Crassamentum nigrum.
>
> R. Lower[174]

Lower continued his pursuit of the problem into the spring, as Ward noted when he visited Oxford in April 1665:

> Dick Lower is answering ye fellow yt wrote against Willis: Dr. Willis and Dick Lower opened a dogg and they first let him blood in ye jugulars to discover whether Arterial blood and venal did differ in colour and constitution.
>
> Mr. ffrancis [Smith, the surgeon] told mee yt hee and Dick Lower found much chyle extravasated.[175]

At about the same time Locke recorded yet another of Lower's experiments on the same question:

> Bloud taken out of ye artery of ye lungs hath its crassamentum nigrum like yt which comes out of ye veins soe yt yt red florid bloud is made only in ye left ventricle of ye heart
>
> R L[176]

To which Locke himself appended a comment of an entirely different import: "wch perhaps is by ye mixture of ye aire with it wch gives it volatilization & colour." After noting some contradictory statements of Helmont and Kerger about whether it was air or fermentation that caused the blood to volatilize, Locke went ahead to make the link between aerial niter and the volatilization of blood that caused it to change color:

> Aer   probably it is ye nitrous salt in ye aier yt gives it [blood] the tincture & volatilizes it, & ye volatile part in circulation being either transmuted into nourishment of ye part, ye remaining bloud in ye veins is lesse spirituous & both in

colour & consistence comes nearer a caput mortuum, & therefore is returnd by ye veins to ye lungues & heart to be new volatilized & soe by succession is made all volatile

J L[177]

Locke then suggested a way to test this theory: distill "ye venall & arteriall bloud of an animall & try whether they give different quantitys of salt." In fact, Locke concluded, it seemed generally probable "yt ye aier volatilizes bodys & takes away theire sulphur by some nitrous particles."[178]

The contrast between Locke and Lower is fascinating. Lower tried again and again to approach the question of blood color through the vivisectional techniques of which he was an acknowledged master. Yet because he had been so closely associated with Willis, he accepted implicitly the Willisian doctrine of a fermentation tied to the left ventricle. Locke, like Lower an alumnus of Peter Stahl's chemistry course, and an even closer reader of Boyle's chemical works, but no vivisector, was free to take the conceptual pieces of the puzzle—aerial niter, volatilization, fermentation, change in color—and allow them to fall into their natural places. By about mid-1665 Locke had concluded that the aerial niter entered the blood in the lungs, volatilized it there through a process of fermentation that changed its color to a florid red, which hue was lost again as the blood circulated through the parts and returned thence via the veins. But he lacked a way to prove it. He could suggest only the chemical adjudication of analyzing venous and arterial blood for differing quantities of volatile salt, an *experimentum crucis* which, even had he performed it, would have turned out at best equivocal. Locke had the hunch, but Lower had the skills. At least for the moment, the two were not to come together.

LOWER'S *VINDICATIO* (1665): A DEFENSE OF OXFORD HARVEIANISM

Boyle's chemical trials on blood, Lower's vivisectional experiments on its color, Locke's speculations about air, fermentation, and the creation of arterial blood—all were part of a broader, emergent conception of a corpuscular and chemical approach to Harveian problems. Just as the Oxonians believed that experimental challenge best revealed nature, so were the foundations of their own Harveianism revealed in response to a challenge upon their doyen, Thomas Willis. In early 1665 the elderly Bristol physician Edmund Meara published a severe quasi-Galenic critique of Willis's ideas on blood and fever.[179] "Dick Lower," as Ward noted, responded. Lower saw Meara's criticisms not just as an affront to Willis, but as an attack upon the entire tradition to which Harvey's work had given rise. During the early months of 1665, most certainly at breakneck speed, Lower wrote out his *Vindicatio* of Wallis's *Diatribae duae*, and it was published by Martyn and Allestry about early May 1665.[180]

Although Lower's book was in format very much a page-by-page refutation of Meara, it displays an overall logic proceeding from Lower's unified conception of physiological processes. As he said in dedicating the book to Boyle, he was defending not just Willis, but the entire new way of philosophizing; in this enterprise, Lower asked Boyle, as its guardian, for his blessing.[181] One would only be following Lower's intentions in analyzing the major themes of the *Vindicatio* synoptically, rather than plodding sequentially from insult to invective. Lower proved himself as much a master of the literary dagger as of the dissecting knife, but there is no need to be bound by his polemical purposes.

The book's *leitmotiv* was the primacy of the blood as a locus of physiological processes. Although this theme was forced upon him by the nature of the debate— Willis believed fevers arose in the blood, while Meara thought they were lodged in the solids—Lower soon made it clear that on this point he was defending not so much Willis as Harvey. To prove the importance of the blood, he cited again and again Harvey's proof, in *De generatione*, that it was the first-generated part of the embryo.[182] The blood was the first formed, as Harvey and Willis correctly believed, because it was the vehicle of the soul; Lower, like those before him, adduced the well-worn scriptural confirmation.[183] He even turned Meara's charge of impiety against Harvey back upon the Irishman; if the Scriptures said that the divine spark was in the blood, how could any good Christian deny it?[184]

Moreover, Lower, in focusing on the importance of the blood, attempted to justify his analysis of blood components by appeal to Harvey. Harvey, Lower said, had given first place to the blood for good Aristotelian reasons: that it was composed of similar and dissimilar parts.[185] Lower then made Harvey an honorary corpuscularian by interpreting this to mean that the arterial blood was a homogeneous substance, while venous blood was a heterogeneous one. This could easily be seen, Lower said, if one cut the carotid artery of a dog; the blood was the same color throughout and had no darker parts, no *crassamentum negrum*. In contrast, blood drawn from the jugular vein and allowed to stand in a bowl acquired a thin florid top and a darker bottom, as it separated into its blood components.[186] All of this disproved Meara's absurd four-humors theory of the blood and justified Harvey's conclusions. Lower then went on, in a lengthy "Digressio de natura sanguinis," to establish a true "anatomy" of the blood, based upon chemical and physiological experiments.[187] These conceptions of the various fractions of the blood and their constituent particles were drawn, Lower said, from a set of lectures that Willis had given recently at Oxford.[188] In the subsequent pages, Lower's and Willis's ideas were so intermixed that it is impossible to differentiate them. Nor is it necessary to do so. Lower and Willis had collaborated for so long that their views on blood as a particulate, fermentable substance had long since fused.

The defense of the primacy of the blood was, in Lower's mind, only part of the defense of all of Harvey's findings concerning the motion and nature of the blood.[189] Meara accepted a kind of circulatory motion, but Lower ridiculed it as not being truly Harveian. He recapitulated how the blood's motion was due to the contraction of the heart and to the arrangement of the cardiac valves. He accused Meara of being an anti-Copernican in the field of medicine, wanting to turn back the tide of recent discoveries in favor of older ideas.[190] According to Lower, this Harveian tradition also included the discovery of the lacteals, the lymphatics, and the thoracic duct, discoveries that complemented Harvey's own of the circulation.[191] Lower did not know that Harvey had never accepted the thoracic duct, so he righteously chided Meara for neglecting both it and the implications it had for a proper explanation of nutrition.[192] Thus Meara still believed that the liver made blood from chyle, a position one could not maintain if one accepted both Harvey's circulation and Pecquet's duct.

Taking another page from *De generatione*, Lower argued through the book that the blood was the seat of heat in the body. Meara's idol, he charged, was some *calidum innatum* that was separate and independent from the fluids of the body. This was false. Insofar as the body had heat, it was contained in the blood.[193] But in attempting to defend Harvey on this point, and Willis as a follower of Harvey, Lower betrayed the fact that he, under the tutelage of Willis, had evolved a system of physiological explanation that was, in many important senses, far removed from the Harveian problems and themes upon which it was based. He accepted totally Willis's idea of the blood as a fermentative liquid, composed of particles of salt, sulphur, and spirit, in a watery vehicle. This fermentation was responsible for the blood being able to convert chyle into itself. The heat in the blood resulted from this same fermentable nature.

The process by which heat was brought out of the blood, *ascension*, was a key concept in Lower's physiological scheme.[194] It took place in the left ventricle of the heart, as the result of contact between the blood and a nitrosulphurous ferment.[195] The heat forced out of the blood in this way was then, by the muscular beat of the heart, distributed throughout the body.[196] The enkindling of the inflammable, sulphurous particles of the blood in this way constituted the *flamma vitalis*, a combustion that was unlike common fire in that it gave off no light or visible flame.[197] As Willis had noted in the *De fermentatione*, Lower said, this ascension was a sudden commotion, a sudden resolving of the particles in the blood.[198]

Lower's ideas on blood color and on the function of respiration were subordinate to this life-giving process of ascension. Because the enkindled blood was distributed to all parts of the body, when it returned to the heart via the veins, its heat was spent. Lower even went so far as to assert that, since the true ascension of the blood did not take place until it reached the left ventricle, the blood in the lungs was the same as that found in the veins. The color and the consistency of the blood, he affirmed, depended upon the flame or ferment of the heart.[199] Lower was later to regret his rash desire for consistency.

Lower's reflections on respiration occurred very much as afterthoughts to his more important ideas on the composition of the blood and the changes it underwent in the heart. Since the blood was, in some sense, a liquid in combustion, it needed the lungs to rid itself of its smokey wastes. This was the function of *expiration*: to rid the blood of vapors and effluvia.[200] He also accepted, in passing, the notion that the blood was mixed by the motion of the lungs during the pulmonary transit. This was why fish had gills as analogues to lungs.[201] He also gave the function of *inspiration* in straightforward terms: "in order that the blood, in its passage [through the lungs], may be impregnated with the nitrous food of the air."[202] The phrase that Lower used in both of his explanations of the function of inspiration, the *aeris pabulum nitrosum*, was the same one that Bathurst had used more than a decade previously in his lectures.

Moreover, Lower said, how much the inspired air conduced to the preservation of the vital fire of the heart could be seen in a simple experiment *in vivorum animalium dissectione*. The experiment he related was exactly the one that his friends, Millington and Needham, had reported to the Royal Society just a few months earlier and which had so impressed Boyle. One waited until the heart of a dissected animal had stopped beating. Then a tube was inserted into the thoracic duct, or into the vena cava, a breath of air was blown through it, and the heart's motion could be restored. The cause of this effect, Lower said, seems to be that in dying animals, the fire of life outlived the motion of the lungs and could be re-enkindled by particles of air. The increased fire in the sinus of the heart stimulated the animal spirits in the fibers of the heart, which excited a contraction and a subsequent pulse.[203]

Such a summary of Lower's physiological ideas looks well past the polemical purposes that occasioned the *Vindicatio*. But that attitude is not unhistorical; Lower's contemporaries did the same. Oldenburg wrote a brief review of the work for the *Philosophical Transactions* of 5 June 1665 that completely neglected the invective. The book was, he said, a "small, but very ingenious and Learned Treatise," in which Lower reported on "many considerable Medical and Anatomical inquiries." Oldenburg cited certain of Lower's discussions as of greatest interest: the primacy of the blood and whether it performed the function of sanguification; whether the motion of the blood, after the heart ceased, argued that life and the pulse rested ultimately in the blood; new experiments to prove that chyle was not transmuted into blood by the liver; the nature of the blood and the difference between venous and arterial blood; and "what the uses of the *Lungs* are in *hot* animals."[204] A man like Oldenburg, conversant with the traditions of research at Oxford, saw past the ephemeral issues of the debate.

One can see at every juncture in the *Vindicatio* the themes that had run through almost two decades of physiological work at Oxford: the close attention paid to Harvey's work, especially the *De generatione*; the importance of the blood and the reasons for the differences between its two species; the treatment of heat as a process linked to the blood and its composition; and the belief that the blood absorbed a nitrous food during the pulmonary transit. The belief in the *pabulum nitrosum* is especially indicative of the shared nature of conceptual frameworks in the Oxford group. Lower's *Vindicatio* of 1665 was the first *published* work since Ent's *Apologia* (1641) that mentioned a nitrous substance in the air that had a respiratory function. Judged by the standards of formal "publication," Lower had put forward a "new" idea in a new explanatory context. Yet he felt he was doing no such thing. The notion was simply part of the conceptual tradition of the social community within which he had come to scientific maturity. The idea, like many others, had been promulgated long before it had been published.

William Harvey *c*. 1627.

By courtesy of the National Portrait Gallery, London.

Nathaniel Highmore.

*By courtesy of the
National Portrait Gallery,
London.*

Robert Boyle's country seat, the manor of Stalbridge, Dorset.

From John Hitchins, *The History and Antiquities of the County of Dorset* (1861–1873). *By courtesy of the
Bodleian Library, Oxford.*

William Petty as a young anatomist.

By courtesy of the National Portrait Gallery, London.

Buckley (Bulkley or Bulkeley) Hall (107 High Street), from the south. William Petty and Matthew Wren lodged here above the shop of John Clarke, Harvey's Oxford apothecary. Petty's club met at Buckley Hall *c.* 1649−1652, and it was here that Thomas Willis, Ralph Bathurst, Henry Clerke, and Petty attempted their dissection of Anne Greene.

By courtesy of the Bodleian Library, Oxford.

John Wilkins.

By courtesy of the Warden and Fellows of Wadham College, Oxford.

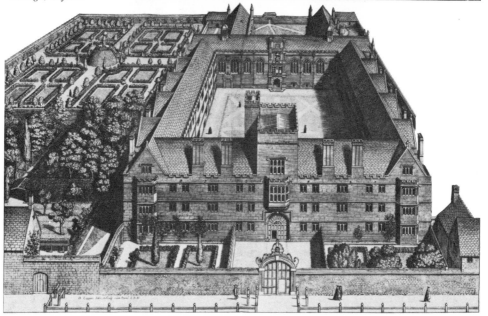

Wadham College, where the Oxford club met under Wilkins's leadership *c.* 1651–1656.

From David Loggan, *Oxonia illustrata* (1675). *By courtesy of the Huntington Library, San Marino, California.*

LEGEND

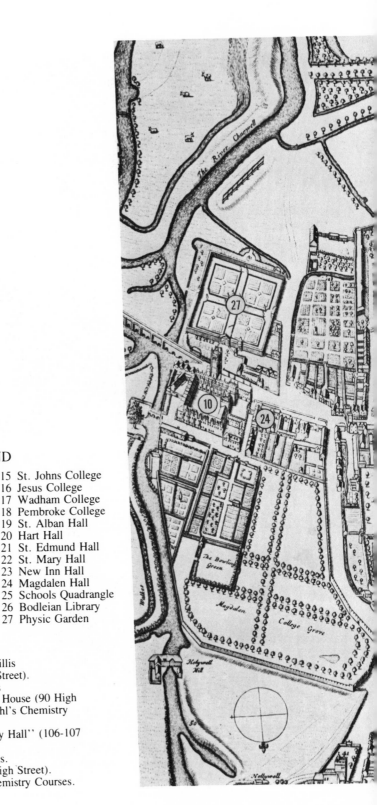

1 University College
2 Balliol College
3 Merton College
4 Exeter College
5 Oriel College
6 Queen's College
7 New College
8 Lincoln College
9 All Souls College
10 Magdalen College
11 Brasenose College
12 Corpus Christi College
13 Christ Church
14 Trinity College
15 St. Johns College
16 Jesus College
17 Wadham College
18 Pembroke College
19 St. Alban Hall
20 Hart Hall
21 St. Edmund Hall
22 St. Mary Hall
23 New Inn Hall
24 Magdalen Hall
25 Schools Quadrangle
26 Bodleian Library
27 Physic Garden

28 "Beam Hall."
   Residence of Thomas Willis
29 "Deep Hall" (88 High Street).
   Robert Boyle's Lodgings
30 Arthur Tillyard's Coffee House (90 High
   Street). Site of Peter Stahl's Chemistry
   Courses.
31 "Buckley" or "Bulkeley Hall" (106-107
   High Street).
   William Petty's Lodgings.
32 "Ram Inn" (113-114 High Street).
   Site of Peter Stahl's Chemistry Courses.

Robert Boyle *c.* 1664.

Drawing by William Faithorne. *By courtesy of the Ashmolean Museum, Oxford.*

High Street, Oxford, looking west. All Souls is on the right. The small gabled house in the left foreground, next to University College, is Deep Hall (88 High Street), where Robert Boyle had his lodgings and where his circle met intermittently *c.* 1657—1668.

By courtesy of the Bodleian Library, Oxford.

Thomas Willis *c*. 1666.

By David Loggan from Willis's *Pathologiae cerebri* (1667). *By courtesy of the Huntington Library, San Marino, California.*

Beam Hall, Thomas Willis's residence *c*. 1648–1667 in St. John Street opposite Merton College. Here Willis practiced medicine and carried out chemical and anatomical investigations in collaboration with Robert Hooke, Richard Lower, Christopher Wren, Thomas Millington and others.

By courtesy of the Bodleian Library, Oxford.

A likeness of Richard Lower.

By courtesy of the National Library of Medicine, Bethesda, Maryland.

Ralph Bathurst, by David Loggan.

By courtesy of the Huntington Library, San Marino, California.

Henry Oldenburg.

By courtesy of the Royal Society, London.

John Locke *c.* 1672, by John Greenhill

By courtesy of the National Portrait Gallery, London.

Gresham College, London, where the Royal Society met after the Restoration.

From John Ward, The Lives of the Professors of Gresham College (1740). By courtesy of the Huntington Library, San Marino, California.

A DISCUSSION AMONG FRIENDS: RESPIRATION WORK AT LONDON, 1667–1669 AND RICHARD LOWER'S *TRACTATUS DE CORDE*

T he year from summer 1664 to late August 1665 had been filled, both at London and at Oxford, with intense speculation and experimentation on the complex cluster of problems concerning blood, respiration, and combustion. Boyle had initiated his study of blood and tested the effect of various chemicals on its color and consistency. Hooke had published the *Micrographia* and elaborated experimentally for the Royal Society his idea that the aerial nitrous substance was necessary for combustion. Hooke, Wren, and Boyle had explored whether "airs" "generated anew," especially "nitrous" ones, could support respiration. In his *Vindicatio*, Lower had upheld fermentation as a heat-generating process that utilized the aerial niter taken in by respiration, but had tied that fermentation, and the resultant color change, firmly to the left ventricle. Locke, in his private speculations, had linked the aerial niter to the fermentative process of volatilization that rendered the blood florid, and may even have suspected that this took place elsewhere than in the left ventricle, but he lacked the physiological technique to substantiate his hunch. Certainly the recent vivisectional evidence—Hooke's pulmonary inflation experiment of November 1664—seemed to support the suspicion of Highmore, Henshaw, and perhaps also Boyle, that the mechanical motion of the lungs, interacting with the spring of the air, was the more important process in respiration. Boyle, Lower, and Needham had continued at Oxford their vivisectional trials on the diaphragm with blowing air into the receptaculum chyli, and had even repeated Hooke's open thorax experiment. More spectacularly, Lower and Boyle had, by the end of August 1665, developed injection techniques until they were on the brink of a successful transfusion.

Much of this momentum was dissipated over the next eighteen months in the confusion wrought by plague and the great London fire, and by movements of the *dramatis personae* to and fro. Lower had indeed returned to Oxford for four months in the spring of 1666 to bring the transfusion experiments to a successful outcome. But he came back again only briefly in late September.[1] There he discussed the protocol for transfusion with Boyle, who proposed no less than sixteen queries that might be answered by the procedure. Would transfusion change the disposition, peculiarities, or habits of the recipient dog? Could diseases or medicaments be so transferred from one dog to another? Would chylous blood sate the recipient dog's hunger, or could a dog even be fed entirely by injected chyle? Could coloring or even breed characteristics be changed by transfusing, for example, a spaniel into a setter? But neither man, it seems, had time to pursue such questions. So Boyle sent Lower's protocol on to the Royal Society so that it might repeat the Oxford transfusion, and a few months later, at Lower's urging, allowed Oldenburg to publish the queries that "they may excite and assist others."[2] Lower, in the meantime, had returned home to Cornwall, where in November 1666 he married the widow possessed of the neighboring manor.

Boyle and his other friends were likewise peripatetic. Beginning in November 1665 the young aristocrat lived in Stanton St. John and in Stoke-Newington, spent a month in London, visited his sister in Essex, stayed at his country house in Chelsea, went back again to Oxford for the autumn, and finally turned up again in London in December 1666. Locke had accompanied a diplomatic mission to Germany in the autumn of 1665, and when he began scientific work again at Oxford in the spring of 1666, he worked with his colleague David Thomas, almost exclusively on medicinal chemistry.[3] Needham visited Oxford in 1665, but otherwise his time was taken up establishing a practice in the Shrewsbury area. Wren left Oxford for Paris in July 1665 and when he returned in March 1666 was increasingly involved not with natural philosophy but with architecture, an avocation that became a profession after the Great Fire of September 1666. Minor members of the group also dispersed. Wren and Millington's colleague at All Souls, George Castle, took his D. M. in June 1665 and departed Oxford soon thereafter. Peter Stahl went to Exeter about 1665. Robert Sharrock found ecclesiastical preferment and left Oxford about the same time. Thomas Sprat had found it earlier, but seems to have left Oxford permanently about 1666. His younger colleague at Wadham, the physician Thomas Guidott, left Oxford about 1666 or 1667 to set up practice in Bath. With so many comings and goings, it is little wonder that few had time to develop the rough answers chalked out over the previous few years.

The process of disintegration was soon reversed. In the year between March 1667 and April 1668, many of the most active members of the Oxford group of a few years before moved to London, there to meet and experiment again. Lower arrived in the capital in early March 1667 to set up his practice. John Locke came in late April to Exeter House, in the Strand, where he was to be personal physician and confidant to Anthony Ashley Cooper, first Baron Ashley (later Earl of Shaftesbury). Sometime that summer, at the urging of Gilbert Sheldon, Archbishop of Canterbury, Willis moved his family into a house in St. Martin's Lane, Westminster, and he too launched his London practice. Boyle was the last to leave

Oxford. He was in London in the winter and spring but returned to Oxford between July 1667 and April 1668. At the end of that month, most of his colleagues gone, Boyle gave up his rooms at John Crosse's and returned permanently to live with his sister, Katherine Lady Ranelagh, in Pall Mall.

HOOKE, LOWER, NEEDHAM AND AN "EXPERIMENTUM CRUCIS" ON RESPIRATION

Lower soon made clear his intention not only to continue contact with his old friends in London, but to resume there his anatomical research. He had passed through Oxford in late February 1667, lifted a last few pints with Anthony Wood and cronies, and visited Locke, still in residence at Christ Church, who gave him a letter to carry to Boyle in London.[4] Once there, Lower not only called on Boyle but also upon his school and college mate, Robert Hooke. It seems that the topics of conversation included the problem of respiration, for Hooke told his Gresham colleagues a few months later "that a friend of his had made many experiments of respiration," which he promised to communicate.[5]

Lower was not the only one of the Oxford club to continue thinking about respiration; John Locke and George Castle did the same. In late 1666 or early 1667, Locke had attempted to sort the problem out for himself privately by writing a brief essay entitled "respirationis usus."[6] Castle aired his views more publicly in the *Chymical Galenist*, a tract against the Helmontian empirics that Castle had begun at All Souls in 1665, but which he completed in March 1667 only after he had settled in Westminster to begin practice.[7] As he explained in the dedicatory epistle to his "honoured and Learned Friend," Thomas Millington, Castle believed he could successfully reconcile the old rules of diagnosis and therapy "to the new Discoveries in Anatomy, and the Democritical and Chymical principles" that had transformed physiology.[8]

As befitted a polemical statement, Castle explicitly stated the Oxford presuppositions that Locke's private essay implicitly assumed—anatomical research and the corpuscular version of chemistry. He praised the men whose work in anatomy was responsible for the "vaste improvement of Physick": Glisson, Ent, Highmore, Wharton, Willis, and especially Harvey, whose circulation of the blood "being laid as a new foundation, the whole Fabrick has been built from the very ground.'"[9] Anatomy was necessary because of the close relation between structure and function, a mechanical determinism that derived ultimately from the atomistic origin of qualities and effects. As Gassendi, Highmore, Willis and especially Boyle had demonstrated, chemistry was likewise explicable rationally by the "atomical Philosophy," a far cry from the meaningless jargon of the Helmontians. He singled out Willis for praise, as a man who, working from anatomy, had explicated chemically and mechanically (and with greater clarity than the Helmontians with their *archei*) the physiological processes such as those at work in fermenting liquids.[10] "Dogs, Pigs and Monkeys," Castle said, "have contributed more to the advancement of Physick" than the vulgar chemists "ever did, or are likely to do."[11]

Castle and Locke also agreed, with some slight differences of emphasis, on what respiration did and did not do. They rejected that it was necessary "upon the score of cooling the blood and Heart."[12] Or, as Locke put it succinctly in disputation style: "an primarius respirationis usus sit refrigeratio cordis? Nego."[13] Rather, its primary purpose was to draw in a "volatile salt" to maintain the fermentation in the heart. Locke especially went on to trace out how air performed essential functions in fermentations and volatilizations, and thus substances that easily caught fire in the open air would not volatilize when denied such access.[14] This volatile acid salt, they agreed, broke up the sulphurous substances in the blood and, by its mixture and fermentation, changed the color. Locke even brought in some recent observations on the air that he had made for Boyle at the Mendip mines in April 1666.[15] Miners would die, not because they lacked cold to temper the heat of the heart but because the air had been exhausted of the agent by which the heart maintained its heat and by which the mass of blood fermented. Locke emphasized again and again how air fostered vital heat in the same way as it maintained combustion; the miners' lanterns faltered at the same rate as the strength of the men who held them.[16]

Although Locke hesitated to designate precisely the identity of this volatile, acid salt, Castle had no such qualms. He believed strongly, as Locke also suspected, that the primary purpose of respiration was to "furnish the Blood with a Nitre most necessary to life." It was this niter that "kept afoot the fermentation of the heart."[17] Such a fermentation produced heat, which was merely a motion of the particles of the blood. Just as Boyle in his *Physiological essays* had shown the variability of different terrestrial niters, Castle noted, so could variations in the quality of the aerial niter cause changes in the fermentation of the blood and hence in the state of man's health. This process began in the ventricles of the heart, where a nitro-sulphurous ferment was lodged that "ferments with the active principles of the Blood and Chyle, and produces heat, flame, motion, and a change of colour, the ordinary effects of Fermentation, as it is well known to *Chymists*." Indeed, it may well be, Castle said, that nitro-sulphurous salts were the "Authors and Causes of all Productions and Generations" in vegetables as well.[18] In support of this complex of ideas, Castle cited a melange of references from Ent, Hooghelande, Gassendi, Willis, and Boyle. Yet, in putting the pieces together, he had seen these predecessors through Willis's eyes, reading them only for the parts that Willis had cited in his "De fermentatione" and "De febribus." More importantly, Castle had elaborated these previous conjectures into the strongest statement so far published about the aerial niter and its *physiological* role in heat, fermentation, and cardiac action.

Yet, if respiration had a primarily chemical function, why was it that Hooke's open thorax experiment of 1664 had seemed to demonstrate the absolute necessity of the reciprocal pulmonary motion? That question was not long in finding an experimental answer. On 2 May 1667, Boyle introduced Lower to the members of the Royal Society, whom he gratified by telling of an experiment he had done at Oxford of breaking the thoracic duct in a dog, thereby apparently causing the animal to starve to death, although well fed.[19] Then, at the next meeting, the following item appeared in the Society's Journal Book:

> The experiments appointed for the next meeting were . . . . 2. To open the thorax of a dog, and to keep him alive with blowing into his lungs with bellows.[20]

Although the experiment was not soon performed, comments over the next few weeks by Boyle and especially by Hooke made it clear that they planned, not just a duplication of the 1664 experiment, but an *experimentum crucis* that would decide between the mechanical theory and the chemical-particulate one favored by the Oxonians. Hooke believed one could determine ''whether it be the supply of fresh air, or the motion of the lungs, that keeps animals alive,'' by repeating the 1664 experiment with the important innovation of ''making an incision in the lungs'' to let out the air as it was pumped in.[21] Whereas some of the Society's members believed that air became unfit for respiration ''by being clogged and entangled with [gross] vapours,'' Hooke held

> that there is a kind of nitrous quality in the air, which makes the refreshment necessary to life, which being spent or entangled, the air becomes unfit.

Hooke cited as a proof of this his previous experiments that mere motion of the air would not keep coals burning in a closed box.[22]

Remembering Hooke's profound distaste for pulmonary inflation when it had been performed in 1664, and his avoidance since that time of any experiments in vivisection, it seems only possible to conclude that he had overcome this aversion through Lower's influence, and that Hooke most likely expected to engage Lower's help in performing the modified experiment. If this was Hooke's hope, he was disappointed. Lower, not yet a Fellow, attended no more meetings of the Society, although his name, and that of Needham did come up in June as possible candidates to be a second curator of experiments, but nothing came of it.[23] Although Hooke was reminded again and again through the spring and summer of 1667 to prepare the open-thorax experiment, and Edmund King even appointed to assist him, it failed to materialize.[24] Hooke finally begged to be excused. King and Peter Balle tried the experiment in July, but it failed, ''the apparatus not being fit.''[25] Hooke, in the meantime, was following his more natural (and mechanical) inclinations: he built an air chamber in which a man could be enclosed, in order to test directly the effects of variable air pressure on human respiration.[26] That Hooke was waiting for help was made clear in a letter he wrote in early September to Boyle at Oxford. ''I hope I shall prevail upon Dr Lower, and for him, so as to get him anatomical curator to the Society. He has most incomparable discoveries

by him on that subject, and a most dexterous hand in dissecting."[27] Lacking
Lower's dexterous hand, Hooke's aversion for vivisection had reasserted itself.

In August of that summer of 1667, while the Royal Society stood adjourned,
there appeared yet another work by a quondam Oxonian that put the question of
respiration into full and sharp relief: Walter Needham's *Disquisitio anatomica de
formato foetu* (London, 1667).[28] Needham, like Lower, dedicated his first book to
Robert Boyle, who had urged him to publish.[29] Its London midwife was most
probably John Wilkins, the man who had proposed the absent Needham for
Fellowship in June and who was almost certainly Needham's sponsor for the
position of anatomical curator at the Council meeting three days before.[30] *De
formato foetu* was published in mid-August, reviewed by the *Philosophical
Transactions* in September, and presented by Wilkins to the Society on 3
October.[31]

Needham's indebtedness to his colleagues was made clear in the first words of
the book. Although his anatomical observations had their origins in work done at
Cambridge over a decade before, he had decided to condemn his findings to
obscurity until inspired by the accomplishments of "nobler talents," especially
Willis and his "immortal book on the brain." He praised the *collegium
anatomicum Oxoniense*: Lower, whose genius had made Needham his devoted
follower; Willis, Boyle, and Millington, whose presence not only added luster to
the proceedings, but who often helped with the cutting; even Needham himself,
who was not negligent in contributing his share.[32] That his compliments were not
just *pro forma* can be seen in the numerous bits of evidence for collaboration that
are spread throughout the book. Needham adduced Boyle's witness for his own
discoveries on the salivary glands, the nasal passages, and some findings on the
anatomy of the calf.[33] At one point Needham illustrated an argument on nutrition
by referring to Lower's experiment of breaking the *receptaculum chyli*[34]—exactly
the experiment Lower had done at Oxford and had mentioned at the meeting at
which Boyle introduced him to the Society. In another chapter Needham appealed
to the witness of Lower, Boyle, Willis, and Millington to corroborate his assertion
that he had discovered the *ductus intestinalis* long before Steno published his
account.[35]

The major portion of the *De formato foetu* was concerned with technical aspects
of the anatomy and physiology of the human fetus: fetal nutrition, the structure and
function of the placenta, the disposition of the chorion, allantois, and amnion, and
their respective fluids, the structure and function of the umbilical vessels, and the
arrangement of the fetal circulation. Throughout this work on a most Harveian
subject, Needham showed himself an apt pupil of the doyen of English anatomists.
He defended Harvey's idea of the nourishing function of the embryonic fluids.[36]
He compared, as Harvey did, the circulatory arrangement of the fetus in the womb
to that of lungless animals with only one ventricle to the heart.[37] Like Harvey's
work, and that of Boyle, Willis, and Lower, Needham's fetal anatomy was
comparative in the best sense of the word. He drew evidence from embryos of the

chick, salmon, skate, and frog, and from the placental systems in the sow, mare, cow, ewe, goat, doe, rabbit, bitch, and finally, the woman. His technical anatomy was first-rate; it was not surpassed for over half a century.

But in certain points of his physiology, especially in the physiology of respiration, Needham's embryological research had led him to conclusions greatly at variance with those of his Oxford colleagues. At the end of his chapter on fetal circulation, he concluded that in fish, the defect of lungs was supplied by the gills, which served the function of mixing the blood.[38] This led him naturally into the subject of respiration in general, the origin of animal heat, and the relation of the heart and lungs to the air in warm-blooded animals—subjects he discussed in a "Digressio de biolychio & ingressu aeris in sanguinem item de sanguificatione."

The hypothesis that blood entered the body to generate vital spirits and to fan the *flammula vitalis* was, he said, of great antiquity. But it was, nonetheless, one that lacked entirely any basis in observation or experiment. No flame of any sort could be observed in the heart. Moreover, the blood itself was not in any way an inflammable liquid.[39] He cited with great respect how the "most learned" Lower argued so accurately for the hypothesis of the blood's ascension in the heart and how it acquired thereby its color and heat. But he begged to differ. The heart was nothing but a muscle; how could it generate such a ferment? More likely, he said, was the notion of the "most learned" Henshaw, who traced the blood's fermentation, not to the action of the air and the heart, but to the reaction between blood and the chyle that was perpetually being added to it.[40] Needham rejected also the even cruder Cartesian notion of the blood "exploding" in the heart, and thus causing a beat. Even if, as some argued, the flame was similar to that of alcohol in its invisibility, the heat of the blood was not sufficient to cause such an inflammation. Moreover, liquids that were enkindled by a process of ventilation were very rare. Lower had said that this flame was not found in fish, yet it was certain that they have life, and sense, and motion. Finally, Needham said, animal spirits were not found in the form of flame, a point he argued from Willis's doctrine of animal spirits, as found in the *Cerebri anatome*.[41]

All of these theories, Needham pointed out, were dependent upon the assumption that air entered the mass of the blood as it was carried across the lungs. But most authors, from Aristotle and Galen in antiquity, to Highmore and Henshaw in recent times, believed only that some *quality* from the air entered into the blood during the pulmonary transit. After reviewing the work of, among others, Boyle, Highmore, and Henshaw on the mechanical motion of the lungs, he adduced, as a strong support of his doubts, the findings of Malpighi that the blood, in passing through the lungs, always remained within capillaries.[42] He concluded from this that it was highly improbable that there existed some way, some path, by which air might mix with the blood. He supported his case by examining in detail the "respiration" of fish, frogs, lizards, and birds. He concluded that although air was in some way "necessary," there was no way of proving that it physically entered into the blood.[43] The true use of respiration, he said, was that, by the constant motion of the lungs, the blood was "comminuted"—mixed and amalgamated—and made fit for a proper circulation. The same function was performed in fish by the constant motion of their gills.[44] Needham, by citing a great deal of observational evidence that supported his own opinion, was challenging those who believed otherwise to adduce experiments to support their theories.

Needham could not have thrown down at a more appropriate moment a more well-argued challenge to the ideas of his fellow members of the *collegium anatomicum Oxoniense*. Coming as it did, directly on the heels of Castle's less systematic rendering of the opposite opinion, it drew the lines of battle clearly. Needham had referred directly on numerous occasions to Lower's ideas and experiments. He obviously had great respect for Lower as an experimentalist, but his own embryological approach to physiological problems had forced him to exactly opposite conclusions. Moreover, Oldenburg's acquaintance with the principals in the debate guaranteed that Needham's views would receive maximum exposure. The review that Oldenburg wrote for the *Philosophical Transactions* was unusually long—eight pages—and contained a precise summary of all of Needham's findings, including (*sans* names) his judgment of his colleagues' opinions on the air and respiration.[45] Lower could scarcely have been given a stronger encouragement to vindicate his own views.

It was a debate soon ended. When the Society reconvened in early October, Hooke was once again appointed to perform the open-thorax experiment at the next meeting, "which experiment might conduce to the discovery of the nature and use of respiration."[46] Hooke, so notified, prepared his apparatus and called in Lower to help. On 10 October 1667 they succeeded.[47] First, they repeated the 1664 experiment of opening the thorax of a dog, exposing the heart and lungs, and keeping the animal alive "by the Reciprocal blowing up of his Lungs with *Bellowes*." The dog remained so for more than an hour, during which time the Society could observe, as it had in 1664, that the animal went into convulsions when its lungs stopped moving and was revived when the bellows were once again plied. Hooke then introduced his new fillip. He attached a second pair of bellows to the first, by a contrivance such that a *continuous* stream of air could be forced into the lungs. He then pricked the pleural membrane "with the slender point of a very sharp pen-knife," so that the "blast of Air" could escape from the inflated lungs as fast as it was pumped in. The lungs thus "were alwayes kept very full, and without any motion." As long as they were kept so inflated, even without motion, the dog lay still, "his Heart beating very regularly." But upon ceasing this "blast" and allowing the lungs to collapse, the dog would immediately go into convulsive fits, but could be "soon reviv'd again by the renewing the fulness of his Lungs with the constant blast of fresh Air." Toward the end of the experiment, Hooke and Lower even cut off a piece of lung and could observe quite clearly "that the Blood did freely circulate and pass throw the Lungs, not only when the Lungs were kept thus constantly extended, but also when they were suffer'd to subside and lie still."

It was a stunningly effective experiment, albeit one that Evelyn—always the proper squire—found "of more cruelty than pleased me."[48] Hooke, with the assistance and encouragement of Lower, had converted pulmonary *inflation* into pulmonary *insufflation*.

The experiment had been done, Hooke said, "because some Eminent Physitians had affirm'd, that the *Motion of the Lungs* was necessary to Life upon the account of promoting the Circulation of the Blood," and that an animal would be suffocated as soon as its lungs had ceased to move. Hooke, for his part, felt the experiment argued that "the *bare* Motion of the Lungs *without fresh Air* contributes nothing to the life of the Animal," and that the cause of death in subsiding lungs was not "the stopping the Circulation of the Blood through the Lungs, but the want of a sufficient supply of fresh Air." In the ensuing discussion, others were not so easily convinced. George Ent, in reflecting on the experiment, would allow only "that it shewed what was not the use of respiration, but not what it was": namely, "that the lungs did not serve to promote by their agitation the motion of the blood."[49] Ent's caution is both amusing and significant. Almost two years before he had opined to Pepys over ale at the Crowne Taverne that it was "not to this day known or concluded on among physicians, nor to be done either," how respiration "is managed by nature or for what use it is."[50] Now he was confronted by a technique designed to confirm an idea that was the lineal descendent of a notion he had once held. But Ent, his intellectual offspring transformed and dressed in the garb of chemistry and the corpuscular philosophy, was reticent to own it. Hooke himself recognized that, although his experiment proved the necessity of fresh air, it was not yet completely determinant. Fresh air could be necessary because it *took something away from* the blood and/or because it *added something to* the blood. To answer this question, Hooke suggested two experiments using lung by-passes. In one he proposed to circulate blood directly from the pulmonary artery to the aorta; in the other he hoped to circulate blood through an open vessel, to see in both cases whether the blood so detoured would serve the animal's needs.[51]

Lower, in the meantime, went on to more, and sometimes more spectacular, physiological experiments. The week after he had helped Hooke with the insufflation experiment, Lower was elected a Fellow and admitted. From then until the middle of March 1668, Lower was almost unflaggingly active in anatomical pursuits and physiological experiments. He demonstrated the Oxford experiment of rendering a dog broken-winded by sectioning the phrenic nerves; the same effect could be produced by piercing the diaphragm.[52] He showed the effects of ligating the vena cava and jugulars, and was later deputed to verify Steno's experiment that an animal could be made insensible by compressing the carotid artery.[53] He did experiments on ripping the *receptaculum chyli* and others showing that foodstuffs passed into the veins only through the thoracic duct.[54] He even showed the cause of blindness in horses by dissecting several of their eyes before the company at Gresham.[55] Many in the Society obviously felt as Oldenburg did when he praised Lower to Boyle at Oxford: "He, in conjunction wth severall others, are like to doe

Philosophy a great service by Anatomicall Discoveries.''[56] To procure such discoveries, the Council dispatched Wilkins to interest Lower in becoming the Society's anatomical curator. Even after Lower demurred, citing his other obligations, the Society showed its gratitude (or its hopes) by ordering that the operator hire ''a small room near Arundelhouse, on the water side, convenient for Dr. LOWER to make anatomical experiments in for the use of the society.''[57]

But the most spectacular of all Lower's work was his collaboration with another *transfuseur*, Edmund King, to perform the first human transfusion in England—the first ever had been done in Paris by Denis in June 1667. Lower induced one Arthur Coga to accept (for a consideration of 20 shillings) this dubious honor, and on 23 November 1667, Coga was given a transfusion of sheep's blood.[58] This first private trial was repeated less than a month later before the entire Society, augmented, as Oldenburg reported to Boyle, by a ''strange crowd both of Forrainers and domesticks.''[59] The fame of transfusion, and thus indirectly that of Lower, was assiduously promoted by Oldenburg in the next few months. He avidly reported news of transfusion to Boyle and Wallis at Oxford, published accounts of both the French and English trials in the *Philosophical Transactions,* and extolled the virtues of the procedure to William Neile in Berkshire, René François de Sluse in Liège, Johannes Hevelius in Danzig, Johann Christoph Beckman in Frankfurt-an-der-Oder, and Francesco Travagino in Venice.[60] Fortunately for Lower's reputation, Denis and Henri Justel also kept Oldenburg abreast of transfusion trials in Paris.[61] When one of the recipients there fell ill from the effects and died in early February 1668, Lower and his English colleagues took warning and ceased their own experiments.

While Lower was thus distracted from respiratory research, Hooke attempted to execute the lung bypass. In November he showed the Society his ''contrivance'' for leading the blood through an open porringer. He was ordered to try it first in private and to arrange a way of regulating the amount of air to which the blood was exposed as it passed from the pulmonary artery to the aorta. In the following weeks Hooke could report only that the experiment had not succeeded ''so well as he wished.''[62] In December Hooke was reminded of his project, and in January 1668 the experiment of ''Circulating the blood, without its passing through the lungs'' appeared again and again on the roster of ''experiments appointed for the next meeting,'' to no avail.[63] Hooke's attempt to demonstrate directly the effect of the air on the blood ran into the recurring problems of technical complexity and declining interest.

Hooke then turned to the other circumstance in which an animal lived without benefit of pulmonary motion: fetal respiration. On 18 December 1667 he joined with Lower and Walter Pope, another old Westminster from Oxford and Hooke's fellow Gresham Professor, in an experiment to see whether the fetus lived *in utero* by having its blood ventilated through the respiration of its dam. They opened a mastiff-bitch about to whelp and, in various ways, tried to ligature or otherwise seal the tracheas of the pups. The whelps failed to live long outside the womb. Hooke concluded, in reporting the results to the Society, that the experiment showed that ''the fetus in the womb has its blood ventilated by the help of the dam.'' The sudden death of the whelps was due, he believed, to the ''want of the ventilation of the blood.''[64] Hooke was vacillating back toward Boyle's opinion that ventilation was the primary function of respiration.

Since Hooke's pulmonary insufflation experiment had seemed to contradict so explicitly Needham's ideas of respiratory function, it was no surprise that the Shrewsbury physician responded to the challenge. Just before Christmas, 1667, Needham got hold of two recent issues of the *Philosophical Transactions,* one that had the review of his book and the other containing Hooke's account of pulmonary insufflation. He considered the experiment, performed some more of his own, and on 10 March 1668 wrote to Oldenburg defending the mechanical theory of respiration.[65] Needham routed the letter via Boyle at Oxford, who read it and passed it on to Oldenburg in early April. In doing so, Boyle suggested that Oldenburg show the letter to Hooke privately before reading it publicly, so that Hooke might consider a reply. They were both, Boyle said, ''such ingenious men, and both members of the *Royal Society,*'' that he hoped ''that they may be brought to agree, without making their opposition éclater.''[66]

The objections Needham voiced in his letter were not, he said, to Hooke's experiment itself, which was ''noble & Handsome,'' but to ''ye logicke of it.'' He remained unconvinced that it proved that air entered the blood. The essence of Needham's own opinion was that the comminution of the blood was done primarily in inspiration, when the pulmonary ''vesicles'' (alveoli) of the lungs were filled and thus pressed against the pulmonary blood vessels. This was exactly what had happened in Hooke's experiment. The impulses of the air as the bellows were worked, and even more the tension of the filled alveoli, pressed upon the vessels of the lungs and forced their contained blood forward into the left ventricle. As for the convulsive motions observed when the lungs were allowed to collapse, Needham could explain this from Willis's *Cerebri anatome.* The pneumonic nerves that Willis had shown to be present in the lungs were irritated by the collapse and sent impulses that were passed through the vagus to the phrenic nerves, where they caused the diaphragm to move. What Hooke had seen as convulsions were merely the normal attempts of the animal to inspire. Nor, if one assumed that the *succus nervosus* came from the blood, could one believe that the deprivation of air would be passed along to the nerves so quickly as to cause convulsions. Indeed, blood was naturally so full of spirit that the lack of a little air could not bring on death so quickly.[67]

Needham went on to cite some experiments that supported his interpretation. If one attached an inflated bladder to the trachea of an animal and forcibly inflated the lungs, ''you shall see ye circulation of ye bloud as well continu'd as if there were a constant repetition of blasts.'' Even more experimentally dubious was ''ye ingress of aire into ye heart.'' In Hooke's inflation experiment Needham found ''yt ye pulse of ye heart is proportionable to ye force of ye bellowes''; by working the bellows more quickly, one could easily double the pulse rate. Yet, if you continue blowing "whilst you open ye left ventricle you shall find no spumosity in ye blood: nor any other signe of ye mixture of ye aire with it." Death was caused, Needham strongly believed, by the obstruction of circulation. If one tied the vena cava above and below the heart and then allowed the lungs to subside, the animal died without any motion: "From whence I gather that it is ye cessation of ye *circuitus sanguineus* yt causeth immediate death, & not ye want of aire." Moreover, a frog, which can live for long periods under water without breathing, does so by first inflating its lungs and then not allowing them to subside. Thus, "in opening of froggs," the lungs "will often leape out of ye thorax & continue rigid unlesse you pricke them." Finally, if, as Hooke believed, the blood circulated as well through subsided lungs as through expanded, why did Harvey go to such trouble to explain how the foramen ovale and ductus arteriosus shunted blood past the compact lungs of the fetus?[68]

As Oldenburg noted in a letter to Boyle, and as was recognized when Needham's letter was read at the Royal Society on 23 April, Needham's objections were "indeed material," and he had suggested "severall good Experiments." The absent Lower was desired to do Needham's experiment of rebreathing with a bladder, with King as his assistant. Boyle, at Oxford, had also had his curiosity piqued by Needham's objections.[69] He had searched among his notes and found the trials on respiration that he and Hooke had done with the airpump at Chelsea in 1662–1663.[70] Thus, when on 30 April 1668 Hooke read his answer to Needham (not now extant), Boyle, attending his first meeting in almost a year, was also able to tell of some previously undisclosed experiments on respiration.[71] Lower was still unavailable, so on 9 May Hooke and King were the ones who carried out Needham's suggested experiment. They filled a large bladder with air and attached it via a brass pipe to the exposed trachea of a dog, thereby forcing it to rebreathe the same air. Within two or three minutes the dog began to struggle violently. By the end of eight minutes it was near death and was resuscitated only when fresh air was pumped into its lungs.[72] They repeated the experiment a fortnight later, this time finding that if they waited until two or three minutes after apparent death, the dog could not be resuscitated. Upon hearing this report, Boyle observed "that it were not amiss to try, whether air might not be made fit for respiration of animals," and intimated that he had made some such experiments.[73] Boyle clearly saw the implication. Needham's mechanical theory had once again been impugned, and the chemical-particulate view seemed all the more likely.

Despite this second experimental defeat, Needham had come within a hair's breadth of a major discovery. He had performed the pulmonary inflation experiment and, while the bellows were working vigorously, had opened the left ventricle. He found what he expected to find: no "spumosity," or other physical signs of air mixing with the blood. Yet had he interested himself a bit more closely in the nature of the blood itself, and had he held a chemical rather than mechanical theory of respiration, he would have seen that such a modification of the pulmonary inflation/insufflation experiment offered a perfect opportunity to examine the properties of the blood *in situ*. Instead, it was Boyle, Hooke, and especially, Lower who seized the opportunity Needham neglected.

FRACASSATI'S EXPERIMENT AND THE FLORIDNESS OF BLOOD

Oldenburg, that irrepressible apologist and scientific magpie, unearthed the important clue. Since 1665 he had, both in his correspondence and in the *Philosophical Transactions,* promoted injection and transfusion, as well as English priorities in these most promising of medical techniques. He was constantly on the lookout for new experiments and new practitioners; witness his role as intermediary between Timothy Clarke and the Danzig injectors, Johannes Fabritius and Michael Behm.[74] Over the summer of 1667—perhaps even as he languished in the Tower of London for maintaining a suspiciously cosmopolitan correspondence in the midst of the second Anglo-Dutch War—Oldenburg found more such references. Injection experiments were buried in the middle of a long *dissertatio epistolica* on the brain by the Bolognese anatomist Carlo Fracassati, which was in turn part of an even larger collection of miscellaneous anatomical letters published by Malpighi and Fracassati in 1665.[75] Oldenburg published extracts from them in the September 1667 issue of the *Philosophical Transactions*.

Oldenburg recounted that Fracassati had injected various strong chemicals into the veins of dogs and observed their effects, sometimes *post mortem*. Aqua fortis (nitric acid) killed the dog "presently" and coagulated the blood in his vessels. Spirit of vitriol (sulphuric acid) did the same, turning the blood both solid and sooty in color. Oil of tartar (potassium tartrate?), on the contrary, killed the injected dog, but the "Spectators were surprised to find his bloud not curdled, but on the contrary more thin and florid than ordinary." This hinted, Fracassati had concluded, that "too great fluidity of the bloud, as well as its Coagulation, may cause death."[76]

Another experiment also bore upon the color of the blood. It was vulgarly maintained, Fracassati said, that when shed blood was left to stand in a dish, the separation of a darker portion below proved that blood contained a melancholy humor. Not true, said Fracassati. The blackish color came "from hence, that the bloud, which is underneath, is not expos'd to the Air." To prove this one need only expose the darker portion to the air, whereupon "it changes colour and becomes of a florid red." *"An Experiment as easie to try, as 'tis curious,"* commented Oldenburg.[77]

The Fracassati experiment showed the Oxonians how to make Columbus's egg stand on end. For over two decades the experiment of letting blood stand and observing its "separation" had been a standard item of discourse among the physiologists. Harvey had cited the experiment in the *De generatione* (1651); Highmore had done the same in the *Corporis humani disquisitio anatomica* (1651). Bathurst had used the example in his "Praelectiones" (1654), Willis in his

Diatribae duae (1659), and Lower in his *Vindicatio* (1665). Now men like Lower and Boyle had to rethink the significance of the simple demonstration they had so often taken for granted. It was a matter of explicit faith among the Oxonians, especially Lower and Willis, that something entered the blood during respiration, but the effect of that agent was conceived as mediated by the process of volatilization, or fermentation, which took place in the left ventricle of the heart. They were ill-prepared to contemplate that the air might have a *direct* effect upon the blood's most obvious property, its color. Lower, for example, both in his letter to Boyle of 1664 and in his *Vindicatio* of 1665, had assumed that the florid crown represented the separation of blood components. To have the observation transposed out of one frame of reference and into another must have been quite a surprise. No one before had had to deal with the possibility that air could have a direct and immediate effect on the blood.

Oldenburg knew that such findings would be of special interest to Boyle. Even before the notice on Fracassati's work appared in the *Philosophical Transactions,* he wrote to Boyle in Oxford informing him that the Malpighi book would be sent on and suggesting that he "order the injecting into ye veins of a Dog" some oil of sulphur and especially some "oyle of Tartar." If in the latter case the dog died, "as 'tis like he will," the animal should be opened, "and his bloud well observed."[78] Boyle read of the Fracassati experiments in early October 1667 and wrote to Oldenburg claiming priority in examining the effects of chemicals on the blood.[79]

He reiterated the results of his Oxford experiments of autumn 1664, that whereas acids such as aquafortis and oil of vitriol coagulated warm blood and made it appear dirty, chemicals "abounding in *Volatil* Salt, such as the *Spirit* of *Sal Armoniack,*" would not only maintain the blood's fluidity, but also "make it look rather more florid than before." Since there had been great interest at Gresham when Boyle reported these results, and since there was "so little difference between the warm Blood of an Animal *out* of his Veins and *in* them," perhaps Fracassati had heard of Boyle's results and had been moved to initiate similar trials. Boyle also suggested that, if the Royal Society did more injection experiments, they might try injecting "spirit of sal armoniac, or of hartshorn alone"; Boyle remembered doing a similar experiment, but had not time to write it out.

A week later, on 26 October 1667, Boyle wrote again from Oxford to say that he had tried Fracassati's "pretty" observation on the florid color of blood, and it held true. The reason given by Fracassati, that blood turned florid when exposed to air, might also "perhaps, be good." Boyle had however "tried some things" that might help to determine whether the explanation was really correct, "of which, if it be desired," Boyle could "easily give an intimation." Appropriately enough for such a subject, he expressed approval of Lower's election to Fellowship, and in a postscript inquired "of your prosecuting at the Society the business of respiration." Boyle was, he had told Oldenburg the week before, interested in the subject and was trying to locate some of his old notebooks.[80]

The subject of "florid" blood then dropped entirely from sight, both at Oxford and London, for six months. When it reemerged, it was under highly suggestive circumstances. On 30 April 1668, Boyle attended his first meeting of the Royal Society since June of the previous year. He had just finished writing up at Oxford the air pump experiments for his *Continuation of new experiments physico-mechanical* and had found and reexamined his vacuum pump experiments on respiration of 1662–1663. In late April he had given up his lodgings with John Crosse and returned permanently to London. His Society colleagues eagerly heard that he had been conducting experiments on airs latent in water and other liquors, on the effects of the absence of air on living creatures, and on the ways of generating and identifying true air. Boyle promised details at future meetings. Hooke was next. He read his answer to Needham's letter and promised experiments bearing upon the Shropshire physician's objections. Hooke then went on to mention Fracassati's experiment. He noted that blood, though of a dark, blackish color, would, "when exposed to the air, become presently very florid." If the florid surface were taken off, and the "subjacent part" exposed again, it too would "acquire the like floridness." Therefore, Hooke said:

> It might be worth the observing by experiment, whether the blood, when from the right ventricle of the heart it passes into the left, coming out of the lungs, it hath not that tincture of floridness, before it enters the great artery; which if it should have, it would be an argument, that some mixture of air with the blood in the lungs might give that floridness.[81]

It seems highly likely that Hooke, who until Boyle's return to London had expressed no interest in the Fracassati experiment, had discussed it with Boyle, who was intensely interested in the chemical and physical properties of the blood. It was Hooke, however, who was much more committed than Boyle to the niter theory of the air and who, in addition, had recently been involved with Lower in doing vivisectional experiments on respiration. Once the Fracassati experiment had been called to mind, Hooke naturally saw what implications it had for their recently devised techniques.

But, as in the case of the original insufflation experiment, this new one proved easier to suggest than to execute. In subsequent weeks what vivisectional work Hooke did was with King—the rebreathing experiments designed to vindicate his own insufflation experiment from Needham's objections. On 28 May Hooke was desired "to try in private the experiment of the floridness of the blood," a reminder that was repeated a fortnight later.[82] He did not follow through. Instead, during July and August he fell back into an old and comfortable line of research, aimed at using the vacuum and compression pumps to prove that the amount of air necessary for respiration was proportional to time, and that in the cases of animals and flames, both lasted longer in compressed air than in ordinary air of the same volume.[83] At one point in the discussion of these results, the logic behind this revival was made explicit. Hooke intimated that he thought alkalis exposed to air "would arrest the volatile salt, which is in the air, and turn it to niter." Daniel

Cox, for his part, was uncertain ''whether the air added that nitrous substance to alkalis, or extricated it out of them.''[84] Whichever of Boyle's two sometime assistants was right, the logic of such compression trials was impeccable. A greater volume of air contained a greater quantity of volatile salt (niter) and, therefore, could support an animal or a fire for a greater length of time. Compressing the air merely increased the density of this volatile salt. Q.E.D.

RICHARD LOWER AND THE
TRACTATUS DE CORDE (1669)

Hooke never carried out the experiment so perceptively suggested at the meeting of 30 April 1668. One reason may well have been that both his anatomical helpmate, Lower, and his patron, Boyle, ceased to cross his path, at least at Society meetings. After late June, Boyle attended only one meeting in August and none the following autumn. Lower's departure from Royal Society affairs was both more precipitate and more permanent. To my knowledge he attended his last meeting on 5 March 1668, weeks before Boyle's final return from Oxford. After five intensive months of anatomical activity on the Society's behalf, Lower stopped abruptly. One is at a loss to explain precisely why. It was no doubt in part due to the very fame his anatomical accomplishments had brought him. Transfusion was the talk of London and Paris. Even though Lower wisely did no more, it seems almost instantaneously to have made a name for him as a society physician. A press of well-to-do patients made it difficult to attend leisurely meetings of the Royal Society on a Thursday afternoon. Much the same thing happened to Willis; after attending a few meetings in October and November 1667, he too became one of the Society's many nominal members. Lower clearly did not abandon the Royal Society because he suddenly ceased anatomical research. Over the winter of 1667–1668, he had teamed with Timothy Clarke to do a series of investigations of testicular structure that subsequently involved the sometimes pugnacious Clarke in yet another priority dispute.[85] In July 1668, four months after Lower had deserted the Society, Pepys met Lower, Clarke and several others at an ale-house in the Strand to look on ''with great pleasure—and to my great information'' as Lower dissected ''several Eyes of sheep and oxen.''[86] Perhaps Lower had decided that between his private research and ever-growing practice, he had no time to perform medical entertainments for gawking virtuosi.

What must have consumed most of his spare moments in 1668 was the composition of his magisterial *Tractatus de corde* (1669). The book had its origins back in the weeks following the insufflation experiment of October 1667, and Lower's subsequent election to Fellowship in the Royal Society. In mid-November, Oldenburg wrote to Boyle that Lower had been making trials of giving a dog dropsy ''by tying up of the *vena cava*.'' These, ''together wth some other Experiments,'' had given Lower ''matter enough to write a Tract.''[87] The contents of the *De corde* also indicate that Lower began writing the book in the autumn of

1667. Again and again throughout the text, he referred to experiments that he had performed before the Royal Society during the brief half-year when he was an intense participant in its activities. The effects of ligatures on the vena cava and arteries,[88] the transfusion of blood into Arthur Coga,[89] the ripping of the receptaculum chyli,[90] and, most especially, the use of the insufflation technique[91]—all of these can be traced back to the anatomical activity in London in the autumn and winter of 1667—1668. Lower did, of course, incorporate into the book research that had been done at Oxford in the period from the Restoration to his departure in mid-1666; he reported many such experiments, including his protocols for animal transfusion. But the intention to write his "Tract" seems to have been formed amid the work he, Hooke, and Clarke carried out for the Society.

Oldenburg, in his capacity as intelligencer for the London virtuosi, noised about Lower's intentions among the Society's foreign correspondents, creating an expectation that Lower was slow to gratify. In April 1668 Henri Justel in Paris asked Oldenburg to send on a copy of "le traitté Anatomique de Mr. lower qui doit estre curieux," apparently assuming from Oldenburg's comments that the book had already been published.[92] Later that month Oldenburg wrote to Pierre Huet in Caen that Lower, "a most dexterous anatomist, is now preparing for early publication a book packed with many anatomical experiments, new for the most part I believe."[93] Early publication proved not to be so early; in September, Oldenburg was still writing to Finch of how Lower's *De corde,* "said to be full also of very curious, & considerable Anatomicall experiments," was "ready for the presse."[94] But it was not until late December that Oldenburg could report to Malpighi that the book was being printed.[95] Promises of imminent publication were still being given in February 1669, and it was not until early March that Oldenburg reported to Huygens that the printing had been completed, and promised to send a copy.[96]

As often had been the case with Boyle's works, publication for Lower seems to have been delayed by his recurring discovery that he had more to put into his book. For example, in discussing the function of the thoracic duct, Lower took pains to refute the opinion of the Dutch anatomist, Louis de Bils, that the lacteals dumped their contents into the lymphatics, rather than into the bloodstream.[97] De Bils's report, asserting this opinion, had only appeared in the *Philosophical Transactions* in October 1668.[98] Such last minute revisions must have been maddening to both printer (James Allestry) and supporter (Oldenburg) alike.

Lower's first book had been dedicated to one Oxford friend of long standing, Boyle; he took the opportunity of his second book to offer it to another such colleague, Thomas Millington. In the dedicatory epistle Lower outlined his reasons for writing.[99] Harvey, he said, in his "magnificent discovery of the circulation," had described the structure of the heart and the motion of the blood in a way whose perfection would seem to leave no room for successors. But, Lower said, pointing to one of Harvey's promises of future works, even the great man realized there was more to say on the structure and movement of the heart, on the velocity and quantity of the circulating blood, and on the reasons for the difference between venous and arterial blood. His own work, Lower said, was only to redeem the unfulfilled promises of "that excellent man," Harvey.[100] Lower's intentions, as expressed in this dedication to his old friend, colleague, and fellow Westminster, were to complete the edifice Harvey had begun.

Lower was by training an anatomist, by inclination, an experimental physiologist, and by temperament, no theorist. The *De corde*, in both its structure and content, reflected these traits of Lower's background and scientific personality. His concern was to examine and explicate the anatomy of the heart, and to establish, by beautifully contrived experiments, a few crucial principles of cardiac action and respiratory function. In doing so he drew extensively upon a store of techniques and conceptions accumulated in almost a decade of scientific research and collaboration at Oxford and London. Yet in all his arguments he avoided, almost studiously, putting forward a general scheme of physiology that might encompass all these individual results. The *De corde* was a book of discoveries, not systems.

Lower's starting point was structure; his first chapter, one of his longest, detailed the anatomy of the heart considered as a muscle. His aim in devoting so much space to *cordis situs & structura* was set forth in the very first pages: to know the nature and qualities of the blood, one must understand the movement of the heart. Since its motion derived solely from its properties as a muscle, structure had to be established before function could be inferred. To do exactly that, he examined the position of the heart, its pericardium and fluid, the origins of the blood vessels, cardiac innervation, the parenchyma or muscular structure, its internal structure (including fibers, valves, and structure of the vena cava and aortic arch), and finally, the anatomy of the fetal circulation.[101] The core of his discussion was the curious arrangement of the cardiac musculature. From specimens of boiled hearts, he traced out how the heart's muscles were arranged on several levels into a whorl, or helical pattern. It followed necessarily from this structural arrangement, he said, that when cardiac muscle shortened, it pulled, as it were, upon itself, and thus could effectively squeeze the walls of the ventricles, one against the other. In demonstrating this unique action of the heart, Lower in a very real sense completed the work Harvey had begun. Harvey had simply described the heart as a muscle and assumed that its contraction would result in the expulsion of the contents of the ventricles. Lower showed *how* this followed necessarily from the arrangement of the cardiac fibers.[102]

The value of this initial exercise in morphology became clear when Lower turned his attention to the first of the important physiological problems with which the *De corde* dealt—the question of the cause of the heartbeat. Harvey had shown in his *De motu cordis,* Lower argued, how the movement of the heart was clearly muscular and how it resulted from a contraction of the cardiac fibers when they became smaller and harder in systole. Yet in spite of this, some "distinguished" and "famous" men, such as Descartes and Hooghelande, had conjectured that the heart moved, not by virtue of its own structure and powers, but rather because it was put into motion by the blood. They made this mistake, he said, ''because they did not pay close enough attention either to the strength of the heart's structure and its great efforts at every systole, or to the rapidity of the blood's movement.''[103] Lower set himself the task of refuting these erroneous, un-Harveian, theories.

The first kind of error about the heart's movement arose from a misunderstanding of chemistry and the nature of the blood. Seeing the ebullition that resulted when some chemical fluids were mixed, Hooghelande had posited the existence of a nitro-sulphurous ferment in the left ventricle that reacted with the blood, causing it to become lighter and swell up and thus to escape into the aorta ''more by its own action than by that of the heart.''[104] To refute this theory, Lower drew upon a decade's experience and reading, and adduced more than a dozen reasons why such a conjecture was wrong.[105]

The first class of reasons was simply observational. Blood was in no way a liquid given to ebullition or effervescence; it was too gentle, inert, and innocuous a fluid to ferment in the violent way Hooghelande and his followers thought it did. Arterial blood was not light or foamy in any way. If collected in a container, it showed the same properties of weight and consistency as venous blood and differed only in color. Moreover, the blood did not enter the heart drop by drop as these anatomists thought, but in a large stream that filled the entire ventricle. And, if the heart was just a passive vessel, why was it so well endowed with fibers and nerves? Besides, pathological cases in which the blood became more watery left the heartbeat unchanged.

Lower brought even more telling logical reasons to bear on this iatro-chemical idea. One knew, he said, that the active phase of heart action was systole, yet the ebullition theory demanded that the heartbeat take place in diastole. Moreover, what ebullition was so regular as the heartbeat? Why was there not a fermentation in the right ventricle as well as in the left? Or, more to the point, why did this effervescence not take place in the auricles, which were the first parts of the body to live and the last to die? Finally, he argued, the blood was simply not in the heart long enough ''for the production of such ebullition.''

But it was with experiments that Lower delivered the *coup de grâce* to the ebullition theory. First of all, he asked, how was this ferment in the left ventricle replenished? It could not remain in the heart after systole, since Lower had already shown that the ventricle contracted completely. To those who would suppose the ferment was replaced via the coronary arteries, he pointed out the imperviousness of the ventricle's inner membrane. But the argument could be clinched more directly; if one forcibly injected dye into the coronary arteries, none of it penetrated into the cardiac sinus. He made the point even more elegantly with an experiment he had performed some years before at Oxford:

> But, to decide experimentally whether or not any ebullition of blood helped the
> blood's movement at all, it occurred to me to see if the heart would continue its

movement undiminished, after I had drawn off the blood, and had replaced it intravenously by an equal volume of other fluids, less liable to become lighter or to froth up.

He drew off through the jugular vein of a dog almost half the animal's total blood volume and then replaced it by injecting an equal amount of beer mixed with wine into the crural vein. He continued the procedure of bleeding and injecting until the fluid coming from the vein had no more color "than the washings of meat, or than claret several times diluted." The heartbeat, meanwhile, became only slowly more feeble, "so that practically the whole of the blood was replaced by the beer before life was replaced by death."[106] Obviously the amount injected was somewhat exaggerated; but Lower's training in injection techniques had provided him with an ideal way of disproving a misleading theory.

When Lower turned from disciple Hooghelande to master Descartes and to the second version of the ebullition theory, his Oxford background served him equally well. Descartes, and some of his English followers, believed that the heart's motion was due to the blood because the substance of the heart itself, containing a "sort of vestal fire," heated the incoming blood, causing it to flow out. Aside from the fact that blood was scarcely a rarefiable liquid, Lower said, the credibility of this theory of effervescence depended upon the heart being the center of heat for the body. In rejecting the premise, Lower was unequivocal: "I am so far from believing the movement of the blood to be dependent on any heating of it within the heart, that I do not think it owes any of its heat to this organ."[107] On the contrary, the heart needed to be heated by the blood flowing in and through it. "The blood is, therefore, entirely responsible both for the heat of the heart itself and for the activity and life of our bodies, which its heat produces."[108] He did allow, however, that since muscles by their activity could have heat, the heart, by virtue of its perpetual motion, might be endowed with a more *constant* heat than any other members of the body.[109]

Lower, in attempting to refute the Cartesian heat version of the ebullition theory, had been led to deny the innate heat of the heart and to place the locus of heat in the blood. He had done much the same in the *Vindicatio* of 1665, as had a long line of Oxonians before him. But the statement in *De corde* was firmer and clearer, shorn of even the weak links to the heart that he had expressed four years earlier. He had succeeded in centering the heat *totally* in the blood. But beyond that point Lower could not, and did not, go. The reason was his long-standing friendship with Willis:

> But we are certainly warmed by a fire that is more than fictitious or metaphorical, and so it would be worth while in the next place to explain at somewhat greater length how the blood itself becomes heated and in its turn provides warmth for the whole of the body. However, since this would be outside the plan of the present work, and I also understand that the learned Dr. Willis is giving the matter some thought in his

book *On the Soul, and Also on the Heating of the Blood*, I should not like to depart so far from professional courtesy as to forestall him in this matter.[110]

From this, and from other hints scattered through the *De corde*, one may conclude that Lower had his own ideas on the origin of heat, perhaps tied to the process of fermentation. But thinking Willis would make the same points in his upcoming book, Lower deferred to his scientific and professional mentor.

After this excursus in criticism, Lower completed his section on the heart by summarizing the heart's true motion, utilizing the principles of muscular structure that he had established in the first chapter.[111] Moreover, in recapitulating the work he and Boyle had done at Oxford in 1664, he demonstrated how the pace of the heart was ultimately controlled by the vagus nerve, a point he proved with both ligature and section experiments.[112] To conclude, he delineated the ways in which a large number of pathological conditions could affect the heart's motion, and, in turn, be affected by its irregularities.[113] In most traditional order, but with most untraditional content, Lower had explicated the anatomy, physiology, and pathology of the heart.

LOWER ON RESPIRATION AND BLOOD COLOR

When he came to present his findings on a related subject, respiration and the color of the blood, Lower followed a mode of argument that represented the logical rather than the chronological development of his ideas. Once again his starting point was the structure and movement of the heart. Those who had hitherto written on the question of the blood's rate of circulation, he asserted, had not adequately considered cardiac anatomy. As a result, they had calculated a circulation rate that was much too low.[114]

To correct these earlier mistakes, Lower assumed that the ventricle contained at least two ounces of blood, the figure Harvey used, and that it expelled its contents completely upon contraction, an assumption Harvey did not make. Thus, given a pulse rate minimally estimated at 2000 beats per hour, one had to conclude that the heart pumped 332 pounds of blood per hour. Since all the blood in the human body rarely exceeded twenty-five pounds, "it will surely follow that the whole of [a] man's blood circulates through the heart six times within a single hour."[115] Despite Lower's arcane arithmetic result[116]—corrected in the *Philosophical Transactions* to the proper figure of thirteen[117]—the point he made was valid and telling; even if one minimized the multiplicands (ventricular capacity and pulse rate) and maximized the divisor (total blood volume), as Lower did, the circulation rate was much faster than any investigator had previously estimated. The only assumption Lower needed, established so well in the first chapter, was that the ventricles expelled their contents completely.

Moreover, Lower said, a circulation rate of this order could be corroborated by other evidence. Could not a man pass two or three pounds of beer in a half-hour? Even assuming this meant only four to six pounds of blood had passed through the kidneys during this period, was this not a large amount of blood passing through one part of the body in a short space of time? Finally, in carefully controlled experiments, animals could be bled completely from arteries in less than five minutes, sure proof of the rapidity of the circulation.[118]

The thrust of Lower's calculations and experiments was clear:

> If, however, this rapid circulation of the blood is accepted (and I think I have given sufficient proof of it), it will be evident that there is not so much *difference between arterial blood and that contained in the veins, as is commonly supposed.*[119]

But having taken his reader this far, Lower was forced to confront his former self—the opinions he had expressed in the *Vindicatio* of 1665. He did so with aplomb. He had been wrong then, he said, because he had "relied more in this matter on the authority and preconceived opinion of the learned Dr. Willis than on my own experience, and confused too far the torch of life with its torch-bearer." He would gladly exchange his "former view for a better one."[120]

The "better view" that he wished to argue was of two parts: (1) the difference between venous and arterial blood was *independent* of any heating of the blood in the heart; and (2) the color of arterial blood must be attributed entirely to the *fresh air* it absorbed in the lungs. The two propositions, which might seem to be encompassed by the second, were viewed by Lower as distinct. He first proved that the color change did *not* occur in the heart, and only then, that it *did* occur in the lungs.

To prove the first, he opened the thorax of a dog, exposed and corked the trachea, and observed that the blood flowing out of the cervical artery was "as completely venous and dark in color, as if it had flown from a wound in the jugular vein." To demonstrate further that the color of the blood was in no way dependent upon its heat, he waited until the dog died and then drove venous blood through the insufflated and perforated lungs. His expectation was confirmed; "the blood was discharged into the dish [from the pulmonary vein] as bright-red in color, as if it were being withdrawn from an artery in a living animal." Finally, Lower pointed out the observational corollary that, if the color of the blood did not depend upon heat, then it was perfectly understandable why arterial blood, when cooled, did not lose its florid color.[121]

The primary proof for the second proposition was given by the same insufflation experiment. If the pulmonary vein were cut, florid blood could be withdrawn. In justifying why he had previously asserted that dark, venous blood flowed in the pulmonary vein, he explained that it had been a legitimate inference from experimental work. He had been misled because he had not been able to keep the animal alive by continuous insufflation, and hence all the air had been forced out of the lungs by the time he was able to grasp and cut the pulmonary vein.[122] Quite naturally, he got venous blood. Lower thanked Hooke carefully for the technique:

> I acknowledge my indebtedness to the very famous Master Robert Hooke for this experiment—by which the lungs are kept constantly dilated for a long time without meanwhile endangering the animal's life—and the opportunity thereby given me to perform this piece of work.[123]

That it was air, and air alone, which was responsible for this change, could, Lower said, be seen when venous blood was exposed to the air; its uppermost part became florid. If this upper part were removed with a knife, the newly exposed venous blood would similarly change color upon contact with the air. The Fracassati experiment had found its logical place in a new synthesis. Indeed, he added, a coagulated cake of venous blood would also change its color when turned over.[124]

Some of Lower's predecessors had come close to perceiving this simple and direct relationship between the air and the color of the blood. Harvey, and perhaps many other anatomists before him, had noted how florid the blood appeared when it was in the lungs, but the Englishman had assumed it was only an artifact. Oldenburg, in his letter to Boyle discussing Hooke's inflation experiment of 1664, had seemed to hint that air might make blood look "florid and sprightly," but the passage is ambiguous. Locke, in his notebook jottings of 1665, had suggested that air "volatilized" the blood in the lungs to make it more florid, but he had taken this idea no further than a speculation. Hooke, had he been able to carry out his lung by-pass in late 1667, exposing the blood directly to the air, would have seen the effect immediately. If Needham, in duplicating Hooke's pulmonary inflation experiment early in 1668, had opened a pulmonary vein instead of the left ventricle, he would have found the blood a florid red. Hooke might have been celebrated as the discoverer of this all-important physiological fact, if he had had the inclination and skill to carry out his brilliant suggestion of 30 April 1668. But instead it was Lower. He combined a deep knowledge of physiological theories, a belief that some part of the air was active in generating heat within the body, and most of all, an unsurpassed skill at vivisection. The last especially, a most Harveian of talents, had led Lower to the most un-Harveian conclusion that air entered the blood in respiration, and caused directly the genuine change in color that Harvey had gone to such lengths to deny.

What substance in the air caused this change? To Lower, the answer was so obvious that he gave it *en passant* while treating the logically derivative question of how this aerial agent entered the blood:

> If you ask me for the paths in the lungs, through which the nitrous spirit of the air reaches the blood, and colours it more deeply, do you in turn show me the little pores by which that nitrous spirit, which exists in snow, passes into the drinks of gourmets and cools their summer wines. For, if glass or metal cannot prevent the passage of

this spirit, how much more easily will it penetrate the looser vessels of the lungs? Finally, if we do not deny the outward passage of fumes and of serous fluids, why may we not concede an inward passage of this nitrous foodstuff into the blood through the same or similar little pores?[125]

Lower concluded his argument by noting how, by an opposite process—the aerial component leaving the blood in the peripheral circulation—one could account perfectly for the darker and blacker appearance of venous blood.

Although Lower had referred just as casually to the *pabulum nitrosum* in the *Vindicatio* of 1665, his passing reference to the *spiritus nitrosus* in the above extract is more than a little puzzling. At all the other points in the argument, he had designated what entered the blood simply as "air." Why should he introduce the *spiritus nitrosus* so briefly and so offhandedly? The answer is to be found in the degree to which Lower, in simply assuming the aerial nitrous substance, was a literal pupil of his Oxford elders. The passage cited above is a virtually unchanged quotation from the closing words of Ralph Bathurst's "Praelectiones" of 1654.[126] Except for some minor changes of tense and case, the only alteration was Lower's addition of the all-important phrase "and colours it more deeply." Lower, as a dutiful student of his mentors' ideas, had read a manuscript of the "Praelectiones" at some time during his training at Oxford, copied out extracts into a notebook, and then used one of them in his writing. Throughout their Oxford careers Bathurst and Willis had been very close, and it was only natural that Lower, as medical apprentice and anatomical assistant to Willis, should have looked also for training and guidance to his teacher's best friend and colleague. No plagiarism was meant on Lower's side, and no umbrage was taken on Bathurst's. For men who had associated with each other for so long, ideas and even exact words were shared generously as common property.

The experiments and conclusions that Lower reported in *De corde* had a significance beyond the spare terms in which he presented them. But Lower himself did not draw out that significance. Through a series of shrewd observations, elegant and ingenious experiments, and forcefully logical arguments, he had completed the dismantling of the Galenic system that Harvey had begun. He had conclusively proved that the heart was a muscle, and that only its muscular contraction moved the blood through the body. He had shown that both modifications of the ebullition hypothesis were untenable. Like many Oxonians before, he asserted the later Harveian proposition that blood was the locus of the body's heat. But only he, among them, dispossessed the heart of its vestigial role as the point at which that heat began. In a sense, his experiments hoisted Galenic physiology by its own petard. The change in color between venous and arterial blood had for so long been taken as *prima facie* evidence of the process of heating that, when Lower proved that the change took place elsewhere, his successors had no choice but to consider the heart simply as the pumping mechanism Harvey had always said it was.

But in addition, he had proven the un-Harveian proposition that the air entered the body and had an active role within it. By demonstrating that the blood color change took place in the lungs, he had provided direct physical evidence, in a way Hooke's experiment of 1667 had not, for the truth his Oxford compatriots had always believed. In proving this point, Lower's order of exposition almost exactly reversed the chronology of his ideas; his thinking had proceeded from the Fracassati experiment, to the insufflation technique, and thence to the necessary lack of difference between arterial and venous blood. The quantitative considerations, first in his exposition, but derivative in fact from his qualitative reasoning, were part of the logic of argument, not the logic of discovery. In this way it extended the role that quantification necessarily had played in Harvey's work. From the same kinds of computations—of ventricular capacity, of pulse, and of total blood—Harvey had produced an estimate of pumping rate that was greater than the rate at which blood was manufactured from the foodstuffs. Hence, he said, the necessity of the circulation. Yet he dared not carry his quantitative argument too far, lest he be brought too abruptly against the question of exactly *why* venous and arterial blood were different. Harvey, in other words, if he had carried his quantitative argument to its logical end, would have had to explain the process by which arterial blood became venous blood, and vice versa. But this would have entailed introducing into physiology exactly that which Harvey had been so careful to exclude—the role of air and the airy vital spirits. Lower, once he knew from the Fracassati experiment why the bloods differed, could press the quantitative argument to its full logical conclusion, even though it had not been the origin of his discovery. But in both cases quantification was largely a *post hoc* reason, one which served to persuade, but not to discover. It was, in a sense, the proverbial whore to biological argument—all things to all men, and only when needed.

COLLABORATION AND RESEARCH

On the most general level of discourse, the events and ideas of 1667–1669 have a simplicity and clarity that has always excited the admiration of scientists who prize brilliantly conceived and elegantly executed experiments. Hooke contrived a way to keep the lungs of an animal continuously inflated. Lower used that technique to prove that when the blood flowing through the insufflated lungs was allowed exposure to fresh air, it changed color from dark to florid red, and when it was allowed no such access, the change in color would not take place. It seems a clear case of one scientist devising a technique, and another using it to explore the physiological properties of the organism.

A rather closer analysis would show these events as theory-testing. Needham proposed that respiration served by its mechanical motion to comminute the blood and promote its circulation, an eminently Harveian kind of concept. Hooke refuted that idea with his insufflation technique. Needham defended the theory and proposed yet more experiments. Hooke performed them, once again refuting Needham. Lower used the insufflation technique to test the theory that florid blood originated in the heart and demonstrated instead that it came from the lungs. He thereby disproved the theory of a cardiac ferment put forward by Willis, which Lower himself had accepted four years earlier.

Both sets of statements are true. Yet such a logic of scientific events rests upon a deeper historical, social, and psychological logic that infuses a bald recitation of sequence, or of cause and effect, with a deeper vitality. The key points of innovation were those of collaboration. Hooke suggested pulmonary inflation the week after Lower was introduced to the Royal Society. He was able to carry it through only with Lower's help. He further conceived the idea of extrapulmonary aeration of the blood, a brilliant notion whose execution was impeded not only by its technical difficulty, but by Lower's distraction with other anatomical work. Hooke suggested the "floridness experiment" the week Boyle returned to London, but never carried it out because this conjunction occurred after Lower had ceased his activity within the Society. Lower's fame in the history of science rests upon carrying out an experiment that was first suggested by Hooke. Equally, Hooke's more limited renown as a physiologist rests upon an experiment conceived under Lower's influence and executed with his assistance. The fine texture of the historical evidence suggests a communication and cooperation expressed when chance brought men together, and founded upon long-standing friendship and interaction. This friendship in turn reached back to common origins at Oxford and, in some cases, even beyond. It was, I think, no accident that Lower, Hooke, Needham, Locke, Millington, Castle, and Pope were all old Westminsters. Busby's boys were always known to be a cliquey lot.

Yet if this coterie of Oxonians had common origins, they were by no means all of a piece. Each emerges with a scientific personality as clearly etched as that of a character in a Restoration play. Lower is cast as the brilliant and sometimes impatient technician, adept at anatomy and vivisection, and proficient at criticizing specific hypotheses in physiology. Hooke is more mercurial, flitting from one experiment to another, mechanically talented, quick to see the connections between disparate ideas, bold in putting forward his speculations, but sometimes disinclined to explore and argue a set of ideas from beginning to end. Boyle, although only a few years older, is the scientific *paterfamilias*. Heir to wealth that frees his time completely for science, he blends perfectly a profound concern for the more philosophical aspects of matter-theory, for chemistry, and for the physiological aspects of anatomy. He is the catalyst to his colleagues' thinking, the man who just happens to be on hand when his friends and collaborators have their best ideas. Yet his rigorously antispeculative philosophy of science and the very breadth of his knowledge and experience lead him to be overly cautious in putting forward general schemes. He is the scientific *éminence grise*, content to encourage speculation but slow to engage in it himself. Needham appears as Lower's twin, a man almost as skilled in anatomy, but one whose close attention to morphology leads him to reject anything that cannot be perceived directly. Castle is the man learned in books rather than things, *au courant* with the corpuscular philosophy, chemistry, and the new physiological theories, but content to restate and to rationalize, rather than explore. Locke is the mirror image of Needham, little interested in exact anatomy, but with a consuming love of chemistry, blended with a naïveté and audacity of speculation unfettered by Boyle's twenty years of sobering experience in the vagaries of the spagyrist's art. Oldenburg is the indefatigable communicator of opinions to all and sundry, the man who presents, summarizes, and ferrets out—the learned gossip that every Caroline play should have.

Each man's characteristics are valuable components of an ongoing scientific tradition. The presence, within a cooperating circle of scientists, of such divergent attitudes and capacities, guarantees that as such elements are needed in the process of scientific change, they are available. The group thus encompasses traits and talents that by their self-contradictory nature are seldom found in a single individual. Rare is the man who is, at the same time, cautious and bold, committed to experimentation, yet able to see beyond it, technically proficient yet speculatively far-ranging, intensely interested in his own scientific problems, yet having time to maintain a vast correspondence with like-minded people. Rarer still is the man who can diffuse his interests equally over many subject fields, thinking as fruitfully in physics as he does in anatomy or chemistry. One man, such as Hooke, might be passionately committed to an idea, such as that of the aerial niter, and be able to experiment upon it in chemical and physical terms, yet be unable to carry through the experiments that would apply that concept to physiology. Another man, like Lower, could be fascinated with anatomical and physiological problems of the heart and lungs, yet have only a passing interest or competence in the chemical properties of the air. In these three facets—matters of scientific temperament, of scientific interest, and of scientific competence—lies the strength of collaborative research. One could, ideally, have all these necessary characteristics in one man; he could be a group unto himself. But more often such traits are parceled out to a number of men who, if fortunate enough to be in contact with each other, exchange the benefits of their differences.

Moreover, the long-standing nature of their collaboration meant that men like Lower, Hooke, Boyle, and Willis shared their experiments and ideas freely. What one finds recorded in letters and in the Journal Book of the Royal Society are merely the tokens of a more far-ranging interaction. Concerns for priority arose only when it seemed that a particular idea or technique would pass beyond the friendly circle of colleagues. Because of this freedom of communication and interaction within the Oxford group, a consensus born of countless discussions and experiments, I find it difficult to engage in the traditional scholarly game of assigning priorities and apportioning praise. Even in such seemingly discrete accomplishments as the insufflation and floridness experiments, innovation seems to lie at the conjunction of men's talents, rather than in each one's individual exertions.

Such a process of interaction seems especially important in *functional* explanation in biology. Not only does physiology, in attempting to explain how an organism works, draw concepts from practically all subject areas of the biological and physical sciences, it also makes use of a wide variety of procedures to confirm those explanations. Explanatory schemes may be set up as possible by observation, adduced as necessary by logic, and proved to be true by experiment. But, unlike many other sciences, the nature of that confirmatory experiment is not easy to predict. It could depend upon a new piece of apparatus, such as Boyle's air pump. It could require a new technique, such as the Hooke-Lower procedure of pulmonary insufflation. It could hinge merely upon seeing the significance of the simple experiment of scraping the top florid layer off venous blood. And, just as in functional explanation the kind of confirmatory experiment is exceedingly difficult to specify in advance, so is it impossible to predict what kinds of interests and training a man should have in order to make his contribution. At one point chemistry is needed; at the next, pneumatics; and at the following, vivisection. The existence of scientific groups guarantees that the proper interests and talents are always available.

NITER, NITER EVERYWHERE 9

The very casual nature of Lower's reference to the *spiritus nitrosus* in the air, the substance that was absorbed into the blood as it passed across the lungs, was a reflection both of his Oxford education and of the rigorously monographic nature of the *Tractatus de corde*. He might have wished to speak of the role the nitrous spirit had in the generation of heat, but there he deferred to Willis. There is no hint that Lower believed it had any other function in the animal body. In this particular instance, silence was eloquent, for in the preceding few years his Oxford mentor, Thomas Willis, had developed theories about one area of physiological action in which nitrous agents were very much involved—muscular contraction.

WILLIS AND THE "EXPLOSION" THEORY OF MUSCLE CONTRACTION

Since antiquity, anatomists had put forward theories to account for how muscles, under the influence of the appropriate nerves, contracted to shorten the distance between the tendons, and thus to produce the complicated movements that made animals so obviously different from plants. These theories rested on the belief that a muscle shortened because it was inflated, thereby increasing its breadth and decreasing its length. Almost all of these classical theories further assumed that this intumescence occurred because the muscles were inflated with animal spirits that had passed down the attached nerve.[1] This seemed reasonable; if one cut its nerve, a muscle could no longer contract.

When, as part of the revamping of physiological explanation in the mid-seventeenth century, anatomists proposed new theories of muscular contraction, they were largely concerned with taking the classical assumption of inflating muscle and putting it on a more modern basis. William Croone, for example, published a small tract in 1664 in which he argued that muscle contraction resulted from the interaction of two sets of spirits, one derived from the nerves and the other from arterial blood. When these two substances were brought together in the space between the muscle fibers, they effervesced like acid and alkali, and expanded. This swelled the width of the muscle and thereby shortened its length.[2] In thus proposing a "chemical" theory of intumescence, Croone was explicitly disagreeing with Walter Charleton, who in 1659 had suggested from a geometrical analysis that muscles might be able to contract without increasing in bulk.[3] Unfortunately, Charleton adduced no experiments to support his more radical contention.

Willis, in the meantime, was developing at Oxford his own corpuscular-chemical theory of muscle contraction, which first appeared in his *magnum opus*, the *Cerebri anatome* of 1664.[4] Willis speculated that since muscular contraction occurred only in a very defined locus, it must be caused by substances brought together there for the first time. His candidates were the particles of animal spirits from the nerve fibers, and saline-sulphurous particles from the arterial blood. These joined together briefly to form a "copula." Under proper stimulation from the nervous system, this copula would break apart and explode like gunpowder, intumify the muscle, and thereby cause it to contract. This occurred in all voluntary muscles, as well as in the heart and diaphragm.[5] Thus Willis could explain the pathological effects of such experiments as he, Lower, and Boyle had performed of ligating or cutting the cardiac branch of the vagus nerve; the heart eventually failed for lack of animal spirits.[6]

Although in most passages explaining his "explosion" theory of muscular motion, Willis was content to call the arterial component merely "sulphureous" or "saline-sulphureous," on two occasions he explicitly said that the blood supplied "nitrosulphureous particles" to this explosive copula. Thus in heavy exercise, large amounts of nitro-sulphurous particles needed to be supplied by the arterial blood, although it was generally the supply of the other component, spirit particles from the nerves, that was exhausted first.[7] Willis's theory was by no means fully developed, but rather dropped into the neurophysiology at various junctures, together with promises that he would discuss the topic further when he came to write explicitly on muscular motion. Thus Willis leaves the reader wondering where these "nitrosulphureous particles" originated—something he failed to explain.

Just before he departed Oxford for London in the late summer of 1667, Willis published his *Pathologiae cerebri*,[8] the long-awaited companion volume to his anatomical treatise of three years before. In this work, widely publicized by Oldenburg,[9] Willis made the "explosion" theory of muscular motion one of the central concepts of his neurology. To understand all kinds of convulsions, he said, one must understand normal muscular motion. This motive power can come from nothing other than the expansion, or explosion, of spirits. Spirituo-saline particles from the nerve and tendon fibers, and "particulas nitrosulphureas" from the arterial blood, were joined together in a copula. When the "instinct of motion" was brought down the nerve, it enkindled the combined niter and sulphur particles; they burst forth, or exploded, blowing up the muscle and causing the ends to be drawn together. Since the animal spirits—the *particulas spirituosalinas*—were not themselves destroyed in this explosion, they remained in the nerve fibers to be coupled again with new nitrosulphurous particles. On the other hand, the raw materials of the explosion, the nitrosulphurous particles, were expended, and had to be supplied anew. Thus muscles were copiously supplied with arterial blood.[10] Although Willis quoted in support of his idea a passage from Gassendi,[11] the French mechanist had not really discussed muscular motion. Willis merely took from him the idea of motive "energy" being supplied *in situ*, and elaborated it according to a corpuscular scheme incorporating the Oxford emphasis on chemical units as a source of heat and activity.

But after having explained his theory in the first few pages, Willis, in typical fashion, devoted the remainder of his long book to showing how all kinds of convulsions and spasms were the result of deranged, or "heterogeneous," explosive copula.[12] Such an extended exercise in hypothetical pathology some-times occasioned references to friends. He tactfully but forcefully disagreed with Highmore on the cause of hysteria. Highmore thought it resulted from blood rushing too copiously to the lungs, whereas Willis believed it was a nervous disorder, created by convulsions of the abdomen and thorax, which in turn were caused by heterogeneous and unduly explosive copula transmitted down from the brain.[13] In discussing the nervous cause of asthma, he cited the results of a postmortem performed by the "learned" Needham on a Staffordshire butcher who had recently died of the disease. He concurred with Needham's judgment that since the lungs and other viscera were sound, the affliction must have been of nervous origin.[14] Such was only one of many cases and postmortems Willis reported to support his notions of a neuropathology based on explosive copula. As in "De fermentatione" and "De febribus," physiological concepts always paid their way in clinical coin.

Given Willis's academic habits, his "explosion" concept of muscular contraction almost certainly found its way into his Oxford teaching during the mid-1660s. In the fragmentary lectures of 1661 recorded by Lower and Locke, Willis had subscribed to the old theory of inflation by animal spirits, and had illustrated it with the inflated bladder experiment done at Petty's club in 1650.[15] As Willis developed his new theory c. 1662–1663, it is hardly likely he would forbear telling his students about it. They could hear about Willis's "explosion" theory in the Schools, and within a few years read about it in his books. As it turns out, Willis had one student—John Mayow—who was at once a fervent follower and a perceptive critic.

JOHN MAYOW'S EARLY LIFE AND THE
TRACTATUS DUO (1668)

Mayow, like Lower, came from the Cornish gentry. Although their homes were separated by only eighteen miles, there is no evidence, or even likelihood, that they knew each other before Oxford. Mayow was ten years younger, born in the manor house of Polgover, in the parish of Morval, Cornwall, in December 1641, and probably educated in the west country. He came up to Wadham in April 1658, in the twilight of Wilkins's Wardenship, and was elected Scholar of the college in September 1659, after Wilkins had already departed to take up the Mastership of Trinity College, Cambridge.[16] Mayow left Wadham a year later, being elected in November 1660, on the recommendation of Henry Coventry, into a Fellowship at All Souls. Coventry was a staunch Royalist who had been put out of his Fellowship there in 1648, reclaimed it after the Restoration, and then resigned it—quite possibly for a monetary consideration, given the record of All Souls for corrupt resignations—in Mayow's favor.

Mayow's subsequent life—social, professional, and scientific—was structured by the institutional and intellectual ambience of All Souls. Because he had been elected into one of the jurist's places within the college, he was obliged by statute to proceed to his degrees, not in arts and medicine, but in civil law. Accordingly, he took his B.C.L. in 1665 and his D.C.L. in 1670. But for him, as for Wren before him, the moribund discipline of civil law was not a bar to scientific pursuits. In these surroundings Mayow had some illustrious fellow-collegians. Wren, although he had resigned his Fellowship in 1662, lived at Wadham and continued to take commons at All Souls until early 1664.[17] George Castle, another frequenter of the gatherings of virtuosi at Boyle's lodgings and at Tillyard's coffeehouse, stayed on until at least late 1665. A more long-term friend was Thomas Millington, who continued to live at All Souls until early 1674.[18]

Mayow's scientific contacts outside of All Souls are more difficult to infer. One was Edward Browne, son of Sir Thomas. Mayow and Browne were both of similar ages and interests. After taking his M.B. at Cambridge in 1663, Browne had studied chemistry and anatomy at Norwich, London, and Paris. They met when Browne spent two years at Merton, 1665–1667, before taking his Oxford D.M. there. Browne was an especially active dissector; during a two-month period in 1664 he dissected various parts of an ox, bull, calf, sheep, dog, monkey, hedgehog, badger, polecat, hare, turkey, rat, and frog.[19] Other dissectors were also in Mayow's orbit. He probably knew of the research activities of Lower and Willis by about 1663, especially since Wren and Millington had been so closely involved in preparing the *Cerebri anatome*. But it is unlikely Mayow was close to them, for his name does not occur in any of the extant correspondence before 1666. John also had a family connection with the Westminsters at Christ Church; the tutor of his younger brother Thomas Mayow, a Student in the House, was Benjamin Woodroffe, an old Westminster, pupil of Stahl, friend of Boyle, and minor virtuoso in Restoration Oxford.[20] Thomas was also at one point a student of John Locke.[21] Yet another of Mayow's acquaintances was certainly Thomas Guidott, three years older, and two years his senior at Wadham.

Mayow's early study of chemistry dates from not later than 1667. On 19 September of that year, John Ward, visiting from Stratford, began a course of "Chymistrie" with William Wildan, an Oxford chemist. His fellow students, as Ward recorded in his notebook, were "Mr. Mayo, Mr. Grosshead, Mr. Plot."[22] This could only be John or Thomas. The former seems almost certain, since Ward included among his jottings a story about Chichele's founding of All Souls and information about the customs involved in proceeding D.C.L.[23] Other anatomical annotations in Ward's notebook may also be from Mayow, since none of the other members of the course was a physician. Ward, for example, noted the position of the recurrent nerves, the existence of a valve in the colon, and that "An anatomicall disquisition [has been] put out by Mr. Dr. Needham de formato foetu."[24] He also recorded a defense of Mayow's colleague at All Souls: "Dr. Wren first invented ye syringing of Liquors into ye veines of doggs, wch some travelers carried beyond seas."[25]

The protocols learned by Ward, Mayow, and Plot in their tutelage with Wildan were quite predictable, including many that Stahl had taught to his students in similar courses in the early 1660s. The students prepared sulphuric and nitric acids. They distilled "spirits" out of various substances, such as wine (alcohol), hartshorn (ammonia), and urine. They learned procedures for purifying metals such as tin, mercury, and silver, and for compounding medicaments, including a large number of emetics. They made the standard explosive compounds, such as "pulvis fulminans" and "aurum fulminans." Their procedures were also a good cross section of available laboratory techniques: destructive and fractional distillation, calcination, purification by sublimation, filtration, alcohol extraction, and precipitation.[26]

Other than these scattered pieces of evidence, we have no certain knowledge of Mayow's scientific associates during his formative years, a lack all the more puzzling because Mayow later showed himself to have been a close reader of Willis, Boyle, and Lower, and well acquainted with such Oxford techniques as injection, transfusion, and the air pump. There are, however, some probable lines of reasoning that suggest his relations to Lower and Boyle.

In September 1666, when Lower wrote from Tremeer to give Boyle the transfusion protocol, he noted that Boyle's letter of June, "after it had lain at Mr. *Crosse*'s about three months, was occasionally found and sent to me in the country by a friend, together with another from Mr. *Mayer*, with a little box of brass pipes."[27] The brass pipes sent by "Mr. *Mayer*" were presumably for Lower to continue his transfusion trials on Cornish dogs. Since there was no "Mayer" matriculated as a student at Oxford at the time, and none of that name among the townspeople, it seems logical to conclude that "Mayer" was yet another of the variant spellings of John Mayow's name. Further, a bad pun later recorded by Anthony Wood seems to suggest that the final syllable in Mayow's name was pronounced as "ow" rather than "o," and thus quite capable of being rendered "Mayer."[28] The same diphthong in Lower's name was pronounced the same way.

A "Mayer" also crops up in connection with Boyle. In 1669 Samuel Colepresse, a friend and correspondent of Boyle who lived near Plymouth, went to study medicine at Leyden. From there he wrote to Oldenburg:

> My most humble service to the Noble Mr. Boyle: when you see him next; and when you light on Mr John Mayer his gent. my service & thankes to him likewise for his kind letter by Mr. John Feake who suites well with this country as yet.[29]

This seems to suggest that "John Mayer" was known to Colepresse as an assistant to Boyle. This is all the more likely because in his *Continuation*, Boyle mentioned on three occasions one "I. M.," whom Boyle calls an "Ingenious young man . . . whom I often imploy about Pneumatical Experiments."[30]

This may well be John Mayow. Colepresse had visited Boyle in Oxford, most likely in the autumn of 1666,[31] and would have had an opportunity then to meet "John Mayer." John Feake, who had carried "John Mayer's" letter to Colepresse, was two years younger than Mayow, likewise a budding physician, and had been at Oxford in the early 1660s.[32] The air pump experiments for the *Continuation* were carried out at Oxford from about August of 1667 to March of 1668, when Mayow was about 25 years old;[33] Hooke had assisted Boyle when he was a neophyte of about the same age. For a number of the experiments Boyle adduced Wallis and Wren as witnesses,[34] and on one occasion "I. M." repeated a suction pump trial for Wallis, Wren, and Millington.[35] Mayow would certainly have known the latter two.

It was right in the midst of these trials that Ward, Mayow, and Plot were going through their chemical course. Ward reminded himself in his notebook: "Remember to see and viewe Mr. Boghils air pump." Some days later, probably in October 1667, he got his wish: "Mr. Boghils Air pump, and ye Torricellian Experiment both seen at Mr. Boghils Lodging." In the next few pages Ward then went on to record information not only about Boyle's chemical apparatus and procedures, but also a number of facts about Boyle's personal life. He noted that Boyle was working on a book about English minerals, "and hath set up an Elaboratory" to experiment with them. He recorded that Boyle never took any "strong drinks or wine," that he ate simple meals of eggs, gruel, bread, butter, and meat, that he had lived in Geneva and traveled in Italy, and that Stubbe had sent him things from Jamaica.[36] The entire pattern of entries suggests that Mayow, as Boyle's occasional assistant, arranged a tour of Boyle's laboratory and told Ward something about his patron. We will, unfortunately, never know for certain whether Mayow was Boyle's informal helpmate at Oxford c. 1666—1668. We may conclude only, on the one hand, that it is congruent with many disparate pieces of evidence, and on the other, that it makes comprehensible many aspects of Mayow's later skills and preoccupations.

One thing is certain: by early 1668 Mayow was well enough versed in the literature, techniques, and concepts of Oxford physiology to see the implications of Hooke's pulmonary insufflation experiment of October 1667. This, more than anything else, seems to have prompted him to write up a brief essay on respiration, to which he added another on rickets, and the two were published as the *Tractatus duo* by Richard Davis, the bookseller across the High Street from All Souls.[37] Mayow's book must have come out at Oxford in the late summer of 1668, just as Lower was delivering the *Tractatus de corde* to the printers in London. Henri Justel soon heard of it in Paris, for he wrote to Oldenburg in mid-October that he understood two books, one on respiration and the other on the heart, were being printed at Oxford.[38] Oldenburg, who had long been interested in respiration, gave the *Tractatus duo* a careful and relatively long summary in the November issue of the *Philosophical Transactions*.[39] A month later Sir Thomas Browne, an attentive reader of the *Transactions*, wrote his son Edward, then in Vienna, that ''Mr. Mayow your freind hath putt out a booke de respiratione et Rachitide.''[40]

The first half of the tract ''De respiratione'' was devoted to the mechanics of inspiration and expiration. Mayow rejected, as Boyle and Bathurst had, both the Aristotelian idea of the lungs attracting air, and the Cartesian notion that the outward pulsion of the chest into the atmospheric plenum mechanically pushed the air into the lungs. He cited the same reason as Boyle: If this were so, why could one draw air from a large flask held to the mouth? And if the Cartesians explained that the pulsion was done by matter subtle enough to penetrate glass, then Mayow would ask in return why this same subtle matter did not keep alive animals sealed in glass containers. Rather, Mayow said, air entered the lungs during inspiration because of its inherent elastic properties, as had been ''splendidly demonstrated'' by the ''noble and learned'' Boyle. The lungs were admirably suited for this, he pointed out, for Malpighi had shown that they consisted of innumerable little vesicles.[41]

How did the lungs carry out this motion? The eminent Highmore had shown, Mayow said, that lungs collapse when punctured, and hence could not move on their own—a fact Mayow, like Boyle before him, had verified for himself in his own vivisections.[42] Rather, Mayow said, the lungs were moved by the actions of the diaphragm and of the intercostal muscles, each of which Mayow explained in turn. In explicating the diaphragm he touched on many of the same subjects previously considered by Willis, and later discussed by Lower.[43] Mayow's consideration of the intercostal muscles was rather more individualistic. From a detailed analysis of their arrangement and statics, he argued the novel point that both the external and internal intercostals served for inspiration, contrary to what had been generally held.[44] In all, Mayow's discussion *de modo respirationis* was a balanced, lucid, and insightful synthesis of recent findings with his own anatomical observations—good exposition, but only his treatment of the inter-costals was really new.

It was when he turned to consider the *use* of respiration that Mayow showed his credentials as a critical scientific thinker. He began by rejecting—"waving" as Oldenburg said—many of the traditional functions. Some thought respiration served to cool the heart, but it seemed instead to promote the incalescence and circulation of the blood. The "most prevalent opinion," Mayow said, was the mechanical theory: that respiration served to transmit the blood from the right to the left ventricle, and/or to churn and mix the separate blood particles (respectively the Needham and Malpighi theories). The strongest evidence for this view was the embryological finding that the unmoving lungs of the fetus had no blood circulating through them. But why, Mayow asked, should nature go to so much trouble just to move the blood from right to left? If we stopped breathing, did our pulse stop? Moreover, the easy passage of liquids through motionless lungs could be verified experimentally. If one injected blood into the pulmonary artery of a dead animal, it would pass easily into the left ventricle of the heart. As for "churning," if this were the use of respiration, any kind of air, even "vitiated," would suffice, a deduction manifestly untrue.[45]

Mayow drew his conclusion in strong terms. "We do not doubt to assert, *that air does not serve for the motion of the lungs, but rather to communicate something to the blood.*" When these necessary and life-sustaining particles were exhausted from the air, it was no longer fit for respiration. This was clearly evident from an experiment "recently done in the presence of the Royal Society": pulmonary insufflation. Mayow then gave a brief but lucid description of Hooke's experiment, emphasizing that although the lungs were kept continuously inflated, and hence unmoving, the animal fared well. Conversely, if one closed the nose and mouth, the lungs might still be inflated, but the animal quickly died. This was clear proof that respiration was neither for the passage of the blood nor for its agitation. Rather, Mayow reiterated again, it was to communicate a vital something directly to the blood, which, being exhausted, the air became useless for respiration. Having spent so long circumscribing the function of inspiration, he dismissed that of expiration in a single sentence: it was to exhale and carry away the vapors of the blood.[46] Mayow was clearly much more interested in what went *in* than in what came *out*.

What was it in the air that was so necessary for life? This was not easy to answer. But in such obscure matters Mayow begged leave to make a conjecture:

> It is very likely that it is the *fine, nitrous particles*, with which the air abounds, that are communicated to the blood through the lungs.

This aerial niter was necessary to every form of life. Plants could not grow in soil until it had been fecundated with the aerial niter; thus they in some sense respired. But it was the place of *nitrum hoc aereum* in the life of animals that demanded closer scrutiny.[47]

In solving this problem, Mayow was once again a student of his Oxford elders. He believed that the aerial niter mixed with the sulphur of the blood to cause a fermentation, and hence an incalescence of the blood. This took place not just in the heart, but in the pulmonary vessels and in all the arteries. But Mayow was most careful, as Lower was a few months later, to reject the idea of a ferment lodged in the left ventricle of the heart. How could this ferment get there in such quantities? How was this done in the fetus, when the blood passed directly from the right ventricle to the aorta? Nor could this fermentation of the blood cause the motion of the heart. Mayow opposed this Cartesian view with the findings of vivisection. The heart expelled blood in systole, he said, not in diastole, as would be required in any ebullition or rarefaction hypothesis. One could fill the beating heart of an animal with any liquid, and it would continue to expel the contents in systole, even after the animal had died. If the heart were completely cut out, and all its blood removed, it could still be made to contract. Mayow, like Lower, concluded his discussion of the motion of the heart with an injection experiment. It could be observed, he said, that if opium or cold water were injected into an animal through the jugular vein, the animal's heartbeat would speed up, not slow down, as was required by the ebullition hypothesis. It seemed, he said, that the heart was nothing other than a muscle, whose sole function consisted in contraction alone, and the expulsion of the blood.[48]

But having linked the aerial niter to the fermentation in the blood, and hence its heating, and having refuted the connection of this fermentation to the heart, Mayow was not content to stop. He, like Boyle before him, was concerned with the question of why death ensued so quickly when an animal was cut off from the air.[49] The answer lay, Mayow believed, in the fact that the aerial niter was necessary not only for the maintenance of heat in the organism, but also for the *contraction* of the muscles. And, since the heart was a muscle, when its supply of this niter was cut off, it ceased beating.

In order to argue this systemic need for nitrous aerial particles, Mayow grafted onto his view of respiration an elaborated version of Willis's theory of muscle contraction. The "illustrious and learned" Willis had shown, Mayow said, that muscular action resulted from the mixture of different kinds of particles, by which a rarefaction and explosion took place, causing inflation of the muscle and thereby contraction. But if this were the case, it could scarcely be supposed that the different kinds of particles had coexisted beforehand in the same liquid. Here he put his finger on the weakness in Willis's scheme. The eminent physician had seemed to imply that the salino-spirituous particles in the animal spirits had been separated out of the arterial blood in the brain, and then recombined with the nitro-sulphurous particles in the same arterial blood to produce the muscular "explosion." Mayow supposed instead that the nitro-saline particles flowed to the muscles with the arterial blood, and were affixed there. The animal spirits were distilled and *altered* in the brain, and thence distributed through the nerves. The mixture of the two in the muscle produced an "explosion," inflation, and resultant contraction of the muscle.[50]

Clearly, Mayow summarized, "the *especial use* of respiration seems to be to

serve the motions of the muscles and particularly of the heart.''[51] In this respect
the heart was no different from any other muscle. When its supply of the *nitrum
illud explosivum* was cut off, its motion ceased. One could live for a certain time
without breathing only because the blood contained, as it were, a store of
impregnated air. This theory of muscle contraction could be confirmed by the
deep respiration and rapid heartbeat that were observed in violent exercise.
Because the muscles needed more nitrous particles, the heart sped up, pumping
more blood through the lungs, to absorb these particles from the air and thereby
supply more niter both to itself and to the other muscles.[52] This principle could
be confirmed, Mayow said, by several experiments. If one waited until both the
heart and the lungs of a vivisected animal had ceased beating, and then blew air
into the heart via a tube fitted into the *vena cava,* the heart's motion could be
reestablished. So it appeared that air was necessary to the continued motion of the
heart. Nor did it much matter *how* the air came into the blood, whether by the
lungs or by any other way.[53] He used this experiment to cap a logically tight
argument. All the parts of the system fitted together with precision. Mayow
showed himself, even at a young age, to be a biological thinker whose critical
abilities were the equal of Lower and Boyle. In fact, it seems likely that this
perception of a new system was Mayow's inspiration in writing the *Tractatus
duo*. The book reported several interesting and original experiments, but its
central conception of an aerial niter posited nothing that would have been
unknown to someone who was acquainted with the Oxford group. Mayow's
major contribution was the logic that tied the elements of his system together.

Mayow had even thought through the embryological objection. If one were to
ask him how the fetus in the womb breathed, he would answer that it lived by
virtue of the air that was supplied to it from the arterial blood of the mother. The
fetus moved and had a heart beat; therefore it too required the aerial niter. In fact,
Mayow said, he supposed that if arterial blood, impregnated with this niter,
could come to the heart instead of venous blood, there would be no need for
respiration. Indeed, this seemed to be confirmed in an experiment in which a dog
was transfused with arterial blood; whereas beforehand the animal struggled and
respired laboriously, after he received the arterial blood he scarcely seemed to
breathe at all. Both of these phenomena—that of the fetus *in utero* and the transfused dog—demonstrated for Mayow this importance of arterial blood as the
carrier of the aerial niter.[54]

The arguments that Mayow presented to the learned world in the autumn of
1668 were, in their component parts, neither new nor startling. The rejection of
the cooling and mechanical functions of respiration, the belief that it served to
communicate an aerial niter to the blood, the notion of a blood fermentation, the
emphasis upon the heart as a muscle, even the effervescent theory of muscular
motion—all were stock conceptual items of the Oxford tradition. But Mayow had
an eye for *system* in physiological theory that his contemporaries did not. He saw
how, by using the central notion of an aerial niter, he could link together

respiration, heat, heartbeat, muscular motion, and embryological evidence. Since he was more concerned with the origins of muscular motion than those of heat, he placed greater emphasis upon the role of the aerial niter in muscular effervescence, than upon its place in blood fermentation. But in a sense this was quite natural. The role of niter in fermentation and heat was, in the absence of chemical experiments on combustion, a logical dead end, whereas its role in muscle contraction led him naturally back to the question of cardiac motion. Mayow, in placing the emphasis upon his niter theory of muscular contraction, was emphasizing the element that bound his system together.

Moreover, in propounding this new system, Mayow drew explicitly upon the conceptual and technical content of the Oxford tradition. Arguments of friendship and propinquity make it most likely that Mayow was exposed to the aerial niter idea not only in the lectures and books of Thomas Willis, but through contact with his fellow-collegians at All Souls. Wren had discussed it in his letters to Brouncker and to Hooke of 1663 and 1665. Castle had used it *inter alia* in his *Chymical Galenist* of 1667. Thomas Millington was also an adherent. In one surviving notebook he saved some amanuensis notes on the role of the "nitrous salt" that fecundated the air plants take in, and the way it was "ye food of ye Lungs."[55] Millington was also a lifelong friend of Bathurst, to such a degree that Bathurst intervened personally to have Millington elected to the Sedleian chair as Willis's successor.[56] All three Westminsters at All Souls provided not only a close and mutually reinforcing source for Mayow's convictions, but another point of access to the extensive series of dissections and vivisections carried out at Oxford in the early and mid-1660s, before Willis, Lower, and Boyle departed for London.

Mayow drew directly upon this heritage of Oxford techniques. His book was precipitated by Hooke's insufflation experiment. The experiment of blowing air into the heart was a slightly changed form of the one that Millington, Needham, and Lower had performed in 1664, and that Lower had reported in his *Vindicatio* of 1665. But Mayow, in drawing upon that Oxford technical tradition, also enriched it with several new experiments. He refuted the mechanical theory of respiration by injecting blood through the motionless lung of a dead animal. He demonstrated the weaknesses of the Cartesian ebullition theory of cardiac motion by his injection of opium and water. And finally, his report that dogs transfused with arterial blood "scarcely seem to respire," albeit most probably an overstatement, was a new and very striking way of using the transfusion technique for physiological ends. The *Tractatus duo* showed Mayow very definitely to be an anatomist and physiologist worthy of his heritage.

It is also important to note what Mayow was not. He was most certainly no chemist nor, in the most general sense of the term, natural philosopher. Although the idea of an aerial niter was almost inherently chemical, not a bit of chemistry found its way into the *Tractatus duo*. As a result, certain notions derivable from a chemical point of view, such as the emphasis upon fermentation as a process that generated heat, got rather short shrift. Moreover, because of his concentration upon physiological problems, the *Tractatus duo* lacked, as did Lower's *Tractatus de corde,* a clear-cut matter theory. Mayow accepted the belief that matter, including his aerial niter, was organized as particles, but he did not go beyond that; his natural philosophy was corpuscular only in the very broadest sense. As a result, Mayow dealt only in passing with the role his aerial niter played in the air; it was simply there. Thus, although it is clear that he had read Boyle closely, and had accepted his concepts of elastic aerial particles, he had no ideas about how the pneumatic properties of the air related to its physiological ones.

Mayow began his scientific career, then, as an anatomist and physiological theorist. He had taken a recurrent theme of the Oxford scientific tradition, the aerial niter, and made it a central concept in a physiological system. But in doing so, he ran up against problems that took him beyond questions of anatomy and physiology: What were the chemical properties of this niter? How did they relate to the heat-generating fermentation in the blood? How did the aerial niter relate to the pneumatic properties of the air? In answering these questions Mayow transformed himself, over the next five years, from anatomist into chemist.

WILLIS ON ANIMAL HEAT AND MUSCULAR MOTION

By the spring of 1669, it must have seemed to an interested onlooker like Oldenburg not only that several sets of physiological debates had been joined, but that some resolution was in sight. Lower's book had come out in late February, Mayow's five months before. They had independently argued that a nitrous aerial substance entered the lungs in respiration, that it generated heat, that this incalescence took place in the blood and not in the left ventricle, and that the heart was nothing but a muscle. Lower had done so at greater length and with deeper insight into anatomy, and had in addition beautifully demonstrated the crucial point that the blood changed color in the lungs.

But Mayow and Lower had come, unknowingly, to disagree on one point— muscle contraction. In 1667 the brilliant Danish anatomist, Nicolaus Steno (Nils Stensen), had argued in his *Elementorum myologiae specimen* that muscle fibers did not, as commonly conceived, run from one tendon to an analogous point in the other. Rather, they were arranged in such a way that the tendons constituted the tops and bottoms of a parallelogram, with the fibers running parallel to the sides, from one tendon to another.[57] Steno's book had arrived in London by February 1668, in time for Lower to make use of his findings.[58] Lower, in his *Tractatus de corde,* elaborated Steno's scheme into a detailed analysis of cardiac muscle. As preface to this, Lower showed that every mammalian muscle was therefore actually biventral, or "two-bellied." Contraction was accomplished, not by an "explosion" and inflation, but by the two bellies moving towards and past each other, like two hands clasping. Thus, Lower asserted, as Steno had not, muscles in contraction got smaller and harder, exactly the opposite of a process of swelling up.[59] In so arguing from anatomical structure to physiological function, Lower had been led implicitly to deny any theory—whether the traditional Galenic or the modified chemical ones of Croone, Willis, and Mayow—that depended upon the muscle swelling. This was what Oldenburg meant when he wrote to his correspondents in early March that *De corde* not only refuted the Cartesian doctrine of an ebullition in the heart, but also "the way of movement in the muscles taught by Mr. Willis, exploding his explosion."[60]

Lower's book set up the question, and his colleagues quickly responded. Two weeks after a copy of *De corde* was presented to the Royal Society, Jonathan Goddard suggested an experiment "for making out the manner of the motion of the muscles."[61] Goddard's experiment was performed a fortnight later, at the meeting of 1 April 1669.[62] He had fashioned a tin sheath, filled with water, into which a man's arm could be inserted. Near the hand a small glass tube was cemented, such that any increase in volume of the sealed arm was detectable as a rise in the water level in the glass tube. With the subject's arm relaxed, the water level was observed to rise with the pulsation of the surface arteries, and to fall between pulses. Then came the experiment. The subject was "ordered to make a contraction or clutching of his fist," and "upon each such contraction, the water in the glass did descend."[63] The experiment seemed to show, Oldenburg recorded, "that in contraction the muscles, upon the whole, were brought into less dimension than in their dilatation."[64] Lower's argument, or at least its conclusion, had been verified experimentally.[65]

Although Goddard's memoir was not published until the eighteenth century, and the experiment itself not until 1677,[66] the results were almost certainly known to Lower and Willis. Willis, especially, had been associated with Goddard in the Oxford club. Moreover, when Goddard turned in his written account of the experiment some months later, Edmund King, who acted as surgeon for Lower and Willis in their London practices, went to the trouble of making his own copy of it.[67]

Goddard's experiment may well have been the reason why, when Willis finally published his tracts on muscular motion and animal heat, the latter was more interesting than the former. In December 1669 Highmore had published a short *Responsio* to Willis's explanation of hysteria and hypochondria.[68] In March 1670 Willis brought out a defense of his own ideas, and appended the short tracts "De sanguinis accensione" and "De motu musculari" to fill out the volume.[69] Oldenburg, as usual, gave it a prompt review in the *Philosophical Transactions*,[70] although one might read into the review's brevity, and the lack of subsequent recommendation in Oldenburg's correspondence, the good Secretary's impatience at the speculative nature of Willis's contributions.

Willis's treatment of muscle was, if longer than his previous hints, less vigorous. He traced out at length his contention that the animal spirits came down the nerves, and were stored in the tendons. In contractions they rushed out into the muscle fibers, where they met the active particles from the blood. The strife and agitation of this meeting caused the fibers to "wrinkle" up, and thus the contraction of the whole muscle took place. He discussed at length the mutual necessity of animal spirits and of blood particles, but refused to hazard a firm identification of their natures. Since it was a question that could not be answered by the senses, he would not "rashly" declare that the animal spirits were spiritous-saline, or whether the particles from the blood were sulphurous or nitrous.[71] Similarly, although he no longer spoke of "swelling" muscles, his explanation failed really to explain how the shortening took place. Obviously the explosion theory had been exploded, and Willis had no real candidate to take its place.

The "De sanguinis accensione" ("On the Enkindling of the Blood"), although better argued and significantly more important, shared some of this tentativeness. Because Lower knew that Willis had his tract in hand, he had forborne discussing the heat of the blood in his *Tractatus de corde*. One rather wonders whether Lower, once he saw the tract in print, felt his deference had been justified.

Willis's starting point was the multitude of explanations that had been given for the warmth that existed in the blood. Some attributed it to an innate heat, he noted, others to a flame in the heart, some to a fermentation in the blood, and others to the enkindling, or ascension, of the blood. That the heat, and hence the soul, were in the blood, he asserted by reference both to the scriptural opinion, and to the "most famous" Harvey's observation that the blood was the first to live and the last to die. But to decide among these opinions, and to prove that the heat in the blood resulted only from its enkindling, he would prove not only that heat could be produced only by ascension, but also that there were many properties in which flame agreed with the life in the blood.[72]

First Willis had to disprove the other putative causes of the blood's heat, especially those that posited an innate heat, a vital cardiac flame, or a sanguinous fermentation. The latter, he said, was impossible; fermentations might produce heat, but only in solids. "No liquors," he said, "either thick or thin, whether they ferment or putrifie, do at any time grow hot."[73] He denied vigorously that the heat of the blood was excited by fermentation. Moreover, he said, neither was there an innate heat nor a vital flame in the heart. The heart was a mere muscle, containing no matter nor tinder for a flame or heat. Its sole function was to circulate the blood through the body, and in doing so it received its heat from the blood.[74]

Having cleared the field of opposing opinions, Willis could proceed to argue his own. He wished to show, he said, that in the case of both fire and the heat in the blood, three things were necessary for proper enkindling: (1) free access to the air, (2) a constant supply of sulphurous food, and (3) a way of expelling the thick and sooty excrements. He believed that if he could show the parallel between combustion and animal heat in these three respects, the identity of the two processes would have been demonstrated.

In arguing the first, the necessity of air, Willis relied, as he had not done explicitly before, upon the idea of an aerial niter. Air was necessary because it provided the "nitrous food (*pabulum nitrosum*) necessarily requisite for the burning of any thing."[75] Indeed, he said, the entire atmosphere was stuffed with these nitrous corpuscles, ready to constitute fire or flame.

If these nitrous particles of the air were excluded from the sulphurous ones, then fire was put out. To prove his point, he cited Boyle's experiments of coals in the vacuum pump; the flame went out, not because it was suffocated with its own smoke, but because it lacked the nitrous food of the air. The death of animals in closed spaces occurred in exactly the same way; the flame in the blood was extinguished from lack of the nitrous food. This point, too, was proved by Boyle's vacuum pump experiments. The animal died sooner when air was drawn away, than when it was not, though one would think that lack of air would create even more space for the absorption of sooty vapors.[76] Willis had taken Boyle's cautiously phrased conclusions of 1660, and uniformly read into them the concept of the aerial niter.

Willis's explanation of the role of the air provided him with yet another opportunity to deny that the heat arose from a fermentation of the blood. Citing Boyle's experiments on coral and acids under reduced pressure, Willis asserted they proved that lower air pressure *increased,* rather than decreased, the fermentation of fermentable or effervescent liquids. In fact, Willis said, Boyle's work showed that the intestine motions of particles of almost everything *except* fire and life were increased when the air was sucked out. "Therefore we may rightly conclude" he said, "that the life of a living creature is either fire, or something analogous to it."[77]

Willis then went on to show how animal heat and fire were analogous in the other two respects. Both required a constant supply of sulphurous food. The lamp received its combustible matter from its fuel, while the blood received its supply from the ingested food. And, in the same way as a flame required constant ventilation, so too did the blood require a constant ventilation, not only in the lungs, but by the pores throughout the body. In making the last point, Willis was especially careful to say that although the two functions of respiration—absorption of nitrous particles and exhalation of wastes—occurred at the same time and in the same place, it was important to keep them distinguished one from the other.[78]

Moreover, he said, the instigation of this process could even be seen. It began in the lungs, where the blood was changed from a dark purple to the florid arterial red, as the "most learned Doctor Lower has observed." It could also be seen in the fact that the crimson top on shed blood would re-form, when it was taken off. This exposure to the aerial niter was only the beginning of an enkindling that continued throughout the arterial system, carrying heat to all the parts.[79]

In fact, he said in summary, the animate body was a great heat engine, a *Thermautomaton*, in which the burning of the sulphurous and nitrous parts took place throughout the body. It was a flame which, like that of a hot iron or rotten wood, produced heat without producing light. But the heat of the blood was, he said, in every essential respect, a process of combustion.[80]

The contrasts between the Willis of the *Diatribae duae* (1659) and that of the "De sanguinis accensione" (1670) could hardly have been greater, both in style and in content. The *Diatribae duae* had been a rambling series of medical essays, displaying a great faith in chemistry and corpuscular matter-theory, but no great acuity of mind. In the "De sanguinis accensione," Willis had a delineated set of physiological propositions to argue, and he did so with admirable brevity and succinctness. The structure and discipline of his format is testimony to the rigor exacted of an author when the subject he treats is one that has been for so long a topic of discussion and debate among his close friends and colleagues.

The content of the essay also reflected a decade of work on respiration. In his emphasis upon the role of the aerial niter, Willis was reacting both to the logic of his subject matter, and to the ideas of his predecessors. He explicitly discussed the origins of animal heat, and dealt with it in terms of the obvious inorganic analogue. In doing so, he drew the analogy so close that it became an identity. The linchpin of this identity was the necessary role of the aerial niter, an idea that appeared with so much prominence most probably because of its championship in the intervening time by Hooke, Castle, Lower, and especially Mayow. One is almost tempted to think that Willis's great emphasis upon the idea was his expression of pique at the fact that Mayow, two decades his junior, had taken an idea that was common property, and appropriated it to private use by publication. In one passing reference to his "De fermentatione" of 1659, Willis noted that, while he had therein explained at length the concept of fire as an intestine motion of sulphurous particles, he had omitted to mention that the constant addition of the nitrous food was necessary to sustain such a process. The implication was that he, Willis, knew of the role of the aerial niter in fire and animal heat, but had just not bothered to mention it.

Willis's resounding rejection of fermentation as a heat-generating process was also, in a subtle way, a recognition of his colleagues' work in the foregoing years. In the *Diatribae duae,* the two concepts of *fermentatio* and *accensio* had shared an ill-defined identity. Willis had treated fermentation as the process by which blood converted chyle into itself, and by which it accomplished the function of nutrition. He had written of ascension as the process that took place in the heart, resulting in the heating and changed color of the blood. But the work of Lower had altered all that. It had forced Willis to concede that whatever process he wanted to call *fermentatio* or *accensio* took place in the blood alone. Since he could find no analogue of a heat-generating fermentation in nature that took place entirely within a liquid, Willis chose to stick with his concept of *accensio,* or enkindling.

Yet in declaring for *accensio* as the common process that united combustion and animal heat, had Willis really clarified the nature of the identity? The answer must be "No." He had argued, as no one had before, concerning the common properties that united the two sets of phenomena. But in doing so he had simply made more explicit and more codified a set of speculations that Oxonians had shared for some years. He had done so under the promptings of the experimental work of Hooke, Boyle, and Lower, and the speculations of Mayow. But his own codification, suggestive though it might be of relationships between animal heat, the blood, the air, and combustion, offered neither new experimental confirmations, nor any distinctly new points of departure for further experiments. As was often the case in his other works, Willis was content to perceive what he saw as the truth, state it, and assert its verity. In this sense, he was still the writer of the *Diatribae duae*.

CLOUDS OF NITER

The publication of Mayow's *Tractatus duo* in 1668, and of Willis's "De sanguinis accensione" in 1670, seem to have precipitated a season of scientific thought in which, wherever one looks, one finds niter as an active corpuscular substance.[81] Moreover, in almost all the instances in which it occurred over the next decade, one can discern lines of intellectual and social descent from Oxford in the 1650s and 1660s. Such patterns do not, I think, represent any conscious proselytizing. The minor figures who used niter in their works in the 1670s seem to have done so merely because it seemed to represent a commonly accepted generalization about nature, one they had heard bandied about in Oxford lectures, clubs, and coffeehouses, and which recurred consistently in the anatomical and chemical works of their more distinguished friends and colleagues. Indeed, as one examines the profusion of "niters" in the 1670s, one can almost draw parallel schemes of social and intellectual affiliation. Those who were recognizably closer to the central personalities of the Oxford group tended to have a clearer and more detailed knowledge of niter and its properties than those whose social links were more tenuous. The structures of scientific perception and social interaction were remarkably congruent.

Henry Stubbe is probably the most idiosyncratic example. When he was put out of his Studentship at Christ Church in 1660, he studied medicine, collected a magnificent library of medical books,[82] and became a great adherent of Willis.[83] He followed this with practice in London and in Jamaica, finally settling in Warwick about 1665. At first he wholeheartedly accepted the corpuscular philosophy, the Willisian doctrine of fermentation in the blood, and considered himself a friend and supporter of Robert Boyle.[84] He even published a few letters in the *Philosophical Transactions* of 1667–1668 describing his experiences in the Caribbean, and noting that tobacco grown on Jamaican soil known for its saltpeter would flash when smoked.[85] Sometime in 1669 Stubbe became convinced that the Royal Society posed a monstrous threat to the Anglican religion, to the universities, and to the traditional practice of medicine. He began a pamphlet war, including sallies against Thomas Sprat's *History of the Royal Society*, in which his Oxford preoccupations came through.

Stubbe directed much of the fire of his *Legends no histories* (1670) against the ineptness of Thomas Henshaw's "History of Saltpeter," which had been printed in Sprat's *History*.[86] Stubbe himself had read extensively on niter, experimented with it, and made many inquiries among the workmen at saltpeter works in Coventry and Warwick; it was an interest that no doubt went back to c. 1657, when he had translated Boyle's essay on saltpeter from English into Latin.[87] Step by step, Stubbe corrected Henshaw's misstatements about saltpeter manufacture, its chemical properties, and its use in gunpowder.[88] In the course of this, it emerged that Stubbe conceived saltpeter as arising from a fermentation in the ground, in which the earth was impregnated with corpuscles from the air, thereby turning volatile and fixed salts from animal excrements into saltpeter.[89] Stubbe poured scorn on Henshaw's idea that there was a universal niter diffused through all of nature, of which the chemical saltpeter was simply a specific manifestation. Not only was this very far from proven, but it was a vague, useless doctrine already published by Glauber.[90] Henshaw would have done better, Stubbe implied, if he had given over such maunderings and applied himself to more precise investigations of saltpeter, as Boyle had.

A more obscure Oxonian, William Clarke, also discussed saltpeter in a book published in the same year. Clarke, a fellow of Oriel, had studied medicine at Oxford in the mid-1660s, returning to practice in his native Bath about 1666. There his colleagues included Thomas Guidott, year-round, and John Mayow, who practiced among the spa-visitors during the summer. Clarke's *Natural history of nitre* (1670) explained, in a somewhat Aristotelian way, the physical, chemical, medicinal, and pyrotechnic uses of saltpeter. Clarke believed that niter was "easily sublim'd into the Air," and conversely that it could be "generated, or separated and drawn from the Air," by heat and dryness. In the air it is "universally dilated," and is therefore called "*Aer Nitrosus,* or Nitrous Air," by "our Modern Philosophers."[91] He updated Sennert with Willis by asserting that thunder and lightning occurred when nitro-sulphurous vapors, or spirits, met in the sky, fermented, and exploded.[92] When he came to discuss "The Use of Nitre to Animals and Vegetables," Clarke overlooked all the problems that had so exercised a careful thinker like Boyle, by cheerfully asserting that nitrous air had three uses in the animal economy: (1) it fed the innate burning flame of the heart; (2) by its cooling properties it allayed and tempered "our natural heat"; and (3) it rendered the humors more fluid, "and so more apt to perform their circular motions, in which Life consists."[93] Clarke, on the fringes of Oxford scientific life, had attributed all the possible functions to the aerial niter, even if two of them were mutually contradictory.

One shudders to think what Stubbe's vitriolic pen could have done to such slip-shod argument.

The Wadhamite Guidott, more identifiably a student of Willis, had rather less grandiose functions for his niter. In a publication in 1669, about the different virtues of the waters at Bath, he accepted that the Cross Bath was more nitrous than the others. This explained, he said, why its temperature was lower. "The greater proportion of Nitre it contains, which being of a cooling nature," would "allay the heat arising from the Sulphur and Bitumen there."[94]

A meteorological niter of a more corpuscular sort came into Ralph Bohun's *Discourse concerning the origins and properties of wind* (1671). Bohun, a fellow of New College, was Ralph Bathurst's nephew, tutor to John Evelyn's son, and a great admirer of Robert Boyle, many of whose works he had in his own library.[95] His meteorological explanations were based upon Boyle's view that air consisted of Gassendian particles with spring and weight, moving in a void. In particular, Bohun believed that the elasticity of the air accounted for many phenomena of weather better than did deductions from the principles of Aristotle or Descartes.[96] Like Boyle, he doubted that nitrous particles were the universal "frigorific," and hence were probably not the cause of cold winds.[97] It seemed more logical that nitro-sulphurous particles exercised their meteorological function when pent up in clouds, from which they broke forth explosively, causing lightning or (if the particles were moving circularly) whirlwinds and typhoons.[98] Bohun's explanations seem to have been widely approved among Boyle's friends, for Oldenburg, in addition to reviewing them in the *Philosophical Transactions*, went out of his way to recommend the *Discourse* to his continental correspondents.[99]

Niter found a slightly different place in the meteorology taught in the late 1670s by Charles Morton to the students at his academy for dissenters at Stoke-Newington, near London. Morton had been at Wadham from 1648 to 1655, where he was "extremely valued by Dr. Wilkins for his mathematical genius."[100] He seems also to have been an admirer of Petty, Wallis, and Boyle.[101] Later at Stoke-Newington, he became close friends with Boyle's sometime assistant, Daniel Cox.[102] In his "Compendium physicae," a manuscript textbook composed for his students and later used for over forty years at Harvard, Morton showed all of these influences.[103] He reiterated the Boylean line that air was an elastic and weighty fluid that was necessary for that kind of motion we call fire, a truth clearly shown when a fire was extinguished in the evacuation produced by a "pneumatic engine."[104] Sulphur and niter were volatilized up into the air as particles, where their opposition caused an attrition or "fermentation." This could increase until it became a flame that broke out as thunder and lightning.[105] Morton even had a mnemonic to help his students remember:

> Nitro-sulphurious fumes from Earth do rise
> And kindled fiery meteors supplyes.

When dissolved by the rain, the aerial niter and sulphur were made fit to enter plants and fertilize them. It even accounted for snow. Water, Morton said, descended as rain, "but meeting with much Nitre it is coagulated thereby." Hail was simply a further degree of the same coagulation, because of a greater quantity of niter in the air:

> While snow from nitre has its Generation
> Hard hail its rise from more coagulation.

Ice then was merely water congealed by the niter of the air. The particles of niter thrust themselves into the pores of the water and obstructed its fluidity. Similarly, the "Dissolution of Ice by warmth is the calling forth these fixing (yet volatile) particles of nitre, and thereby leaving the water to its natural fluidity."[106]

The aerial niter travelled to other parts of the colonies as well. Thomas Trapham, Jr., a physician of Magdalen whose surgeon father helped Willis in dissections,[107] settled in Jamaica. In a tract of 1679 he described the air of the island as abounding in "nitrous parts." Thus it came that great quantities of saltpeter were imported into Europe from the West Indies, and why the tropical rains were so fruitful there. "That all Air hath nitrous parts, though some far more then others," Trapham felt was no difficult belief. Jamaica was such a place. Because of the heat of the sun volatilizing the niter in the earth and the nearness of the surrounding sea, Trapham found the Jamaican air "to be very nitrous, thence penetrating and thence cooling," a true febrifuge. Accordingly, Jamaicans suffered only rarely from continued fevers, and had instead intermittent ones (such as malaria).[108]

A related link between the aerial niter and disease was seen by Nathaniel Hodges, a man connected with the major Oxford figures in many ways. He was an old Westminster, and thereby a friend of Millington and Needham.[109] Before coming to London at the Restoration to set up his practice, his medical life at Christ Church in the 1650s had brought him into friendship with Willis. He was a great exponent of learned chemistry, of the new anatomical discoveries,[110] and especially of Willis's notions of fermentation as a process generating heat, which he both defended in public,[111] and recapitulated in his private notebooks.[112] Hodges kept in touch with his Christ Church friends, at least through the 1660s; he solicited Willis's opinions on difficult cases,[113] and gave John Ward encouragement in his medical studies.[114] In March 1662 Ward visited Hodges, recorded his advice on materia medica, and noted that his bookshelf contained a chemical work "by Gunter Billichius with an appendix of fermentation from wch Dr. Willis doubtless had some of his notions of fermentation."[115]

Thus when Hodges, as one of the few physicians to remain in London during the great plague of 1665,[116] came in the late 1660s[117] to write up his description of the epidemic and its causes, he reached into the conceptual grab bag of his Oxford training, and plucked out the aerial niter. The pestilence arose, he said in his *Loimologia* (1671), from a poisonous aura resulting from the corruption of the nitro-aerial spirit.[118] This nitro-aerial spirit transpired as particles up into the air from the earth. There it resided in the pores of the aerial particles, where it preserved the life, not only of plants, but of animals and even of man himself. Hodges saw no objections to the doctrine that this nitro-aerial spirit maintained the state of the blood and animal juices unimpaired, and the whole body in a vigorous and healthful condition.[119] But this nitro-aerial spirit was also capable of corruption, whereby the conformation of its particles was changed, and it became harmful, reacting with the saline nature of the blood, corroding and dissolving its fibers.[120] This new hypothesis, Hodges believed, not only explained the epidemiology of plague, but also the effects the disease had on the blood. While he was clearly little interested in the physiological function of the nitro-aerial spirit, and the way it maintained bodily health, Hodges was willing to use the notion as a convenient explanation of pathological states.

A more passing medical use of the aerial niter was made by William Cole, of Worcester, yet another physician trained at Oxford in the early 1660s. His book on animal secretions of 1674 accepted Boyle's ideas of the particulate air having weight and elasticity,[121] used Willis's ideas of a chemical/corpuscular fermentation,[122] and frequently cited the work of Lower.[123] In tracing the transformations of the chyle, Cole asserted that it abounded with nitrous-aerial particles whose elasticity helped the heart transform the chyle into blood.[124] Cole must have assumed his readers knew these were taken in through the lungs, for he does not even bother to specify their origins.

Even in those cases during the early 1670s in which the aerial niter cropped up in books not written by Oxonians, a closer look shows that sensitivity to the concept, especially the aerial niter's possible role in respiration, is traceable back to the Oxford medical milieu in the 1660s.

In his immensely popular *Pseudodoxia epidemica*, Sir Thomas Browne had previously, in examining the myth that the chameleon lives on air, included a rather standard Galenic discussion of respiration. When he came to revise that chapter for the sixth edition of 1672, he dropped a passage from Hippocrates on the cold qualities of the air, and substituted for it a sentence stating that the air entered the lungs, "that by its nitrous Spirit" it could "affect the heart, and several ways qualifie the blood."[125] Otherwise, the discussion of respiration remained completely traditional. Browne had obviously been impressed enough by the views of his son Edward's "freind," John Mayow, to incorporate his idea of an aerial nitrous substance—drawn either from his own reading of the *Tractatus duo*, or from Oldenburg's review in the *Philosophical Transactions*.

An even more interesting example of action at a distance is that of Malachi Thruston's *De respirationis usu primario, diatriba* (1670). Thruston had been educated as a physician at Cambridge in the 1650s and 1660s, where he was a fellow of Gonville and Caius College.[126] Both his background and the overall structure of his physiological views would suggest that he had little connection with either the Oxford virtuosi or the chemical view of respiration. He argued that the fundamental attribute of the blood was its continual movement, for which it had to maintain fluidity. This was achieved by respiration, in which the more pure, subtle, and penetrating parts of the air interacted with the blood to maintain it in a proper temperament. Animal heat was generated in the heart by a process of fermentation, and was thence communicated to the blood. Thruston cited the "very ingenious experiments" of Hooke and Lower that proved blood acquired a florid color when exposed to air in the lungs, but seems not to have fully appreciated Lower's conjoint argument against a ferment in the heart.[127] The overall emphasis of Thruston's book was that respiration promoted the fluidity of the blood, and hence its due circulation, a central idea he shared with his fellow Cantabrigian, Walter Needham.

It was only when Thruston came to speak in more detail about the nature of the air that some conceptual congruences to Oxford notions begin to appear.[128] Air had, he said, several properties. It was elastic, as Boyle's experiments had shown. It was laden with many nitrous particles (*nitrosis particulis*), as the experiments and reasonings of Ent, Gassendi, Digby and others had shown. Some believed, he noted, that the nitrosity of the air served a function by its frigidity, others believed it contributed to heat, while yet others believed it was the cause of rarity, or conversely, density. Although he hinted at one point that these nitrous particles conserved the heat in the blood,[129] Thruston was unwilling to decide between the putative effects, and referred his readers to the fourth book of Descartes's *Principia* for a full discussion—by no means the case, since Descartes there had dealt only with the properties of solid saltpeter.[130] Thruston then went on to assert that the density or rarity of the air had to be within certain limits in order to be fit for respiration, and that when the most penetrating part of the air was absorbed by the lungs, the respired air in this way lost part of its elasticity.[131]

Why did Thruston mention the nitrous aerial particles, and then not really use them in his theory of respiration? The answer is to be found in the history of the work's composition. The first half of the book was presented as a thesis at Cambridge in 1664. It must have been composed no earlier than the autumn of that year, since Thruston referred to books by Willis, Henshaw, and Croone, all of which were published in 1664. More significantly, he mentioned in the same thesis the fact that Needham and Boyle, working in Oxford, were able to reactivate the heartbeat of a dead dog by blowing air into the thoracic duct.[132] He used this experiment, as well as a report of Croone's 1664 work at the Royal Society of reviving choked chickens by blowing into their lungs, to argue the point that air conduced to the motion of the blood, and hence to life.

It seems clear that during the early 1660s, while Needham was sampling the academic wares at Oxford, he was in touch with his old friends back at Cambridge. The vague idea of an aerial nitrous substance most probably came to

Thruston's attention in this way, but without the other concepts that were tied to it as it had developed within the Oxford tradition. Indeed, Needham, while he could very easily have passed on the unit-idea "niter-in-the-air," would have been very unlikely, given his own opinions on the function of respiration, to pass on the entire cluster of ideas concerning the physiology of the blood and lungs. Moreover, the evidence for a correspondence between Thruston and Needham is quite strong; Needham contributed a laudatory letter to preface the *De respirationis usu*.[133]

During the years that intervened between the presentation of the thesis in 1664 and its publication in 1670, Thruston continued to have the kinds of contacts which would have kept him aware of the aerial-niter idea. At some time before he took his M.D. at Cambridge in 1665, he met John Wilkins—possibly during Wilkins's tenure of the Mastership of Trinity, when the Oxonian also met Walter Needham. Wilkins proposed Thruston as a Candidate for the Royal Society in May 1665 and he was elected at the next meeting; these were the very meetings at which Wilkins elaborated his suggestion of injecting blood from one dog into another.[134] Later in 1665 Thruston moved to Exeter, where he set up a practice, and where, as in Cambridge or London, he did not lack for contacts with Oxonians. Seth Ward had been made Bishop of Exeter in 1662, and the two virtuosi struck up a friendship. Wilkins was made a canon and preceptor in the same cathedral in 1667. In May 1668 Oxford's perennial teacher of chemistry, Peter Stahl, gave a course at Exeter, similar to the ones given at Oxford,[135] and Thruston was one of the attendants. His manuscript notes from the courses,[136] in a neat, distinctive hand, showed that he, no less than his contemporaries at Oxford, welcomed an opportunity to use chemistry to make more concrete the concepts of natural philosophy. In the fall of that same year, Wilkins related to the Royal Society a series of experiments on transfusion and kidney surgery that he had seen Thruston perform at Exeter.[137] This activity as a provincial natural philosopher came to a culmination in February 1670, when Thruston published, with the active encouragement of both Ward and Wilkins, a revised and expanded version of his Cambridge thesis *De respirationis usu*.[138] Thruston's book was naturally accorded first-class treatment. Within a few weeks Oldenburg had given it a careful review in the *Philosophical Transactions*, as well as recommending it privately in his correspondence with Malpighi, Travagino and Gornia in Italy, and even to Winthrop in Boston.[139] With such backing, it was perfectly natural that Thruston should be especially aware of the findings of Hooke and Lower, and that the Oxford niter should enter into his discussion *en passant*; he was close enough to be aware of such ideas, but not enough to be committed to them as part of a complete system of physiological explanation.

The aerial niter has claimed the attention of historians of science very largely because it has been seen by some as simply another name for oxygen. Historians have even gone so far as to claim for Mayow, most especially on the basis of his *Tractatus quinque* (1674), the honor of having discovered oxygen a full century before Lavoisier. If the profusion of aerial niters in the late 1660s and through the 1670s demonstrates anything, it is that the simple term "aerial niter," or "nitrous aerial particles," or "nitro-aerial spirit," cannot always and easily be equated with oxygen. We do indeed know that saltpeter fixes oxygen from the air, that oxygen is taken in by respiration, that it oxidizes foodstuffs to produce heat and waste products, that oxygen ultimately provides the energy for muscle contraction, and that similar oxidative reactions produce heat and light in ordinary combustion. But do we now think of thunder and lightning, or snow and hail, as originating from oxygen in the air? Or of oxygen being, of itself, a fertilizer for plants? Or of fevers cooled by oxygen, or plague caused by corrupted oxygen? Or of oxygen contributing to the fluidity of the blood? Of course not. Yet all of these represent legitimate attempts to establish a connection—a "transdiction" if you will—between the observable chemical properties of a solid, and the less definable effects of air on brute matter, plants, and animals.

The belief of so many Oxford people in the possibilities of such a transdiction probably had several roots. At the most general level, practical experience in laboratory chemistry, such as that obtainable in the courses of such "chemists" as Stahl and Wildan, accustomed men to thinking of qualities and effects as arising from chemical properties and interactions. Did the air cool falling rain or fevered brows? Perhaps this was due to the refrigerative properties of niter. Was air necessary for both fire and the incalescence of the blood? Perhaps the heat-sustaining properties of niter were responsible for both. Did arterial blood seem necessary for muscular contraction? Perhaps this was due to the explosive capacities of the nitrous particles absorbed into arterial blood in respiration. Such kinds of thinking were almost certainly encouraged more specifically by Willis's use of "fermentation" and "nitro-sulphureous particles" in club gatherings in the 1650s, and especially in his Sedleian lectures during the years 1660–1667, before he taught by deputy. This may well account for the similarities of phraseology in such diverse characters as Clarke, Bohun, Morton, Hodges, and to a lesser extent Mayow himself, men who are not known to have been associated closely with the research activities of such central Oxford figures as Bathurst, Willis, Boyle, Hooke, Lower, and Needham.

The particular brilliance of Mayow, and on a more speculative level, Willis, lay in perceiving not only that the properties of an aerial nitrous substance could solve important problems in physiology, but that in doing so one could draw more general conclusions about the active component in the air. Mayow was led to define the chemical characteristics of "oxygen" only after physiological questions—animal heat, muscular motion, and fetal respiration—had enabled him to shear away from the aerial niter those putative properties that were not essential to its physiological role. He, like Julius Robert Mayer two centuries later, took the physiological route to the discovery of an important principle in the physical sciences. In turn, Mayow could do this only because the general nature of those nitrous properties, both chemical and physiological, had been for so long a topic of concern in the milieu within which he came to intellectual maturity.

FIRE AND LIFE: JOHN MAYOW'S *TRACTATUS QUINQUE* (1674) AND A GENERAL PHYSIOLOGY OF ACTIVE PARTICLES

Scientific ideas have life cycles of their own, the mutual outcome of their own internal logic and of the interest and creativity of the men who put them forward. The two and a half years from the publication of Needham's *De formato foetu* to the appearance of Willis's tracts on animal heat and muscular motion—from September 1667 to March 1670—had been an intensive period of speculation and experimentation. It was succeeded by one of equal length in which such studies, especially at the Royal Society, took on a fragmented character greatly contrasting with the focused investigations of the late 1660s.

Participants were lacking. Lower and Willis had passed entirely out of the Society, each busy with his London practice. Boyle continued to live in town, but attended less than a dozen meetings. Hooke, although his post required him to be in constant attendance, was distracted by his blossoming career as an architect, and initiated few new lines of experimentation; in November 1670 he was rebuked by the Council for neglect of his duties as Curator of Experiments.[1] The attendance of Castle, Charleton, Clarke, Locke, Pope, Wilkins, and Wren is recorded for only a handful of occasions each during the early 1670s. Needham moved to London in the winter of 1671, was officially admitted to the Society, and thereafter became a regular participant, but his contributions were confined largely to communicating letters on natural history from his relative, John Templer of Northamptonshire.[2]

Oldenburg continued to give heat and respiration prominence in the *Philosophical Transactions*, but to little effect. In August and September 1670 he published the air pump experiments on respiration that Boyle had done at Chelsea in 1662–1663.[3] But they elicited no response at the Royal Society, most likely because Boyle was himself entirely absent from meetings for the following nine months. Oldenburg also printed communications by Lorenzo Bellini and by Marcello Malpighi on the structure of the lungs,[4] but these had little applicability to the aspects of respiration that had been mooted in England. He even summarized an eccentric English tract on respiration[5] that proposed an absurd theory of the role of the diaphragm, and argued that there were other uses to respiration

> than the Cooling of the Heart, the Fanning of the Blood, the Discharge of steams, the Conveyance of a Nitrous aliment, the Comminution and subdueing of the Blood, and its intimate commixture with the Chyle, and the promotion of the Blood from one ventricle of the heart to the other . . .[6]

Having impugned all parties in the recent debates on respiration, it is little wonder that the author chose to publish anonymously.

Hooke continued to think about respiration, but most of his comments at the Royal Society were in passing response to some external stimulus. In February 1670, a week after Thruston's *De respirationis usu* had been presented to the Society, Hooke promised "an anatomical experiment concerning the lungs." At the following meeting he reported that the experiment had not succeeded in private, and promised to try it again; he never did.[7] In June 1672 Needham produced a letter from Templer on the structure of the lungs that emphasized the complex ramifications of the alveoli and blood vessels.[8] This occasioned a discussion of respiration, whose purpose Hooke thought to be "that by the air something essential to life might be conveyed into the blood; and something that was noisome to it, be discharged back into the air." He further suggested that colored wax injected into the arteries might reveal valves that helped distribute the air to all parts of the blood. This was recommended to the "physicians of the Society," but nothing ever came of it.[9]

Hooke did, however, come up with another mechanical wonder. In February and March 1671 he built and used an evacuation chamber big enough to accomodate a man. The subject crouched inside a small barrel attached to a pump and containing a pressure gauge. This barrel was in turn immersed in the water of a larger barrel, in order to insure a reasonable seal. The apparatus was designed, Hooke reported, "to find what change the rarefaction of the air would produce in man, as in respiration, heat, &c."[10]

In the first trials Hooke was his own guinea pig. Some air was withdrawn, and he remained in the chamber for a quarter of an hour without "any inconvenience." He felt, however, that "a man could not endure much more than the evacuation of a fourth part of the air." A week later Hooke had done more trials, and reported that one-tenth of the air had been withdrawn, with "no other inconvenience but that of some pain in his ears."[11]

The most important experiment took place privately at Gresham on 13 March 1671, with Brouncker in attendance. This time the maximum evacuation attained was one-quarter, and the time Hooke spent in the barrel "above a quarter of an hour." Once again he suffered no ill effects other than some pain in his ears, and temporary deafness. Hooke had, at the Society's instructions, "taken a candle with him into this vessel." Brouncker and Hooke reported that although the "engine" had not been evacuated fully one-quarter for the entire duration, "the candle went out long before he felt any of that inconvenience in his ears."[12]

The experiment is particularly significant because it represents the first recorded time that an animal (in this case Hooke) and fire had been put under reduced pressure together at the same time. Previous trials, like those of Boyle in *New experiments* (1660), had separately put animals and flames under reduced pressure. Other experiments, such as those done at the Society in 1663 and 1665, had put animals and flames in the same closed containers, but not under reduced pressure. Hooke had proved for rarefied air what had long been known for air at atmospheric pressure: that fire was extinguished long before life was incommoded. Once again it seemed that if fire and animals consumed the same substance from the air, experiment could not prove it. Inexplicably, after Hooke had done these few experiments with his air chamber, he did no more.

THOMAS WILLIS AND THE "FLAMEY SOUL"

Such scruples about experimental verifiability little bothered Thomas Willis, whose literary productivity in London, as at Oxford, was unaffected by a busy life as a clinician. In the winter of 1672 he brought out at Oxford the last of his trilogy on the brain—the massive *De anima brutorum*.[13] Most of the book, a classic in the literature of physiological psychology, was a highly elaborated version of his Christ Church lectures of 1661 on the senses and on neurological disorders.[14] As in his two previous neurophysiological works, Willis gave detailed corpuscular explanations of nerve function and dysfunction, buttressed by many cases and postmortem results, some of them going back to his Oxford days.[15] He prefaced these considerations with more than a hundred pages of comparative anatomy, including some of the first detailed descriptions of such invertebrates as the earthworm, oyster, and lobster.[16] In carrying out these dissections Willis had, as at Oxford, recruited some helpers. His surgeon, Edmund King, did much of the work, helped by John Masters, yet another one of Willis's medical apprentices recruited out of Christ Church.[17] He addressed the book to his Oxford colleagues who, he hoped, would give it the same approval accorded his lectures some years before.[18] As with almost all of Willis's books, Oldenburg gave it a laudatory review in the *Philosophical Transactions*, and recommended it to his foreign correspondents.[19]

Bodily activities, Willis believed, were coordinated by a "corporeal soul," or "soul of brutes," which existed by itself in lower animals, and coexisted with the rational soul in man. This corporeal soul had two parts. The "vital soul" was a "flame" in the blood. The "sensitive soul" consisted of animal spirits diffused through the brain and nervous system. Although Willis adopted a modification of the Aristotelian terminology, both parts of his corporeal soul were perfectly naturalistic; both consisted of the movements of chemical particles in appropriate directions to accomplish specific ends. He was, he felt, simply following all the modern philosophers, especially Gassendi, in interpreting the animal soul as atomistic.[20]

Nitrous aerial particles came into the discussion because the vital soul in the blood was nothing other than a kind of flame, fed by sulphurous matter from the food, and nitrous particles from the air. He reiterated his now familiar definition of fire as a swift motion of contiguous particles, which could exist in a vehement form in ordinary flame, and in a more subdued form in the "flame" of the blood itself.[21] Willis then went on to trace, as he had not done before, how this process was found throughout the scale of animate nature. Boyle had shown that insects needed air; Willis demonstrated how, lacking lungs, they therefore had spiracles to draw in the nitrous food from the air.[22] The beating of an oyster's heart circulated its blood through its gills to pick up the nitrous corpuscles in the water.[23] The lobster did the same, as could be proved by injecting ink into its heart.[24] The earthworm had holes on its back to take the place of lungs in bringing in nitrous particles.[25] Injections likewise demonstrated how in fish the heart pumped blood through the gills to take up the nitrous particles in the water. Thus if fish were put in too little water, or in water from which the nitrous particles had been driven off, they perished for lack of nitrous food. Before they died, they put their mouths up to the surface in an attempt to breathe the naked air.[26] Reptiles and amphibians also needed the *pabulum nitrosum* of the air, although because the "flame" in their blood was so moderate as to seem cold to human senses, they needed much less of it than warm-blooded animals. By the time Willis reached mammals and birds, he was saying nothing he had not already said in 1670, or intimated in earlier works.[27] But he had succeeded, via some ingenious dissections, in showing that the respiration of nitrous particles was consonant with the anatomical arrangements of a large number of animate forms.

Willis also suggested, although rather in passing, a more general physical use for aerial nitrous particles. In discussing sense perception, he hypothesized that the circumambient air consisted of many kinds of corpuscles. Motion of one class of corpuscles affected only one sensory modality. The motions of saline corpuscles, for example, repercussed from vibrating objects, moved the eardrum and were perceived as sound. Other effluvia in the air were perceived as smells. The effects could move independently of each other, just as waves on the surface of a pond could cross and emerge unchanged. This accounted for why we cannot hear smells, or smell sounds.[28] In this scheme, Willis seems to have believed that light was transmitted as a wave disturbance in the "luminous or nitrous particles" in the air. Since fire was nothing but a vigorous interaction of the sulphurous particles of the flammable body with the nitrous particles from the air, the sulphurous particles breaking forth from the fire agitated the nitrous particles diffused through the air, and this agitation was propagated out from the flame to be perceived as light. It was, in its own way, a very neat scheme. Since every kind of effect was transmitted by a different class of particle in the air, one could explain, for example, why certain effects traveled more rapidly than others. Sonorific (saline) particles propagated effects more slowly than luminous (nitrous) particles. Thus one saw the flash of a gun before one heard its report.[29]

With the exception of this ingeniously Cartesian use of nitrous particles to transmit light, Willis's ideas in *De anima brutorum* were, if more greatly elaborated and more firmly based on anatomy, also quite predictable. The corpuscular concept of the "flamey soul in the blood," fed by sulphurous substances within and nitrous ones without, went back to Oxford in the 1650s. It had been assumed in the *Diatribae duae* of 1659, and had become progressively explicit in Willis's four subsequent works. The emphasis changed over the years, responding to the ideas and discoveries of Boyle, Hooke, Lower, Mayow, and Goddard, but the essential line of reasoning remained the same—one rooted in his early chemistry and contacts with Bathurst.

BOYLE AND HOOKE ON FLAME AND AIR

Willis's repeated speculations on a *flamma vitalis* in the blood prompted his very antispeculative friend, Robert Boyle, to bring forward some earlier experiments by which the notion might be tested. The intermediary was Oldenburg. As a sequel to Boyle's respiration experiments published in 1670, Oldenburg asked Boyle to gather what experiments he had done on the relation of flame and air, especially as they might relate to the necessity of air to the "vital flame," a topic on which, as Boyle put it, "divers of our Learned men have spent both Thoughts and Discourses in inquiring and disputing."[30] Although most of these experiments were assembled by late 1671 or early 1672, it seems to have been the publication of Willis's *De anima brutorum* in February 1672 that brought them to the printer. Oldenburg wrote to Magalotti on 13 June 1672 praising Willis's book, and noting that Boyle had "made ready an experimental discourse" on "the relation between flame and air."[31] He reported to Huygens in September that the work was being printed, and in November that it would be out soon.[32] But, as usual, Boyle kept adding new items (mostly on unrelated topics in hydrostatics),[33] so that *Flame and air* did not actually come out until early March of 1673, although it continued to carry the date of 1672 on the title page. Oldenburg reviewed it promptly in the *Philosophical Transactions*[34] and recommended it to Swammerdam in Holland, Malpighi and Gornia in Italy, and Bartholin in Denmark.[35]

Both sets of Boyle's experiments, on flame and on the *flamma vitalis*, were carried out largely with the vacuum-pump. The instrument used followed Boyle's second design, and was first constructed about 1666, incorporating new desiderata by Boyle, and improvements suggested by others, especially by the

The *I.* Plate

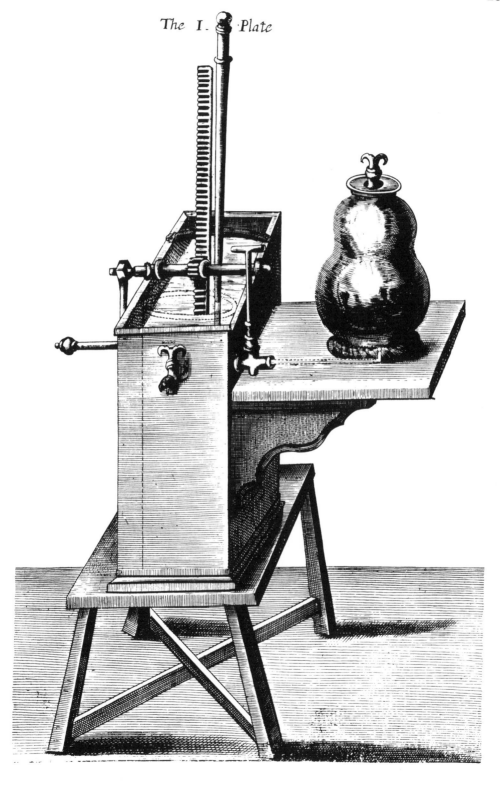

"ingenious Mr *Hook*."[36] The piston was placed under water to improve the seal. The outlet pipe was embedded in a flat iron plate, to which the receiver was cemented directly. This level base not only enabled objects to be more easily kept within the receiver, it also permitted the elimination of the stopcock on the receiver itself, which had always been a source of leakage (see p. 251). It was this instrument that Boyle had used at Oxford in 1667–1668 for the experiments published in his *Continuation* (1669). Indeed, the experiments on fire might well have been done at the same time and place, since in the preface to that work he mentioned "two Clusters of Pneumatical trials, the one about Respiration, and the other about Fire and Flame," that he had foreborne to publish at that time because his "Notes and Observations" were "out of the way."[37] He seems to have found the respiration experiments in packing up to leave Oxford in April 1668; perhaps those on fire surfaced in the same cleanup.

Although the experiments on fire are divided somewhat artificially into three "titles" on producing, preserving, and propagating fire without air, the results are best considered as a whole. Boyle had easily confirmed his earlier finding that an ordinary lamp, even one whose alcohol was impregnated with metal (and whose flame might therefore be more durable) would last less than two minutes after evacuation began.[38] Sulphur—"Brimstone"—could not be ignited in a vacuum, even when dropped upon a red-hot iron. It would melt, bubble, and sometimes even sublime, but never take fire. Brimstone already burning when the evacuation began would also go out, although it seemed to last longer than a lamp flame. Conversely, when air was readmitted into the receiver, the hot sulphur would once again burst into a blue flame.[39] Such experiments, Boyle remarked disingenuously, had the difficulty that the pump drew out great quantities of sulphurous smoke "that was offensive enough both to the eyes and nostrils."[40] There was another experiment that must have been easily as offensive, and even more interesting. Boyle fashioned what we can now identify as a hydrogen sulfide generator, lit the "inflammable steams" as they emerged at the mouth of the vial, and put it into the receiver. He observed the pretty result that at the first evacuation, the flame appeared five or six times as large, which Boyle (no doubt rightly) attributed to the increased bubbling under reduced pressure. After the third exsuction this flame too died from lack of air.[41]

But Boyle's most interesting experiments were those in which he tested the capability of gunpowder, which Hooke had argued contained its own supply of aerial nitrous particles, to explode without atmospheric air. His first trials had been in the summer of 1660, in which he had attempted to fire off gunpowder in a vacuum by concentrating the rays of a burning lens upon it. The attempt failed, so Boyle tried the same procedure with *aurum fulminans*, which was popularly supposed to be of the same nature as gunpowder. The *aurum fulminans* obligingly went off under the sun's rays. It did the same when a mere quarter of a grain was dropped onto a hot iron in an evacuated receiver.[42]

When Boyle came to repeat the gunpowder experiments some years later he encountered similar difficulties. Gunpowder dropped on a hot iron in a vacuum would flame, but not explode. It did the same when dropped on coals. Boyle even arranged to fire off a very small flintlock pistol in his evacuated receiver, but got no flash until he let the air back in.[43] By using a thinner glass receiver, he finally succeeded in igniting a few grains with a burning glass. But although the individual grains under the light would explode, the rest of the powder in the train would not. In other words, it seemed that flame could not propagate itself through a vacuum, a result Boyle confirmed using an instrument that tested the power of gunpowder. He could ignite the powder in the touch-hole using his lenses, but the flame would not be propagated through the hole into the main chamber.[44]

Boyle's only real success in propagating the flame of gunpowder "without air" came in an experiment burning the explosive under water. He stuffed gunpowder into a goose quill, lit it, and plunged it into water, where it continued to burn. Although some might interpret this to mean that gunpowder could burn without air, Boyle was cautious. He pointed out that air lurked in the pores of the water, and between the grains of gunpowder itself. It may even be, Boyle said, that saltpeter stored aerial particles. Thus the gunpowder burning under water could perhaps get air from the dissipated niter itself. To test this hypothesis, Boyle had, he said, tried to "reproduce" niter in a vacuum, so as to preclude the possibility that in its formation, the saltpeter had captured aerial particles. There were, he said, other conjectures he could make about the inclusion of air in saltpeter, but he would forebear.[45]

Boyle's discussion of this experiment is especially interesting because he seemed to believe that air was *physically* enclosed in saltpeter, rather than being *chemically* combined in it. The distinction was admittedly fine, but it seems almost as if Boyle, conceiving of the air as his little springs, was loath to think of them entering into real chemical combinations such as might "fix" them into solid saltpeter. In this sense he continued to differ from Hooke, who had since the early 1660s consistently conceived of the nitrous aerial substance as having a real and chemical existence in the air, whose powers were continued unchanged when it was "fixed" in saltpeter.

Boyle's "New Experiments About the Relation betwixt Air and the Flamma Vitalis of Animals," a subsection of *Flame and air*, were conducted with equal care. He first compared the effects of air at atmospheric pressure on an animal and flame in a sealed container—thereby repeating the Royal Society experiments of 1663 and 1665. The general results were the same. A greenfinch and an alcohol lamp (or two) were put together into a ten-quart container. The flame went out within two minutes, while the animal, even if kept confined for five or six times as long, showed no ill effects. A mouse and a lamp, as well as a mouse and a candle, showed the same pattern.[46]

Boyle then put his finch and a lighted candle in the receiver together, plied the pump, and noted the effects of evacuation. The results turned out to be comparable to those in air at normal pressure:

> . . . we found, that the Flame began more quickly to decay, and the Bird to be much more discomposed than in the former Experiments; but still the Animal outlived the Flame, though not without Convulsive motions.[47]

Boyle then varied the experiment. He pumped out a receiver containing the bird and a taper; once again the bird showed some symptoms, but outlived the flame. He tried it again with his bird and a piece of charcoal. If they evacuated until the coal could no longer be revived by admitted air, the bird became "very sick indeed, but yet capable of a quick Recovery."[48]

We now know that there are some perfectly good reasons for Boyle's results, as well as those of the earlier experiments at the Royal Society. Although it is true in qualitative terms that both combustion and respiration utilize oxygen from the air, and give out carbon dioxide as a gaseous product, flames and animals greatly differ in their sensitivity to either low oxygen content or high carbon dioxide content. Beginning at a normal atmospheric oxygen content of 21 percent, a candle will be extinguished when the oxygen level has fallen to 14.8 percent, and the carbon dioxide risen to 4.1 percent. By way of contrast, the oxygen content has to fall to almost 3.5 percent, and the carbon dioxide rise to 11.5 percent, before an animal such as a mouse will be dead. Such a great differential sensitivity to oxygen tension at atmospheric pressure accounts for why Boyle, and his Royal Society colleagues before him, always found that in a sealed container the animal outlived the flame.[49]

The same differential sensitivity applies to an animal and a flame that are jointly subjected to reduced pressure. In the extreme circumstance that the receiver is emptied very rapidly, the animal and the flame will indeed be extinguished at the same time. But Boyle's air pumps were not capable of producing such near-instantaneous vacuums. The slower the receiver is evacuated, the longer the animal will outlive the flame. In modern trials duplicating what must have been approximately the evacuation rate of Boyle's pumps, such that a candle was extinguished in slightly less than a minute, a mouse lives almost five and a half minutes, exactly the ratio that Boyle found.[50] Similarly, when Hooke took the candle into the barrel with him in 1671, it went out before he was greatly inconvenienced, precisely because its tolerance for a lower oxygen tension was so much less than that of Hooke himself. Thus, although Boyle's initial air-pump experiments at Oxford in 1659, conducted *separately* on flame and on animals, had seemed to show a new and striking common dependence of both fire and life upon air, careful prosecution of those experiments by putting a flame and an animal *together* in the same receiver, whether evacuated or not, seemed to disprove that relationship.

In drawing his conclusions, Boyle was therefore even more cautious than he had been some years earlier. He saw several possible ways to interpret the results. The survival of the animals could indicate (1) "That the Common flame and the Vital flame are maintained by distinct substances or parts of the Air." Or it might mean (2): "that common Flame making a great waste of the Aereal substance, they both need to keep them alive, cannot so easily as the other find matter to prey upon, and so expires, whilst there yet remains enough to keep alive the more temperate Vital flame." And there was always the possibility (3) "that both these causes, concurr to the *Phaenomenon.*" Boyle refused to decide between the alternatives.[51] His reticence seems to have arisen, not so much from his doubts about the existence of some Willisian flamelike process in the body that needed *access* to air, but from his disinclination to believe that a vital flame actually *consumed* something from the air. Thus he went on to record experiments that purported to show that glowworms needed air to give off light,[52] to demonstrate his old point that animals closer to the womb were more easily killed by the lack of air,[53] to show that silk worms or gnat larvae could not metamorphose in a vacuum,[54] and even to support Willis explicitly by adducing some examples of explosions caused by mixtures of liquids.[55] He was, Boyle himself confessed, a man who was much beholden to the air for many discoveries, so that he was always disposed to give it its due.[56] But Boyle was also a man both scrupulously respectful of the limits of experimental inference, and one who tended to conceive the role of the air in physical rather than chemical terms. Therefore he would not, he said, give air "a larger Jurisdiction than I find nature to have assigned it."[57]

Thus, although Boyle could demonstrate the great importance of air in the production and maintenance of flame—just as he had previously, in his countless air pump experiments on respiration, demonstrated the necessity, especially of warm-blooded animals, for constant access to the air—he could not construe the results of joint experiments on flame and life to support an absolute equation between the two. We now know that if only he had kept his animals in the receivers just a little longer after the flame went out, he might have been able to quantify the differential sensitivity of a flame and an animal to oxygen content. As it was, his experimental technique, and perhaps his presuppositions, did not lead him in that direction.

Boyle's experiments on flame and air had repercussions at the Royal Society even before they appeared in the booksellers' shops. At the meeting of 13 November 1672, just as the tracts were going through the press, Hooke suggested "that it were worth trying, whether air be consumed, or increased by burning"; he was desired to design some experiments to determine the question. Boyle, attending one of his few meetings that year, further suggested that one might try "to make air of finer bodies than ordinary, such as are distilled liquors," to see whether they followed the pressure/volume relation (Boyle's Law) known to hold for common air. "He moved also, that it might be examined, whether, in making salt-petre by art, there is any air intercepted and compressed."[58] All three suggested lines of investigation had been prompted by Boyle's work then in press. Over the next eight months the first two were pursued by Hooke with some tenacity, and with surprising results.

The question of volume change in combustion was seemingly clarified within the next few meetings. After a few initial failures[59]—duly noted in Hooke's diary as "Experiment of fire and air"[60]—he reported to the Society on 4 December 1672 that, using unspecified apparatus, he had found that burning "neither increased nor decreased" the volume of the ambient air.[61] Once again, it must have seemed that any theory that required a substance to be consumed from the air in combustion was on shaky experimental foundations.

Hooke took up the question again in mid-February 1673, just as Boyle's tracts on flame and air were finally coming out. As in November, the first trials miscarried because of a failure of the apparatus.[62] Hooke saw the first glimmerings of hope at the Royal Society meeting of 5 March. The Journal Book for that day records that "the apparatus failing again," Hooke "was ordered to fit it with care."[63] He must have done so later that afternoon, for he recorded in his diary that night: "Shewd fire Experiment. Air decreased."[64] Over the next fortnight Hooke continued to tinker with his apparatus. On Tuesday, 11 March, he was "at home all day trying the fire experiment but could not make it succeed." The next day he was "Preparing Experiment for Arundell [House, where the Society met] of fire."[65] But this too must have failed, for the experiment does not even appear in the Society's minutes that day.[66]

Finally, on 18 March 1673, Hooke found what he was looking for—decrease in volume caused by combustion. He jotted triumphantly in his diary: "tryd the Experiment of fire by the help of burning glasse and found air decrease. writ Lecture upon that Experiment."[67] The next day at the Society:

> *March* 19. Mr. HOOKE read a discourse of his, giving an account of the success of this experiment, which, he said, he had made, about the increase or diminution of air by burning; which was, that the air was diminished one twentieth part.

The Society encouraged him to continue "to prosecute these experiments."[68]

Although Hooke was clearly pleased, his diary entry for that day conveys little further information about his apparatus: "Tryd Experiment of Burning, found air decrease. read a Lecture thereof at Arundell. Sir R. Moray president and 18 others."[69] It seems likely, however, that Hooke's experimental technique must have involved collecting the vitiated air over water. The carbon dioxide produced by combustion was dissolved, thereby giving a decrease in volume after the enclosed gases had returned to room temperature. Hooke's figure of 1/20 is in fact remarkably good, indicating that he must have succeeded in dissolving almost all of the carbon dioxide produced by the flame before it was extinguished. Hooke had, working from his own assumptions about the consumption of aerial nitrous particles in combustion, arrived at both the technique and the result that were to play so crucial a part in Mayow's argument.

A month later, and once again coinciding with Boyle's reappearance at the Society, Hooke began the next step in the program: collecting "airs" from other sources. On 23 April 1673 they put aquafortis (nitric acid) and pulverized oyster shells (calcium carbonate) in a flask, tied a flaccid bladder over the mouth, and collected the gas in the inflated bladder. This was then put aside until the next meeting, "to see, whether these exhalations would prove permanent air." Boyle, for one, felt they would, since he "had frequently made such kind of experiments, and thereby produced true air, which lasted for several months together."[70] Two weeks later the bladder was still full, so it was suggested that various trials be performed upon this very same gas. Would it serve to support a burning candle, and if so, for a shorter or longer time? Could air produced in such a way be made fit for respiration? Had the aquafortis lost weight in generating this air?[71]

Over the following month some fascinating answers to these questions were discovered. Hooke first found that mixing the original gas with common air *shortened* the amount of time the mixture would support a burning candle.[72] This was a suggestive result, but not accurate enough to be conclusive. So Hooke designed some new apparatus. On 28 May 1673 he presented his revamped experiment. He first generated a quantity of carbon dioxide using aquafortis and oyster shells. After determining that a candle would burn for at most forty-five seconds in a container filled with "common air," he plunged it into the same receiver filled with the "generated" air.[73] The results were remarkable:

> . . . the same candle put in the vessel filled with the factitious air would not burn in it, but only an inch beneath the mouth of the glass, where the outward common air had some communication with the produced air; for, being put lower, it went out immediately upon several trials. It was observed, that the candle being gone out near the orifice, it would catch the flame again, when hastily drawn up close to the top. Besides, it was taken notice of, that when this factitious air was driven out of the vessell, the flaming candle held over it was presently blown out by it.[74]

Hooke's entry in his diary that night was shorter but equally to the point: "at Arundell shewd Experiment of a damp air which quenched flame made out of A[qua] F[ortis] corroding oyster shells."[75] The following week Hooke tried the same experiment with "air produced out of bottled ale," and found that it too would not support combustion. It was proposed that "something might be thought upon for correcting this air, so as to make a candle burn or animals live in it," which Hooke thought he might be able to do "by precipitation."[76]

Such experiments, although not printed at the time, brought into clear contrast two views of the air in its relation to flame. On one side was Boyle's essentially physical concept of the air. Any air was "true air" if it met the criterion of elasticity. Moreover, since flame could neither be supported *nor* propagated without air, Boyle seemed inclined to believe that the entire air played some part in this process. Even the role of saltpeter as a surrogate might be related to the air that it had "intercepted and compressed." On the other side, Hooke strongly believed that only a portion of the common air was involved in combustion, a theory he must have felt was supported both by the decrease observed in burning, and by the failure of "generated air" to support combustion at all. Unfortunately, this latent opposition between the two views was not brought to resolution. A few weeks after Hooke had done his experiment on ale gas, the Society adjourned for almost five months. When it reassembled the following October, the mercurial Hooke had gone on to other projects. It was rather John Mayow, working now in isolation at Oxford and at Bath, who had pursued the aerial niter to its logical conclusion.

JOHN MAYOW AND HIS *TRACTATUS QUINQUE* (1674): THE CHEMISTRY OF NITRO-AERIAL PARTICLES

The outlines of Mayow's life between 1668 and 1674 are shadowy. He was in residence at All Souls most of the time during 1668, 1669, 1670, and 1671, absent during most of 1672 and 1673, and in residence again during most of 1674.[77] In order not to lose his college seniority, Mayow continued to occupy his jurist's fellowship, serving as the college's Dean of Law in 1667−1668, and taking his D. C. L. in July 1670. He never took a medical degree. Anthony Wood, in recording Mayow's D. C. L., noted that he "was now, and after, a profess'd physician."[78] Among his medical colleagues at Oxford, Mayow seems to have been especially close to Millington, whom Mayow called both "learned and ingenious," and who gave Mayow a particularly interesting case to include in his *Tractatus quinque*.[79]

Wood also recorded that Mayow "became noted for his practice" in medicine, "especially in the summer-time, in the city of Bath."[80] Bath was just then beginning the rise to fashion that in the eighteenth century was to make it the most fashionable spa in Europe, and Mayow seems to have been one of those physicians who built their careers tending the well-to-do as they took the waters. This squares with the literary evidence; the *Tractatus quinque* contains not only some detailed comments on the city of Bath, and on the qualities of the Cross, Hot, and King's Baths, but also a chemical theory of how the springs there came to be so hot.[81]

It was in these two locales of Oxford and Bath, that Mayow performed his chemical experiments, and wrote them up in a greatly expanded version of his *Tractatus duo* of 1668.[82] The major addition was the first tract "On niter and the nitro-aerial spirit" ("De sal nitro, et spiritu nitro-aereo"), occupying 265 pages, more than the remaining tracts put together. Rewritten versions of "De respiratione" and "De rachitide," together with two other short tracts "De respiratione foetus in utero, et ovo," and "De motu musculari, et spiritibus animalibus," filled out the now lengthy work. Although it was published with the date 1674, the book seems to have been completed by the early spring of 1673. Thus Mayow noted in "De sal nitro" that since that treatise was written,

Boyle's "recently published experiments" on flame and air had "come into our hands," occasioning Mayow to intercalate a paragraph discussing Boyle's experiments on the difficulty of propagating the explosion of gunpowder *in vacuo*.[83] The work was given the Oxford University imprimatur dated 17 July 1673,[84] and published by the Sheldonian Press there. It was eventually distributed by John Crossley, an Oxford bookseller who had published previous items in the new sciences. It appears not to have been sold in London until mid-1674, appearing in the Term Catalogue there for 6 July, and being reviewed in the *Philosophical Transactions* of 20 July 1674.[85]

Although the work was divided into tracts, Mayow clearly conceived it in a single vision. If the *Tractatus duo* of 1668 showed Mayow as a competent anatomist and physiologist, the *Tractatus quinque* showed him, in addition, as a chemist and natural philosopher whose speculations ranged over almost all the phenomena of nature. The *esprit de système* that had given form and forcefulness to the *Tractatus duo*, created in the *Tractatus quinque* a work that was part genius and part fantasy. But in both, it was the natural outcome of Mayow's interpretation of the tradition within which he had been raised, and whose chemical and physiological problems he was attempting to solve.

It was a book with two overarching themes. In the first, Mayow developed in great chemical and physical detail the idea of active nitro-aerial particles and their ubiquitous function in the properties and transformations of matter. In the second theme, Mayow used the characteristics of his nitro-aerial particles to fill in, more forcefully and with greater clarity, the physiological system he had sketched out in the "De respiratione" of 1668.

The very opening sentence of his book reflected his Oxford background. It was obvious, he said, that the air surrounding us, although invisible and almost ineffable, "is impregnated with a universal salt of a nitro-saline nature, that is, with a *vital, igneous*, and highly *fermentative spirit*."[86] For Mayow, the question was not whether this substance existed in the air, but how its properties might be described, demonstrated by experiments, and used to explain natural phenomena.

His starting point, as Boyle's had been almost two decades before, was the nature of saltpeter. Mayow asserted that niter was composed of at least two parts, a fiery acid salt, and an alkali salt. This he proved not only by observing the distillation products of niter (acid spirit [nitric acid] and fixed salt), but also by citing Boyle's experiment of the "redintegration"—reconstitution—of niter. Charcoal could be added to melted niter until it was completely deflagrated, and then, by pouring spirit of niter (nitric acid) upon the remains, the niter could be recovered. Finally, Mayow said, this compound nature of saltpeter could be seen in the way niter was generated in the earth. It was formed, he said, echoing such writers as Hooke, Castle, Clarke, and Stubbe, by the combination of its "most volatile and subtle part" from the air, with the fixed salt component from the earth.[87]

Mayow's next step, once he had demonstrated the compound nature of niter, was to explore the properties of its *aerial* component. He rejected out of hand the notion that the aerial part of saltpeter might be a volatile form of the entire niter. As an alternative, he had once believed, he said, that the aerial part of niter might be the acid spirit of niter (nitric acid). But he was obliged to give up this idea because such a spirit was too heavy, and, more significantly, too corrosive, to exist in the air *per se*. In order to lead his reader to the proper identity of this aerial substance, Mayow engaged in a neat bit of dialectic. First of all, he said, citing the *New experiments physico-mechanical* (1660), Boyle's work had shown that there was something in the air necessary to flame. Mayow gave these particles the neutral interim name of "igneo-aerial." Moreover, these igneo-aerial particles constituted only *part* of the atmosphere, since much air was left in a closed container even after a flame had gone out. Now to saltpeter. Mayow first refuted Willis's contention, made in "De fermentatione" (1659) that niter contained an enclosed sulphurous particle. Having disposed of this error, which might make the activity of saltpeter depend on sulphur, one was now free to consider that it was only by means of these same igneo-aerial particles that gunpowder burned. This was indeed the case, Mayow said, reporting experiments showing that gunpowder could burn under water, where it had no access to the igneo-aerial particles of the atmosphere. Mayow could even, in the intercalated passage referring to *Flame and air*, use the igneo-aerial particles to explain why Boyle was not able to propagate the explosion of gunpowder. The igneo-aerial particles in saltpeter enabled the individual corns to explode, but the lack of igneo-aerial particles in the surrounding vacuum interrupted the continuity necessary for the propagation of the flame.[88]

Mayow could now close the circle. He had already demonstrated that a part of niter was derived from the air; now he could assert that this part was the very same igneo-particles that enabled gunpowder to burn in the absence of air.

> From what has been already said, it is, I think, to some extent proved that nitre contains in itself the igneo-aerial particles required for the production of flame. Wherefore, since some part of nitre is derived from the air and igneo-aerial particles exist in it, it seems we should affirm the proposition *that the aerial part of nitre is nothing else than its igneo-aerial particles.*[89]

One could further affirm that since nitric acid was derived from the volatile part of saltpeter, and not from its fixed salt, these same igneo-aerial particles were contained in nitric acid, probably joined with some other component that was liquid.

> But now since the aerial part of nitre exists in its acid spirit [nitric acid], but not in the fixed salt, which, as we have already shown, forms the rest of the nitre, we may conclude that *the igneo-aerial particles of nitre*, which are identical with its aerial part, *are hidden in the spirit of nitre, and constitute its aerial part.*[90]

It was true lawyer's logic: a part of the air supports combustion by its igneo-aerial particles; saltpeter contains igneous particles that support combustion without air;

saltpeter is compounded of an aerial component and a fixed salt component; nitric acid distilled from saltpeter leaves behind its fixed salt; therefore the essential component in all four processes is the same particle.

In this way Mayow could account for the properties of flame, of air, of niter, and of nitric acid, all by referring their action back to the same igneo-aerial particles. And, since he had now proved these igneo-aerial particles to be found in nitric acid and in saltpeter, as well as in the air, he felt justified in giving them their proper name:

> With regard then to the aerial part of nitrous spirit [nitric acid], we maintain that it is nothing else than the igneo-aerial particles which are quite necessary for the production of any flame. Wherefore, let me henceforth call the fiery particles, which occur also in air, *nitro-aerial particles* or *nitro-aerial spirit.*[91]

As we shall see, Mayow was quite clever in declaring two completely equivalent names for his active substance, because it allowed his very language to plead his case for him. When talking about particulate or corpuscular interactions, he would refer to his "nitro-aerial particles"; when speaking of the substance's chemical properties, he spoke of it as the "nitro-aerial spirit." The judicious use of each term was calculated to appeal to the mental images a physicist or chemist found congenial.

Mayow's next step was to establish the more precise role of the nitro-aerial particles in combustion. He did not mince words: "On this point my opinion is that the *form of flame is chiefly due to the nitro-aerial spirit set in motion.*"[92] In support of his contention, he rejected firmly the opinion of some "recent philosophers," that fire could be produced by the violent agitation of *any* kind of particles. Nor were sulphurous particles, which were always observed to be necessary for fire, the essential component of flame. Their function was the secondary one of throwing the "nitro-aerial particles into a state of rapid and fiery commotion."[93] Mayow went on to argue, from a number of chemical experiments, how the characteristics of fire—subtle, sharp, and caustic— proceeded from the properties of the nitro-aerial spirit. Moreover, he said, only by assuming that fire was composed of the nitro-aerial particles in rapid motion could one explain, for example, why a poker held in the fire became hot. The rapidly moving nitro-aerial particles entered the iron, and heated it, while the sulphurous particles did not, and were collected on the outside. Finally, in an almost prescient remark, he noted *en passant* that the absorption of nitro-aerial particles accounted for the increase of weight that was observed when antimony was calcined. In fact, he said, the role of niter in many practical laboratory operations was exactly that—to effect the calcination in the best possible way, by contributing the nitro-aerial particles whose motion was the essence of flame.[94]

Mayow was rapidly developing his nitro-aerial particle into a *primum mobile* of the physical world, and as such, had to deal with a conception which had occupied much the same sort of place in the world schemes of his Oxford predecessors—fermentation.[95] Mayow accomplished this by the simple process of encirclement. He argued that fermentation, wherever found, was nothing but the intestine motion of nitro-aerial particles when they came into contact with sulphurous and salt particles. The process of combustion was, in turn, seen as simply one kind of fermentation. He asserted that fire was "nothing else than an exceedingly impetuous fermentation of nitro-aerial and sulphureous particles in mutual agitation."[96] This was shown by the fact that all fermentations produced heat. They did so because they all depended, to a greater or lesser degree, upon the agitation of the nitro-aerial and sulphurous particles.[97] The motion of nitro-aerial particles at the base of all heat-producing reactions made them all, in a sense, fermentations. This view was to be most helpful in explaining the origins of animal heat.

Once begun, Mayow saw that phenomenon after phenomenon could be explained by the use of his nitro-aerial particles. Could heat be generated by friction? Then it must be because matter had nitro-aerial particles fixed like wedges into its pores. This same property could explain why niter cooled and frozen water expanded; the nitro-aerial particles entered into the substance of water and expanded it. He even used the idea of interstitial nitro-aerial particles to explain the resiliency and elasticity of bodies.[98] Was fire a property of the nitro-aerial particles? Then equally light must be. He developed, to a much greater extent than Willis's previous hints, a quasi-Cartesian theory in which the nitro-aerial particles were the medium through which the impulse of light was propagated.[99] The propagation and shape of flame could be explained in the same way.[100] By many of the same kinds of extensions, the nitro-aerial particles could also explain the nature of lightning and thunder, sea spouts, the heating of certain salts when mixed together, and the origins of heat in the underground springs at Bath.[101] Mayow had found his lever, now he was trying to move the world. His speculations in these areas, as opposed to his work on combustion and on respiration, were founded on far fewer observations, and no directly pertinent experiments. They were the exuberant fantasies of a man who thought he had found the key to the universe.

But if Mayow, in attempting to make his nitro-aerial particles the cornerstone of a system of natural philosophy, was led to propose increasingly more speculative mechanisms of nature, in one field—the explanation of elasticity—he hit upon the very area that both pneumatic chemists and physiologists had contrived to avoid. Of all the Oxonians, only Boyle had concerned himself during the 1650s and 1660s with the empirical stumbling block to all theories that posited that a nitrous substance was removed from the air in combustion and/or respiration. *Why did the air retain its normal pressure, or elasticity*? Mayow, once he had used his nitro-aerial particles to explain the resiliency of solid matter, had an answer; it did not. He cited, very rightly, the fact that although Boyle's experiments established the *fact* of the air's elasticity, they did not explain whence this elasticity derived. Mayow had the answer. Predictably enough, it was by means of nitro-aerial particles.

Mayow's starting point was a simple experiment that in other forms had been done by Bacon and Van Helmont, but never with the same technique, intentions, or results. Mayow put a burning candle into an inverted cupping glass, placed the glass in a trough of water, equalized the inside and outside water levels, and then let the candle burn until it went out. The inside water level would rise, thus proving that the enclosed air had lost some of its elasticity. He repeated the experiment under more exact conditions in which the combustible matter (camphor) was ignited, after the apparatus had been set up, by means of a burning lens. He found that the elasticity of the air was reduced by one-thirtieth (1/30). This loss of elasticity came about, he said, because the fire struck out the nitro-aerial particles from the air (see p. 264).[102]

We now know that Mayow's dramatic result was an experimental error. A burning candle gives off carbon dioxide and steam to replace, almost volume for volume, the oxygen it consumes. The steam condensed and the highly soluble carbon dioxide dissolved in Mayow's collecting trough, thereby reducing the volume, or "elasticity," of the original air. A similar "error," due to similar apparatus, almost certainly accounts for Hooke's slightly later observation that air was decreased in volume one-twentieth (1/20) by supporting combustion. Both results are within the same order of magnitude. Both accord well with the quantity of carbon dioxide released when a fire is extinguished in such circumstances, and agree within the limits of true experimental error in such rough-and-ready trials. Both experiments, although conducted independently, sprang from the common assumption of a nitrous substance that constituted a small part of the air, and which alone supported, and was consumed in, combustion. And in both cases it was a fruitful error, revealing the nature of the underlying process more truly than if the experiment had been conducted "properly." Boyle, with his finely honed instincts for the physical properties of the air, preferred always to estimate its elasticity by the more precise means of an enclosed mercury gauge. But it was just this technique, unbiased by the assumption of consumable aerial nitrous particles, that would never have revealed the existence of an oxygenlike substance in the air.

NITRO-AERIAL PARTICLES IN PHYSIOLOGY

The experiment of a candle over water provided Mayow with just the link he needed to convert what had been, up until that point in his argument, a physico-chemical theory, into a physiological one. Boyle, in his respiration experiments carried out in 1662−1663 and published in 1670, had carefully determined that an animal could die in a sealed container without the enclosed air losing any of its elasticity. We now know that, as in combustion, respiration

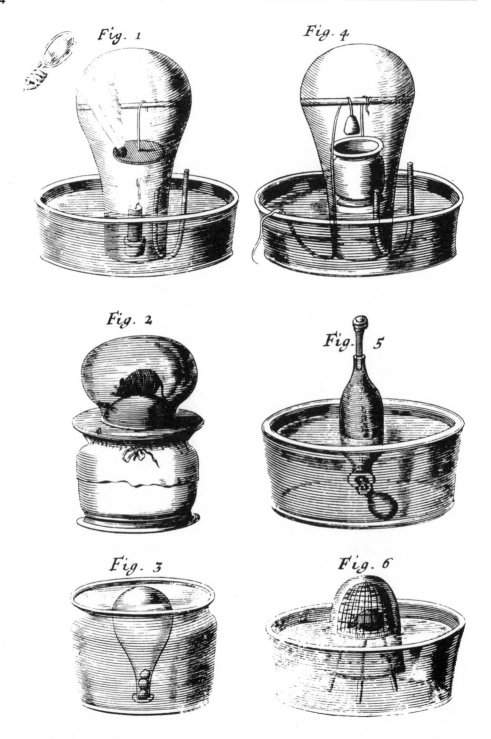

John Mayow's pneumatic experiments on combustion and respiration, *Tractatus quinque* (1674)

replaces the oxygen in the air almost volume for volume with carbon dioxide; most naturally Boyle would not have observed any loss in "elasticity" of the vitiated air. Mayow, although he never attempted to explain away Boyle's results directly, could now bring forward several respiration experiments in which a decrease in volume, and hence a loss in elasticity, were demonstrable.

If a bladder were stretched tightly across the top of a large jar, and moistened, and a small animal, such as a mouse, were put in a bell jar on top of the bladder, within a short time the bell jar would stick to the bladder. Within a longer time, the bladder would positively bulge into the bell jar. Mayow thought that the explanation of the phenomenon was clear; the elasticity of the air within the bell jar had been reduced, and hence could not resist the pressure of the outside air.[103]

Mayow had an even better experiment up his sleeve. To prove more clearly that respiration reduced the elasticity of the air, and moreover to estimate in what proportion the volume of the air was reduced, he replaced his candle flame with an animal. He put the animal into a cage that was suspended over the water in a bowl. He lowered an inverted glass over it, and equalized the external and internal water levels. He then raised the glass slightly, so that the internal water level would be more easily visible, and marked its place with pieces of paper on the outside of the glass. "Soon," he said, "you will see the water sensibly rising into the cavity of the glass," even though the animal's heat and exhalations "might be expected rather to produce an opposite effect." When the animal had died, Mayow marked the water level again. He then righted the glass and compared the amounts of liquid it held when filled to the two marks, for that was the extent to which "the air is lessened as to its elastic force and volume by the breathing of the animal." "In fact," Mayow said, "I have ascertained from experiments with various animals that the air is reduced in volume by about one fourteenth [1/14] by the breathing of the animals."[104]

Mayow drew the physiological conclusion of his experiments in very strong terms:

> From what has been said it is quite certain that animals in breathing draw from the air certain vital particles which are also elastic. So that there should be no doubt at all that an aerial something absolutely necessary to life enters the blood of animals by means of respiration.[105]

In equally strong terms, he rejected the opinion (of Malpighi, Henshaw, Needham, and Thruston, although unnamed by Mayow) that air did not enter the animal, and that respiration served merely to churn and divide the particles of the blood. In that case, *any* air should suffice, which it did not. Moreover, one had no reason, Mayow said, to deny the entrance of air into the blood, even though we could not see the ducts by which this was accomplished. In fact, he noted, if the lungs were boiled and dissected, inspection with a microscope showed an infinite number of minute openings that might be the ducts through which air was conducted into the blood.[106]

Mayow could then bring these two sets of experiments, the physical and the biological, together:

> Hence it is manifest that air is deprived of its elastic force by the breathing of animals very much in the same way as by the burning of flame. And indeed we must believe that animals and fire draw particles of the same kind from the air, as is further confirmed by the following experiment.[107]

He then related the experiment of putting both a lamp and an animal in the same container over water. The lamp went out first, and the animal died soon after. More significant, though, was the fact that the animal lived a little over *half as long* as it might have in the same circumstances if the lamp had not been present.[108] Mayow had done exactly the experiment tried at the Royal Society in 1663 and 1665, and reported by Boyle in *Flame and air* in 1673. But because his trough and cupping-glass experiments had shown that animal respiration could "consume" more of the vital aerial substance than fire, he waited to see exactly how much longer the animal would last. His patience was rewarded by observing that the animal's survival span was greatly decreased by including the flame.

Mayow could explain the fact that the flame went out first by positing that fire needed a more "abundant and rapid stream of nitro-aerial particles" than did an animal. The difference between the two reductions in elasticity, between one-thirtieth (1/30) and one-fourteenth (1/14), was consonant with this difference.[109] The logic was impeccable. Fire uses nitro-aerial particles and reduces elasticity; respiration reduces elasticity; therefore there is a probability that respiration uses nitro-aerial particles. Fire makes air significantly less fit for respiration; therefore it is *certain* that respiration uses nitro-aerial particles.

Mayow accounted for this relationship between nitro-aerial particles and elasticity of the air by an atomic-structuralist explanation. He assumed that aerial particles were bent bodies, to which the nitro-aerial particles were fixed. Mayow seems to have visualized an aerial particle in this manner:

The particle of air was, as it were, wrapped around the nitro-aerial particle.[110] Part of the elasticity of the air came from the restorative force of the air particle, and part from the action of the enclosed nitro-aerial particle. Flame caused air to lose its elasticity by removing these nitro-aerial particles:

> Thus we must suppose that the sulphurous particles of fire, when thrown into violent agitation, approach all the particles of air which are nearest them, and impinge on the nitro aerial particles which the air contains and by their collision drive them forcibly out, and that at last from these, violently ejected and in vehement commotion, fire is produced, as will be shown more fully below.[111]

This twofold origin of elasticity of air accounted for why, after a flame had gone out or an animal had died in a confined space, the air retained a portion of its normal volume. Only the component of spring derived from the nitro-aerial particles had been removed.

Mayow's brilliant exposition of the relationship between combustion, respiration, air pressure, and nitro-aerial particles, as well as the experiments with which he demonstrated his ideas, is an unsurpassed example of rigorous logic applied to an almost-truth. Because of his structuralist interpretation of the role nitro-aerial particles played in elasticity in general, Mayow *knew* they had to have something to do with the elasticity of vitiated air. Because he knew that had to be so, he focused upon the one set of experimental parameters that, we can say in retrospect, would give him those answers: the situation in which the carbon dioxide produced by oxidation would be absorbed in the water trough. Moreover, since he approached the experiment knowing the relationship he sought to demonstrate, he was not put off by the fact that the lamp and the animal did not expire at the *same* time. Wilkins, Hooke, Goddard, and Boyle had done the same kinds of experiments a decade before. But for them, the fact that the animal lived on after the flame went out meant that there was at best a doubtful similarity between combustion and respiration. Mayow, since his highly developed theory of the nitro-aerial particles told him there had to be a relationship, could treat exactly the same evidence in a different way. To him it was simply an indication of *differential* needs for his nitro-aerial particles. Most importantly, since his assumptions had led him to look for a loss of elasticity, he had conducted measurements upon the differential use of nitro-aerial particles (one-thirtieth [1/30] versus one-fourteenth [1/14]) that could then be brought in to shore up his contention that combustion and respiration were the same process, even though his predecessors had judged similar evidence to have told them otherwise. It was a magnificent vindication of the logic of theory-directed experimentation.

Mayow's experimental proof that something was absorbed by the blood in its passage through the lungs brought him to the point from which, in a sense, he had begun. His work had started in the late 1660s with the question of why respiration was necessary, and in answering that question he had been led to develop a theoretical construct—the nitro-aerial spirit—that turned out be the almost universal *explicans*. But doing so had taken him far from his original questions, and much of the remainder of the *Tractatus quinque* was involved in getting back to his physiological base. He did so in two ways. First, he wrote into the "De sal nitro et spiritu nitro-aereo" a chapter dealing with his more sophisticated ideas on the absorption of the nitro-aerial particles into the blood, and on the function they served there. Second, he rewrote "De respiratione," and parsed out into two subsidiary treatises the ideas on muscular motion and fetal respiration that had occupied only a few lines of the 1668 tract.

Mayow began his new consideration of the physiological role of the nitro-aerial particles with an odd question, and an even odder experiment. Where, he asked, were the nitro-aerial particles separated from the aerial particles—in the lungs or in the blood? He had come to believe the latter, and explained his belief by way of an experiment. One arranged a small vial of nitric acid inside an inverted cupping-glass, which stood over a trough of water. One further arranged the apparatus so that bits of iron could be lowered into the acid by means of a string running to the outside. When the iron was dipped into the acid, a reaction ensued, and the water rose up inside the glass, up to a limit of about one-fourth of the volume of the air. If, however, one continued the reaction, the water level would begin to fall again, a result that clearly baffled Mayow, since it meant that "opposite effects seem to be produced from the same cause."[112] Although he had to adduce a tortured explanation to account for the anomalous second half of the experiment, the initial findings demonstrated what he wanted to prove. It showed that a *fermentation* could reduce the elasticity of the air, and hence, by implication, strike out its nitro-aerial particles (see p. 264).

We now know that Mayow's crucial experiment was another "error." Although the reaction between iron and nitric acid is complicated by several secondary byproducts, it involves primarily the evolution of nitric oxide. This reacts with the oxygen in the air to produce nitrogen dioxide, a gas extremely soluble in water, which therefore rises to take its place. The diminution in volume—in Mayow's case 25 percent—is essentially a test of the oxygen content of the air, and was used as such by the late-eighteenth century pneumatic chemists.

Mayow's experiment, and the conclusion he drew from it, is however more than simply a "precursor" of Priestley's method of assessing the "goodness" of air, as is commonly assumed in histories of chemistry. Mayow would not have dropped so odd an experiment into his discussion of blood without intent. To him it proved that a fermentation taking place in a liquid could remove the nitro-aerial particles from the air. Mayow had found just the reaction he needed, to prove by close analogy that fermentation in the blood was capable of removing the nitro-aerial particles:

> After this experiment we must suppose that air when breathed by animals loses its elastic force in the following manner. For I assume, in the first place, that the mass of the blood is a liquid conspicuously in a state of fermentation, as will be shown below. Since then through the action of the lungs aerial particles are mixed intimately and in the minutest parts with its fermenting particles, it comes to pass that the aerial particles have their elastic force diminished by the particles of the blood in the same way as by the vapours of fermentation in the aforesaid glass.[113]

Indeed, he said, it was probable that the elastic component removed in this way was the nitro-aerial spirit, since the respired air was unfit for sustaining life.

Having thus established, in a beautifully contrived experiment, the link between nitro-aerial particles and fermentation in the blood, he went on to draw the conclusions for which he had long been preparing the reader. First, he asserted that the nitro-aerial spirit, mixing with the saline-sulphurous particles of the blood, excited in it a vital fermentation, one that was analogous to all those observed in nature, and which resulted from a motion of the nitro-aerial particles. Moreover, this might be seen in the experiment of the "illustrious

Lower,'' who had shown that the change in the color of the blood took place in the lungs. Just for good measure Mayow also mentioned, as Lower had done in exactly the same context, the Fracassati experiment.[114]

In addition, Mayow adduced two interesting and original experiments. If shed venous blood that had been allowed to stand for some time were put into Boyle's air pump, and then subjected to evacuation, *only the florid blood on the surface would effervesce.* But if arterial blood, entirely florid, were subjected to a vacuum, the whole mass would expand and rise ''in an almost infinite number of bubbles.''[115] Mayow's observation that it was largely the arterial blood that gave off a gas under reduced pressure is particularly significant. Lower's experiment had demonstrated that the change in color took place in the lungs, and that the change required access to fresh air. But he had not, strictly speaking, proved that something *entered* arterial blood. Logically, the result could as well have been explained by positing that something *left* the venous blood. Mayow's air pump experiment enabled one to decide between the two equivalent but opposite explanations. The presence of nitro-aerial particles in the arterial blood also explained why the venous blood was darker; the nitro-aerial particles had been separated at the periphery, and the fermentation of the blood had died down.[116]

Neither the experiments nor the conclusions drawn from them are to be found anywhere in Boyle's printed works. We know further that almost a decade after Boyle first built his air pump, he noted that men had great difficulty in making such an engine, and that he had heard of but ''one or two Engines that were brought to be fit to Work, and of but one or two New Experiments, that had been added by the Ingenious Owners of Them.''[117] It seems therefore all the more likely that the *new* air pump experiments reported by Mayow in the *Tractatus quinque,* of which there were other original ones in addition to the foregoing,[118] were done with Boyle's pump at Oxford, and may date back to when ''John Mayer'' was assisting Boyle with his experiments there in 1666–1668.

The absorption of nitro-aerial particles into the bloodstream during respiration, with the concomitant change in the color of the blood, thus provided an explanation for the origin of heat in the body. Mayow had previously argued that all substances, or at least all thick ones, grew warm while fermenting, and that fermentation was due to the motion of the nitro-aerial particles. He could now conclude that animal heat arose from a fermentation in the blood, caused by the saline-sulphurous particles putting the nitro-aerial particles into motion. Via the long route of practical and speculative chemistry, conjoined with physiological experimentation, Mayow had arrived at a response to the question that had gone unanswered since Oxonians had declared heat no longer to be innate. How was the heat in the organism generated? The answer was fermentation, in the very broad yet precise way Mayow had defined it.

Yet in declaring, as he did, for this mechanism, he had to deal with the opposite opinions of the man whose writings and teaching had made fermentation an accepted concept among the Oxonians—Thomas Willis. He was not unaware, Mayow said, that the "learned Dr. Willis," in his treatise "De accensione sanguinis," had argued that the heat of the blood was not due to its fermentation, indeed, that liquids never acquired heat in fermenting. Mayow asserted in response that most fermenting liquids did indeed grow somewhat warm, and the blood even more so, since it was so rich in saline-sulphurous and nitro-aerial particles. Willis had pointed to Boyle's experiments on the effervescing of liquids in vacuum, and said that these proved that fermenting liquids did better without air. Mayow denied that these experiments had any bearing on this case. The effervescing that was observed by Boyle was due, he said, simply to the expansion of the dissolved air, not to any fermentation. Therefore these experiments could not be adduced to disprove a fermentation in the blood.[119] After refuting a number of Willis's other arguments, Mayow vehemently rejected the idea of a vital flame, or enkindling of the blood:

> From what has been already said, it is I think in some degree made out that the fermentation of the blood, and hence also its heat, arises from nitro-aerial particles fermenting with its saline-sulphureous particles; so that we do not need to have recourse to an imaginary Vital Flame that by its continual burning warms the mass of the blood . . . [120]

Neither the blood, nor any part of the body, he concluded, was in any way constituted to produce and conserve fire or flame, as these were commonly conceived.

Mayow, in his ideas about the role of the nitro-aerial particles in the blood, and their place in the fermentation that caused animal heat, had in a sense outflanked the opposition. He had shrewdly seen that the difficulty of linking combustion and animal heat, of seeing the analogy between the two, lay in trying to think of one in terms of the other. One had to conceive of a fire breathing or an animal burning. We do not today think of an "analogy" between combustion and metabolism, in the sense that we explain one with reference to the other. We explain both by reference to a third, and more inclusive concept, oxidation. Mayow too, saw that exactly this kind of logical move was needed. He proposed that fermentation was that third entity, in terms of which the other two processes had to be considered. And since, in turn, he had defined fermentation as the motion of his nitro-aerial particles, he had a most logical way of explaining the necessity of air (or an air surrogate) for both processes. Mayow had seen the logical function of Willis's idea, in a way Willis himself had not.

The remaining portion of the *Tractatus quinque* was devoted to examining at greater length two other themes concerning the use of the aerial-niter concept that Mayow had adumbrated in the *Tractatus duo:* the respiration of the fetus, and the cause of muscular motion. Mayow also took the opportunity to rewrite the "De respiratione," inserting changes that brought it in line with his new ideas. For example, on all the occasions in which he had spoken of an *accensio* of the blood

(Willis's term), he inserted the word fermentation, thereby strengthening the links that bound blood to the process of heat generation. Whenever in 1668 he had referred to an aerial niter, or an explosive niter, he substituted the term nitro-aerial particle. He now had a much more unified conception of physiological processes, and he wanted the new edition of the "De respiratione" to reflect it.[121]

The "De respiratione foetus in utero, et ovo" made exactly the point that the reader had been led to expect; Mayow solved the Harveian problem of how the fetus respired by positing that it subsisted on a supply of nitro-aerial particles transferred to it from the mother via the placenta and umbilical circulation. After arguing that all the other putative functions for the umbilical circulation were either incorrect (e.g., cooling the blood) or secondary (e.g., nutrition),[122] Mayow set forth his own conception in very clear terms.

> These observations premised, we maintain that the blood of the embryo, conveyed by the umbilical arteries to the placenta or uterine carunculae, brings not only nutritious juice, but along with this a portion of nitro-aerial particles to the foetus for its support; so that it seems that the blood of the infant is impregnated with nitro-aerial particles by its circulation in the umbilical vessels, quite in the same way as in the pulmonary vessels. And therefore I think that the placenta should no longer be called a uterine liver but rather a uterine lung.[123]

He appended, by way of confirmation of this, his observation of 1668 that a transfusion of arterial blood seemed to obviate the need for respiration.[124] Boyle, in 1660, with a very vague notion of an aerial substance, had suggested that the fetus *in utero* might exercise some "obscure" respiration; for Mayow, with his strong conception of nitro-aerial particles, the solution of this longstanding embryological problem simply "fell out" as a necessary consequence.

Finally, one of the strong points of his treatise of 1668 had been Mayow's exposition of the systemic use of the aerial niter to power muscular contraction. He made the same point in 1674, albeit with changed emphasis. He greatly expanded his consideration of muscular motion into a separate treatise, "De motu musculari, et spiritibus animalibus," but retained the same outline of conceptions. Muscle contraction resulted, he said, from an interaction of the saline-sulphurous particles from the blood with the nitro-aerial particles that constituted the animal's spirits flowing in through the nerves.[125] Mayow did, however, change his ideas of muscular motion in two important ways. In the first change he reversed the sources of his interacting particles. He had implied in 1668 that the nitro-aerial particles were supplied by the arterial blood, whereas by 1674 he wished to argue that they came from the nerves. Whereas in 1668 he had believed that this interaction produced an effervescence that inflated the muscle, and hence shortened it, by 1674 he had seriously considered Lower's criticism— that the volume of a muscle did not increase in contraction—and had proposed a new idea. In the same way as heat could cause contraction of a gut string, so could the motion of the nitro-aerial particles—which was by definition heat—cause the contraction of the individual fibrils of a muscle, and hence of its whole mass.[126] In changing these mechanisms, Mayow was thereby bringing his ideas of muscular motion into line with the changed emphasis of his entire physiological system. In 1668 the integrating theme was the necessity of nitro-aerial particles for the "explosion" that powered muscles. In 1674 the integrating theme was the necessity of nitro-aerial particles for all aspects of *heat* (metabolism) in the body, of which muscular contraction was but one manifestation.

The ideas that Mayow argued in the *Tractatus quinque* were logical end points, both of his own physiological opinions of 1668, and of the chemical tradition that had informed the idea of an aerial nitrous substance since its first evocation at Oxford in the early 1650s. But in carrying forward these two lines of intellectual development, Mayow the physiologist had ended up as Mayow the chemist and natural philosopher. Only by giving the notion of an aerial niter a fully corpuscular form, and following out logically the implications of its atomistic nature, could Mayow have knit together so many diverse lines of observation, experiment, and speculation. Only by judicious use of ingeniously designed experiments, such as his "proof" that both combustion and respiration absorbed nitro-aerial particles, could Mayow make the link between combustion and metabolism in the strong way that he did. This, in turn, provided him with the conceptual basis for uniting the two processes under the term fermentation, and assigning to his beloved nitro-aerial particles a crucial role in both.

Chemistry played a vital role in this transformation, both the speculative corpuscular chemistry of Boyle, and the practical chemistry taught in dozens of informal classes in the university town. And it was in this area, rather than in the anatomy and physiology with which Mayow had first been concerned, that he did his most original thinking. He displayed the same kind of virtuosity with reagents, cupping glasses, and pneumatic troughs, that Lower had shown with a dissecting knife.

If John Mayow brought to a logical end many of the diverse threads of thought that had run through the work of the Oxford community since the 1640s, he did so in a way that—while sharing some of the philosophical, scientific, and personal hallmarks of the Oxford group—was peculiarly his own.

Although he often quoted Descartes, and used some Cartesian explanations in his physical theory, the units that comprised Mayow's system were, in their format, a straightforward application of the Gassendian scheme of atoms and molecules to the problem of chemical composition. Even his theory of the nitro-aerial particle giving elasticity to the enclosing aerial particle was an idea that was well within the "structuralist" tradition. Yet even if this part of his corpuscular philosophy was quite orthodox and "mechanical," another part was less so. Mayow clearly assumed that ultimate particles had distinct properties,

properties that were not derived from their magnitude or figure. In this sense, in his belief in the real distinctiveness of saline-sulphurous corpuscles, or nitro-aerial particles, he was closer to the corpuscular philosophy of Boyle, and even more so to that of Willis, than to the French mechanists. He was an atomist, but more a chemical atomist than one who looked to matter's mechanical properties for an explanation of change.

Mayow was clearly much more at home with the work and conceptions of some of his fellow Oxonians than with others. In both the *Tractatus duo* and the *Tractatus quinque*, Boyle and Lower were mentioned in the highest of terms. In the latter work Boyle is named more than any other single author, and almost as many times as all the Continental writers put together. Hooke and Millington were also cited approvingly in the *Tractatus quinque*. But Mayow's attitude towards Willis seems to have undergone a change between 1668 and 1674. In his first book Mayow spoke well of Willis and mentioned his theories on several occasions. Indeed, the overall shape of Mayow's physiology, especially in its emphasis on the active role of an aerial nitrous agent, was molded by the Willisian approach so popular among Oxford physicians in the late 1650s and through the 1660s. But by 1674, after Willis had declared so forcefully that fermentation was not a source of heat, and that there was a vital flame in the blood, Mayow's references to Willis, while remaining formally courteous, were almost invariably critical. He even went so far as to write into the revised edition of the "De respiratione" a number of additional comments, mostly critical, on Willis's theories, which had appeared in books published even before 1668. Mayow, in coming to his own ideas about the nature of the aerial nitrous substance, had most definitely parted ways with Willis.

Technically, Mayow was very much the heir of the Oxford tradition. His extensive use of chemical evidence interpreted in a corpuscular way, the experiments he did with the air pump, his research on the chemical and physical properties of gases—all had strong precedents within the work done at Oxford in the previous two decades. If anything, Mayow was unusual in the degree to which he took up Boyle's interest in gases, a question that intrigued few of the anatomists such as Lower or Willis. Although Mayow's initial work had been in anatomy and physiology, his original contributions in vivisection were slight compared to his work in chemistry.

As should be most manifestly clear, Mayow shared completely in the Oxford tradition of concern for certain areas of the biological sciences: the nature of the air, its relation to respiration, the physiological functions of the blood, the origins of heat, and the applicability of embryological evidence. In all of these areas, he made major contributions.

Yet in spite of this—the degree to which Mayow was heir to the natural philosophy, techniques, and conceptual concerns of over a quarter-century of work—he remains, somehow, unconnected. Part of the reason was chronological. The generation of scientists who lived and worked at Commonwealth and

Restoration Oxford was one composed of men born in the period 1615—1635, who had come to scientific maturity even before Mayow arrived in Oxford. In turn, many of them had left Oxford before his ideas were fully developed. Mayow, born in 1641, was moving into science just as many of the members of this core Oxford generation were moving out. One could make the case that much of the novelty of his ideas depended upon the fresh attitudes with which he approached old problems. But his membership in a different generation had, nonetheless, the effect of isolating him temporally.

To this must be added a degree of geographical isolation. Mayow did his best work at an Oxford bereft of many of the minds that had graced its golden age. His other pied-à-terre was not London and the Royal Society, but Bath. And although the spa town was one of the more interesting of the provincial centers, it by no means had a first-class community of virtuosi.

But the primary point of discontinuity was methodological. The research of almost all the Oxonians had expressed a consensus of caution best enunciated in the numerous works of Robert Boyle. Experimentation and the testing of limited hypotheses were to be the order of the day. Progress was only to be had by rejecting the systematizing of the ancients, and of those moderns such as Descartes, who, while rejecting the contents of scholastic philosophy, had succumbed to its spirit. Mayow, especially in his *Tractatus quinque*, had violated that unspoken injunction; he had constructed a system. It does the historian little good to protest that only with this inclination could he have put together all the disparate parts of his conceptual patrimony. He integrated them into a system of relations whose similarities to modern ideas of oxygen, combustion, and respiration have made Mayow an object of fascination to generations of historians of chemistry. In a sense, posterity has rewarded John Mayow for those very qualities of which his contemporaries were suspicious.

THE
DECLINE OF
THE OXFORD
TRADITION

On 10 July 1674, just a few days after the announcement of John Mayow's *Tractatus quinque* had appeared in the London term catalogs, Oldenburg dashed off a newsy letter to Boyle that included the following:

> I hear some very Learn'd and knowing men speake very slightingly of ye Quinque Tractatus of J.M.; and a particular friend of yrs and mine told me yesterday, yt as farr as he had read him, he would shew any impartial and considering man more errors than one in every page.[1]

At a distance of three centuries, Mayow's errors in chemistry seem no more egregious than those of his contemporaries, and his insight and experimental skill considerably greater. But for his colleagues, Mayow's deviance from accepted opinions may well have been distracting.

Oldenburg reflected this ambiguity in the review that appeared in the *Philosophical Transactions* less than a fortnight later.[2] He was clearly fascinated with the book, for his review spans thirteen pages, more space than Oldenburg had given any book since he founded the periodical. Such attention stemmed both from Oldenburg's long-standing concern to promote studies on air, niter, and respiration, and from his knowledge that Boyle had a scientific stake and (if the identification of "I. M." be correct) a personal involvement with the issues and the author. Yet Oldenburg also knew that Mayow had made many bold assertions and sweeping judgments that he, as a bystander but no chemist or physiologist, lacked the expertise to assess. Moreover, Mayow had indulged in the kind of system-building that Oldenburg, in reviewing such books in the past, had dismissed with mildly contemptuous brevity.

The review itself was both fair and detailed almost to a fault. Oldenburg took over eight pages recapitulating "De sal nitro" alone, following Mayow's arguments step by step: the nature of niter and nitric acid; the necessity of air to fire; fire as nitro-aerial particles in motion; the role of nitro-aerial particles in fermentation; the generation of niter in the earth; how nitro-aerial particles caused rigidity and elasticity, especially of the air; how this elasticity was impaired by combustion and respiration; the role of nitro-aerial particles in animal fermentation and incalescence; the propagation of fire via nitro-aerial particles; their role in waterspouts, light, colors, thunder, lightning, slaked lime, and thermal springs.[3] Oldenburg pointed out Mayow's indebtedness to Boyle for some of his experiments, and to Descartes for some of his cosmological conceptions, but no more so than Mayow had already acknowledged himself.[4] Passing over "De respiratione" and "De rachitide," since they had already been reviewed as *Tractatus duo* in 1668, Oldenburg detailed Mayow's views on fetal respiration, and emphasized how Mayow thought he had solved "that difficult *Quaere*," why the extrauterine fetus could live for several hours after birth as long as its membranes and placenta were intact.[5] Oldenburg closed with a summary of Mayow's views on the role of nitro-aerial particles in muscular contraction. In all, it was an eminently fair review, done in Oldenburg's usual way, and at greater-than-usual length because of his interest in the subject.

What it did not include was Mayow's evidence. Neither the protocols nor the results of Mayow's brilliantly conceived experiments found their way into the *Philosophical Transactions*. In part this was simply Oldenburg's style of reviewing. In part it may also reflect Oldenburg's awareness of the doubts already cast on Mayow's chemical skills. But in large part the nature of the review derived from the structure of the book itself. Boyle had always put his experimental findings first, and tucked his theories into comments or "Digressions"; this style is reflected in the very titles of Boyle's books, which often contain the words "Experiments," "New Experiments," "Experimental," "Natural History," or even "Natural Experimental History." Mayow, on the other hand, had integrated his own experiments into a complete system, and Oldenburg drew the reasonable conclusion that the system transcended the evidence. In a sense he was right.

The author himself visited London from Oxford a few months after the review appeared. In late October, Mayow and Hooke talked over coffee at Child's, one of Hooke's favorite haunts.[6] Two weeks later, on 11 November 1674, Mayow visited Hooke at Gresham where, as Hooke recorded in his diary, they talked about a subject of mutual interest:

> Dr. Mayow Here. I gave him motion of θ [Robert Hooke, *An attempt to prove the motion of the earth from observations* (London, 1674)]. Discoursd of Air.[7]

Later that afternoon Hooke bought Mayow's *Tractatus quinque* ("Mayow. 2s.6d"),[8] and must have read it immediately, because the next day at the Royal

Society, after Wallis presented a paper on gravity, Hooke turned the subsequent discussion to elasticity:

> It being, among other particulars in discourse upon the reading of Dr. WALLIS's paper, remarked, that the explication of the cause of springiness would contribute very much to illustrate the nature of air, Mr. HOOKE said, that he had considered that subject, and particularly to make a springy body out of a body not springy.[9]

Or, as Hooke himself recorded in his diary: "Neer 40 of the Society met I promisd Discourse of Spring of body of Air &c."[10] This episode had a famous outcome. Hooke looked back over his notes and found some ideas and experiments on spring he had put down c. 1660. He then worked out apparatus to demonstrate the relationship between applied force and elasticity.[11] In 1675 Hooke showed these experiments to the King at Whitehall. In 1676 he published an anagram embodying his discoveries,[12] and in 1678 revealed the nature of that relationship: "Ut tensio sic vis"—known ever since as "Hooke's Law,"[13] which, like "Boyle's Law," is dutifully learned by every schoolboy. Mayow's concepts of the role of nitro-aerial particles in elasticity may not have convinced Hooke, but they seem to have prompted him to vindicate his own ideas in a uniquely fruitful way.

Hooke continued to see Mayow in the following months of December 1674 and January and February 1675, mostly at Garaway's coffeehouse.[14] At one of these gatherings Mayow told of "Experiments of Weighing," unfortunately not further specified in Hooke's diary.[15] On 8 December 1674 Hooke "Cald at Dr. Mayow," and two days later was his host at a Royal Society meeting at which Nehemiah Grew read a discourse on the nature of chemical mixtures; as Hooke noted it in his diary: "Cole, Mayow, &c., present at dispute of the figure of salt."[16]

By mid-February 1675 Mayow had returned from London to All Souls, and to his practice. His patients at Bath and in the west country seem increasingly to have absorbed his attention in the succeeding years. In mid-1676 "Dr. *Mayow* of *Bath*" was consulted by one Anthony Williamson of Cornwall, who, in swallowing musket balls to cure himself of colic, had lodged one in his lungs. Mayow prescribed hanging upside down, and breathing various fumes to force the bullet out by coughing, but this therapy was unsuccessful in expelling the object.[17] Other than this one oddity, we know nothing of Mayow's professional life. In 1677 and 1678 he was at All Souls only a few weeks each year, and resigned his fellowship altogether in early 1679.[18]

Mayow visited London only intermittently in those years. Hooke had coffee with him at Garaway's in March 1676 and again in November 1677.[19] In late January and early February 1678 Mayow once again attended several meetings of the Society, at one of which a letter describing the outcome of the Williamson case was read; Mayow commented that Millington had treated a similar one.[20] A fortnight later, on 7 February 1678, Hooke proposed Mayow for fellowship in the Society, an honor to which he was elected *in absentia* at the anniversary meeting the following autumn, but to which he was never admitted.[21]

Mayow's last days were touched with unhappiness of his own making. The Oxford antiquary and gossip, Thomas Hearne, noted in his diary in 1706:

> Dr. Mayow of All Souls College was a very ingenious man, & an Excellent Scholar; but by resolving to marry a Wife of a Great Fortune fatally miscarried to his unspeakable and insuperable Grief. For when he was one Year at ye Bath, and happening to lodge at ye same House with an Irish Lady and her Daughter, as was pretended, who went for vast Fortunes, when ye old Lady feigning an occasion went to London, he courted the young one and quickly married her: but she proving nothing like ye Fortune yt he imagin'd he was so confounded with it yt he did not long enjoy his life.[22]

This would be in the spring of 1679, when he resigned his fellowship at All Souls, and moved from Bath to London. There he lived "in an apothecary's house, bearing the sign of the Anchor,"[23] in York Street, Covent Garden, a few doors from Walter Charleton and a block from Richard Lower in Bow Street.[24] Mayow died there, aged 37, about 30 September 1679, as Anthony Wood recorded, "having been married a little before, not altogether to his content."[25] Mayow had not even visited his coffeehouse friend in those months. Hooke heard the news only in early November, when he visited Boyle for the first time since mid-September; "Mayow dead a month since," he wrote in his diary.[26]

FRAGMENTATION, DEATHS, DISTRACTION, AND DISPERSION

Nineteenth- and early twentieth-century writers were genuinely perplexed that the lines of investigation worked out by Boyle, Hooke, Lower and especially Mayow, on such important topics as air, blood, respiration, and combustion, failed to be followed up.[27] The idea of particles in the air having the chemical and physiological properties of oxygen seemed, on first glance, to have been obliterated by phlogiston theory in chemistry, and by iatromechanics in physiology, not to reemerge again until the 1770s. Nitrous aerial particles, that substance that was oxygen and many other things besides, seemed to have disappeared from the globe.

We know now, thanks largely to the investigations of Partington in the history of chemistry, and Hall in the history of physiology, that such an image—of the English physiologists crying in the desert—is false. Before his "rediscovery" in the late 1790s as an unappreciated forerunner of Lavoisier, Mayow was read closely, and referred to, by many eighteenth-century chemists.[28] His views were likewise discussed by many early eighteenth-century physiologists, including

Boerhaave and Haller, only to be dismissed for reasons that seemed at the time to be perfectly scientific.[29]

Even if one looks at the dissemination before 1710 of Mayow's ideas—admittedly the most extreme statement of the Oxford approach to physiological problems—one finds no ignorance of his views throughout the Continent. The *Tractatus duo* was reprinted twice at Leyden, in 1671 and 1708. The *Tractatus quinque* was reprinted at the Hague in 1681, and translated into Dutch in 1683. Parts of it were included in Manget's widely circulated *Bibliotheca anatomica* of 1685 and 1699.[30] In the Low Countries Mayow was cited by Diemerbroeck (1672), Herfeld (1683), Verheyen (1693 and 1710), and Barchusen (1703); in France by Lemery (1675), Mariotte (1676), Littre (1689) and Viridet (1691); and in Italy by Barberius (1680) and Mistichelli (1709).[31] Mayow's work attracted a great deal of attention in German-speaking Europe, being used by Pechlin (1676), Ettmuller (1676 and 1685), Neukranz (c. 1682), König (1682 and 1686), Bohn (1685), Bernoulli (1690), and Morhof (1692).[32]

Curiously enough, several of the Germans who were most attentive to Mayow's work had an English, and specifically Oxford, *Verbindung*. Johann Bohn had visited London in the early 1660s, and both he and König corresponded with the Royal Society. Michael Ettmuller was a young man of twenty-four when he spent a year at Oxford in 1668.[33] There he was influenced by Boyle, and was acquainted at the very least with Wallis, who passed along letters to him from Oldenburg in London.[34] In Ettmuller's lectures as professor of medicine at Leipzig in the early 1680s he especially cited the work of Mayow and Boyle as showing that the air was necessary for both fire and life, and that if it were altered or lacking, both fire and life would be extinguished. Ettmuller also cited Mayow's theory that the nitro-aerial spirit mixed with the blood, where it interacted with the salino-sulphurous particles, and thereby excited a vital fermentation that was necessary for animal life.[35] His student, Zacharia Neukranz, presented under Ettmuller's presidency a thesis in which Mayow's views were discussed in detail, and subjected to some penetrating criticisms.[36] Daniel Georg Morhof visited England c. 1660, and again in 1670, where he became acquainted with Boyle; he is even said to have studied briefly at Oxford. When he returned to the Continent, he translated into Latin and published one of Boyle's tracts, complete with a laudatory poem.[37] Later, in his widely read and often reprinted *Polyhistor literarius* (1688–1692), Morhof praised Mayow's *Tractatus quinque,* and gave long summaries of the Englishman's theories of combustion, the nitro-aerial spirit, the origin of cold, the heat of Bath-water, and the origins of whirlwinds, while rejecting Mayow's theory that light was transmitted by nitro-aerial particles.[38]

Within England proper, the published references to Mayow's work tended likewise to come out of Oxford, and from among his friends in particular. Thomas Guidott, fellow Wadhamite and spa-practitioner, brought out in 1676 a *Discourse of Bathe,* in which he examined Mayow's opinions on the waters there. In general, Guidott accepted many of Mayow's views. He disagreed, however, with Mayow's assertion that the waters of Bath contained no niter. He quoted Mayow at length,[39] and then spent the next dozen pages recounting the chemical evidence that proved Mayow had been deceived.[40] Guidott even complained a bit peevishly that Mayow had incorporated some of Guidott's ideas on niter, without acknowledgment; "Mayow hath ploughed with my Heifer," Guidott said.[41]

Another Oxford friend of Mayow, Edward Browne, was a bit more chary of using nitro-aerial particles. In his lectures before the Company of Surgeons in 1675–1678,[42] Browne cited approvingly Lower's arguments on the structure and motion of the heart.[43] He went on to discuss the structure of the lungs, explaining how the pulmonary artery divaricates to encompass numerous "little bladders of the lungs."

> Although this be an artery yet the blood in it is not so florid as in the veyne of the lungs which some attribute to the mixture of nitrous particles brought into the lungs by inspiration and mixed with the blood which returned to the heart by the veyne and makes it more florid.[44]

Browne further noted that injections made into the pulmonary artery did "not so easily" return by the vein as in other parts of the body. He attributed this to the "minute small ramifications of these vessells," and also to the propensity of the blood "to froath in the lungs." In this passage across the lungs, Browne believed, the blood from the pulmonary artery also threw off its "fuliginous parts."[45] In all, it seems that while Browne took notice of the Oxford ideas of nitrous particles, he inclined more towards the opinion of his fellow Cantabrigians Needham, Croone, and Thruston, that respiration served largely the mechanical and depurative functions.

When Robert Plot, Mayow's fellow-student in Wildan's chemical course, came to recount the scientific glories of the university town in his *Natural history of Oxford-shire* (1677), he did not neglect the Oxford physiologists. Plot defended Joyliffe's priority in discovering the lymphatics, and praised Highmore's anatomical discoveries and defense of Harvey.[46] He recapitulated Willis's doctrine of fermentation, which Plot himself had "embraced" as the "most universally known and received" explanation of natural phenomena.[47] Plot likewise praised at length the discoveries of Willis in cerebral anatomy, on the soul of brutes, and on the anatomical basis of pharmacological action. But it was "too difficult a task," he said in reference to Willis, "to give a just account how far *Physick, Anatomy, Chymistry,* and *Philosophy,* stand indebted to *him* for their *Improvements.*"[48] Plot noted many of Wren's instruments, including one "of *Respiration,* for straining the *breath* from thick vapors," to make it suitable for rebreathing. But in our knowledge of the air, he said, nothing has assisted us more than "the *Pneumatick Engine*" invented at Oxford by Boyle, "that miracle of ingenuity," with the "concurrent help of that exquisite contriver, Mr. *Robert*

Hook.''[49] Plot then went on to defend Wren's priority in injection experiments, and noted that Lower's successful transfusion had arisen out of such trials. He recapitulated Lower's discoveries in *De corde*, especially his conclusions on the speed of the circulation of the blood, "and the cause he assigns of the *florid colour* of it when emitted."[50] The "ingenious *John Mayow*" had taught us that "the Air is impregnated with a *Nitro-aerial Spirit*," whose motion was "*Fire* it self," and that all fermentations depended on this spirit. These and "many other *Phaenomena of Nature*," Mayow had "ingeniously deduced from his *Nitro-aerial principles,* and confirmed them by *Experiments.*"[51]

Other Oxonians, although not known to have been friends of Mayow, likewise used the respiratory aerial niter. Henry Mundy, educated at Oxford in the late 1640s and early 1650s, and a practitioner at nearby Abington, summarized Mayow's views in his book on the "vital air" in 1680. William Coward and Bernard Connor, Oxford educated in the late 1670s, both recapitulated Mayow's ideas in books published in 1695.[52] In all, it seems that the idea of nitro-aerial particles hung in the Oxford air long after Willis, Lower, and Mayow had left.

In London the idea of an aerial nitrous substance lingered on in the lectures and sporadic experimental work of Hooke. One manuscript composed in the mid-1670s, the "Philosophical Algebra," shows the distinct influence of Mayow. Hooke posed his ideas in the form of queries on the air. Did not, he asked, nature generate heat according to the same principles in such diverse phenomena as the sun, fire, motion, fermentation, hot springs, quicklime slaked, and animals? Were they not all due to a dissolving, or "menstruum" action, of which the "Nitrousness" of the ambient air was the cause?[53] Was not the "heat in Animals" caused by the intestine motion of body fluids

> and more especially by the uniting of the Volatile Salt of the Air with the Blood in the Lungs, which is done by a kind of Corrosion or Fermentation, which to me I confess seems somewhat more than probable . . . [?]

Hooke then marshalled the evidence for this view: (1) animals that breathe great amounts of air were very hot; (2) exercise that caused the blood to circulate more, to make the animal breathe more, also increased the heat and copiousness of the expired vapors; and (3) animals that moved little, and whose blood circulated slowly, were cold, and breathed seldom. Hooke even suggested that the "Ebullition of Steams into the Lungs, which are carried out with the Breath by Expiration," resulted from a mixing of the blood with "the Salt in the Air."[54]

Hooke and Wren rang changes on the same theme in a series of discussions, both in private and at the Royal Society, in the winter and spring of 1678. Hooke's revived interest seems to have been occasioned by the confluence of reading Ettmuller's book on respiration,[55] renewed studies on air pressure at the Royal Society,[56] a visit to the Society by Mayow,[57] and the fact that Wren composed a theory of respiration and muscular motion that he had delivered to Boyle.[58] The resulting discussions at Gresham in late April and early May touched the same old themes—fetal respiration, the pulmonary insufflation and rebreathing experiments, animals in the vacuum pump, and the role of air in muscle contraction.[59] Thomas Henshaw defended his dead brother Nathaniel's old theory that respiration served to promote the circulation of the blood, which Hooke patiently demolished in order to lead the Society to the conclusion "that the principal use of respiration is for the mixing the nitrous parts of the air with the blood; which part of the air being once spent and separated from it, the remaining part thereof is altogether useless for that purpose."[60] A year later, in January-March 1679, there was a reprise of the aerial niter, this time in its chemical aspects. Hooke repeated many of his now-standard combustion experiments (candles in vitiated air, coals heated in closed containers, melted saltpeter with coals thrown in, etc.) to demonstrate that the nitrous substance in saltpeter and in air was responsible for combustion.[61]

Hooke returned to the same themes in his "Lectures of Light" and "A discourse of the Nature of Comets," read at Gresham in 1680–1682.[62] His treatment was the *Micrographia* revisited and expanded: how the aerial niter, acting as a menstruum for sulphurous bodies, created the fire of the sun, the luminosity of comets, thunder and lightning, fire on earth,[63] and how the heat of an animal depended on respiration, which was in turn dependent not on the . motion of the lungs, but only on access to the nitrous substances contained in fresh air.[64] Hooke had found his variant of the Oxford niter, and was sticking with it. Even in 1690 the mention of niter brought forth the old topic in coffeehouse discussion: "With Sr. Christ. Wren at Mans: theory of Niter air flame," Hooke wrote in a diary fragment.[65]

What is significant about these recurring echoes at London, Oxford, and Bath, is not just the tenacity with which a concept holds in the minds of men, but also its failure to elicit any further experimentation. The ideas of air, blood, heat, niter, respiration, and fermentation that had been so eagerly scrutinized over the two decades that separated Bathurst's "Praelectiones" from Mayow's *Tractatus quinque* no longer moved men, either individually or collectively, to ingenious designs of apparatus and experiments.

Deaths exacted part of the toll. John Wilkins spent less and less time at the Royal Society after his promotion to the Bishopric of Chester in 1668, and died in London in 1672; Lower attended him in his final days, and Needham did the postmortem.[66] "Chester is a kill-bishop," Anthony Wood moralized.[67] Timothy Clarke died in the same year, his long-promised treatise on injections surviving not even in manuscript. Peter Stahl, the chemist of the Oxford club, died unnoticed in London sometime in the early 1670s. George Castle died in 1673, and his position as Physician to the Charterhouse devolved through the old-boy network to his fellow Westminster, Needham. Malachi Thruston was removed from the list of sympathetic provincial virtuosi at about the same time; he went insane, and had to be cared for by his family in Somerset until his death in 1701. Thomas Willis, never active in cooperative work after he came to London, except privately with Masters and King, died in 1675 and was given a scandalously extravagant funeral at Westminster Abbey that befitted a man who was both a fashionable physician and a scientific investigator known throughout Europe. Henry Oldenburg, who since his Oxford sojourn of 1656−1657 had so fruitfully rubbed one man's mind against another, died in 1677 mourned by all except Hooke, with whom he had feuded bitterly for the previous two years. Mayow died obscurely in 1679.

Just as devastating to the Oxford research tradition was the less dramatic withdrawal of quondam anatomists, chemists, and physiologists from cooperative scientific work, and often from investigations altogether. One by one, they went their separate ways, distracted from experiment and even discussion by professional duties or personal inclinations.

In the early 1670s Richard Lower continued to revise his *Tractatus de corde* for successive editions.[68] These were always duly noted by Oldenburg in his correspondence and in the *Philosophical Transactions*.[69] Lower even continued to present copies to Christ Church and to the Bodleian. But he had long since become too busy a practitioner to carry on the work of his younger days. In the few years after Willis's death, he became London's premier physician, "and no man's name was more cried up at court than his."[70] He moved to successively more fashionable addresses[71] until the outbreak of the "Popish Plot" in 1678, when he followed his political convictions and sided with the Whigs. This lost him his following in the strongly Tory court.[72] If one may credit the distorted testimony of two physicians he snubbed, Lower became increasingly imperious with some of his fellow physicians. He found their ignorance and incompetence inexcusable; they found his arrogance, based upon his reputation as an anatomist, insufferable.[73] He continued on good terms, however, with his old and respected friends, such as Needham, Millington, and King, and often consulted with them.[74] But cooperative practice never led to cooperative science. After the Glorious Revolution of 1688, of which he understandably approved,[75] Lower advised the Crown well on the organization of naval medical services.[76] He died in his house in King Street, Covent Garden, soon thereafter, in January 1691.[77]

Of the Westminsters in the Oxford group who were physicians, only Walter Needham stayed intermittently active in the Royal Society, and that only until early 1678. He contributed several papers on anatomy and physiology to the *Philosophical Transactions,* served c. 1673−1675 as anatomical lecturer to the Company of Surgeons, and even wrote a book, "Institutiones medicae," that never saw print.[78] He too died in 1691, three months after Lower, and not a quarter of a mile away. But Needham, alone among those active at Oxford in the 1660s, had not accepted the supposition of an aerial nitrous substance. One of the few men who might have carried on the Oxford anatomical traditions, he had his own rather Cambridge view of its contents. Needham never did anything again on respiration.

Wren and Locke had rather different distractions. After his appointment as

Royal Surveyor in 1669, Christopher Wren became almost totally immersed in the task of rebuilding London, thereby creating for himself a reputation as the English architect par excellence. Although in his sporadic appearances at the Royal Society he sometimes referred to his earlier anatomical investigations,[79] and was even moved in 1678 to write up a theory of respiration and muscular motion (now lost), he did nothing new in the sciences that had fascinated him in his younger days. John Locke moved rapidly away from science and medicine in the late 1660s, to become embroiled in the Whig politics of his patron, the Earl of Shaftesbury. He continued to read intermittently in medicine, to jot down items in his notebooks, to talk with Boyle, and even to correspond occasionally with the Royal Society when political expediency dictated a trip to the Continent. Nitrous aerial particles cropped up occasionally in his commonplace books,[80] and even in a primer on the "Elements of Natural Philosophy" composed in the 1690s for a tutee,[81] but it was no longer a subject of active interest. Locke had long since committed himself to political theory and epistemology, of which he became as quintessentially English a representative as Wren of architecture.

Thomas Millington moved to London in 1676, and rose rapidly to eminence as a clinician. He delivered the Harveian oration in 1679,[82] was knighted, and before his death in 1704 had become a respected and powerful president of the College of Physicians. Millington's Westminster classmate Nathaniel Hodges had worse luck. His London practice fell on hard times, and he died in debtors' prison in 1688. Other Oxonians—Robert Sharrock, David Thomas, William Petty, Nathaniel Highmore, Thomas Guidott—lived out the rest of their lives largely in the provinces.

That left, besides Hooke, only Boyle still engaged in science. He was of little help. After attending the Royal Society with decreasing frequency in the early 1670s, he stopped altogether in early 1674,[83] ironically only a few months before Mayow's book appeared. From then until the end of his life, I can find his presence at the Society recorded less than half a dozen times, most of which were to present copies of his books.[84] In 1680 he was elected President, but declined by letter.[85] Boyle continued to send his books to the Society, while Hooke in his frequent visits to Pall Mall no doubt kept him informed of the proceedings, but Boyle's only continuing influence there was a very indirect one through his assistants, Denis Papin and Frederick Slare, who occasionally reported experiments done in his laboratory. Boyle, who in the 1650s and 1660s had been so prominent a catalytic presence among his colleagues at Oxford and London, withdrew completely.

He carried on investigations, largely chemical, but few of them bore upon the subjects he had so brilliantly illuminated in the 1650s and 1660s. His *Second continuation* of the air pump experiments, reporting experiments carried out with Papin's help largely c. 1676–1679, had nothing new about respiration.[86] His *Memoirs* (1684) on human blood, gathered together from old notes at Locke's request, shed no new light on the relation of air and blood.[87] Nothing in his surviving correspondence bears upon respiration, heat, or related problems. In his book on the natural history of the air, edited and brought out posthumously by Locke in 1692,[88] Boyle recounted many phenomena relating to the pressure of the air, and reiterated his concept of air composed primarily of "Elastical Particles" like slender wires or curls of hair, compressed by their own weight and therefore exerting force, or spring, in all directions.[89] Boyle relegated the aerial niter to a title "Of Salte in the Air," in which it was only one of many kinds of saline particles that most likely impregnated the air. Moreover, Boyle said, he had done many experiments with saltpeter, and never found it to volatilize easily. He had even "caused Earth to be dug up in an old Pigeon-House, because that is accounted the most nitrous sort of Earth," and distilled it, but could not detect "the volatile saline Parts" that some "learned Men conceive the Air to be stored with."[90] Therefore he had to confess:

> I have not been hitherto convinc'd of all that is wont to be delivered about the Plenty and Quality of the Nitre in the Air: For I have not found, that those that build so much upon this volatile Nitre, have made out by any competent Experiment, that there is such a volatile Nitre, abounding in the Air.[91]

Boyle died at his house in Pall Mall on 30 December 1691, apparently still convinced that his physical conception of the air, so brilliantly supported by all his air pump experiments, was the only one really consonant with experience.

If the approach of the Oxford group to Harveian problems was not to be continued in London, nor was it to be in Oxford itself. No investigators arose there to take the place of those departed in the 1660s. The continuity of master and pupil was broken. Ralph Bathurst reverted to a church career not long after the Restoration; he was made a royal chaplain in 1663, Dean of Wells in 1670, and refused the bishopric of Bristol in 1691 because it would have interfered with his activities as President of Trinity. In 1664 he wrote to Seth Ward that he had considered publishing his lectures on respiration, but had decided against it, "now the genius of this nation is so active in philosophical disquisitions that *nihil modicum placebit.*"[92] Bathurst's lifelong friend, Willis, had carried out his teaching duties well in the 1660s, but upon leaving Oxford in 1667, he left the lecturing duties to a deputy whose qualifications were more clerical than scientific.[93] Millington succeeded to Willis's Sedleian professorship in 1675, and taught for only about a year before he also assigned his duties to a series of deputies. Wren, too, taught by deputy for the last few years before he resigned his Savilian professorship in 1673. Wallis held on to the other Savilian chair until his death in 1703, but his interest in chemical and medical questions had always depended upon his contacts with Oxford physicians.

From 1677 to 1690, Robert Plot, with the encouragement of such old hands as Wallis and Bathurst, ran an "Oxford Philosophical Society." It met first in college rooms, and in the newly-constructed Ashmolean Museum from 1683, when formal minutes began to be taken.[94] Some of the items on the agenda are familiar—dissections, air pump experiments, injections, and even an occasional discussion of niter. What was lacking, despite some good work done in comparative anatomy, was well-conceived experimentation. The meetings often consisted merely of discussion and the exhibition of some natural curiosity. The only real trial bearing upon respiration, in which William Musgrave excised a section out of the lungs of a dog, was merely a repetition of one done two decades before by "ye famous Dr. Lower." Musgrave had heard of it from Robert Fry, an Oxford surgeon who had assisted Lower in the original experiment.[95] This bit of scientific *déjà vu* is indicative of the failure of the medical sciences at Oxford to continue into the late-seventeenth century the work begun so brilliantly decades before.

A research tradition does not alone consist of ideas and results, printed and standing on the bookshelf. It must also have momentum, a continuity of interest and direction that defines and maintains investigations on a given set of problems, and continues, in the process, to throw up new ones for solution. By the mid-1670s, the Harveian research tradition stemming from Oxford had lost that momentum. Deaths, and men moving out into other areas of interest, had so impoverished the community that was the carrier of that tradition, that it could no longer continue in the work of the few that were left. The failure was not a matter of Mayow's book being so complete a synthesis as to discourage further research: far from it. The impediment lay rather in the dissolution of the uniquely large and diverse group that had maintained *inter alia* a continuous discourse on certain selected problems of physiology, and which had, between the members of the group, generated a completely new system of physiology. There were no longer men to carry on the ideas.

Or rather there *were* men, institutions, or ideas aplenty. They were just different men, working out different ideas, in institutions of different membership. William Musgrave and Edward Tyson, active in anatomical research at Oxford and later in London in the late 1670s and through the 1680s, were dissectors easily as talented as Lower. But they focused their attention largely on the anatomy of particular animals; Tyson is most remembered for his anatomy of the chimpanzee. The Royal Society, after an institutional crisis in the mid-1670s to some degree precipitated by the Hooke-Oldenburg feud, flourished again in the late 1670s and through the 1680s, meeting regularly and filling journal book after journal book with its proceedings. It even included the participation of some peripheral Oxford virtuosi from the 1650s, such as Sir Joseph Williamson and Sir Robert Southwell, now returning to an old avocation after careers as civil servants. The Oxford Philosophical Society of the 1680s was, in terms of numbers and correspondence, much larger and more active than its counterpart clubs of the 1650s and early 1660s. But the peculiar constellation of personalities and conceptual challenges had broken up. Harvey's discoveries that seemed to blaze so enticing a trail in the 1650s had by the 1680s passed so completely into the accepted body of physiology as to pose few problems.

Memories of bygone intellectual excitements are short, especially when there are few around to remember. Hooke was a voice from the past when, at the Royal Society meeting of 20 December 1682, the gathering was reminded of his

> experiment of blowing into the lungs of a dog, which keeps the animal alive, not by raising the lungs to make the blood circulate, but perfecting the blood by the nitrous particles of the air, which have that efficacy, as to make the blood come out of the lungs an arterial blood, which entered into them a venal.[96]

Nothing followed from this comment. The concerns of the 1660s and early 1670s had so far passed out of corporate memory that in 1686, when Abraham Hill brought into the Society an extract on the pulmonary circulation from Michael Servetus's long-lost work of 1553, "it was ordered, that the journal-book be produced and examined concerning the experiments made to prove the arterial blood made in the lungs."[97] Significantly, the search was never made.

IDEAS AND SOCIAL INTERACTION

To begin a narrative with Harvey, and to end with Mayow, is to sandwich a half-century of medical science between two almost totally contrasting physiologists. In a very real sense, there was a greater conceptual distance between Mayow and Harvey than between Harvey and Vesalius. The difference lay in the fact that Harvey, while he had proposed a new description of the heart's motions and the blood's course, had accepted, as recent scholars have emphasized, almost all of the Aristotelian natural philosophy, and many of the Galenic notions of physiological process. Mayow's work culminated the development of a new system of physiological explanation, in which many of the major functions of the organism—heat, movement, and interchange with the air—were explicated in terms of corpuscular matter-theory and physico-chemical processes.

The path from Harvey's *descriptions* to Mayow's *explanations* was, as I have shown in the foregoing narrative, exceedingly complex. Conceptually, the major theme that linked the two was the circulation, and the emphasis upon the properties and functions of the blood that the circulation implied. But the community of anatomists and physiologists at Commonwealth and Restoration Oxford were Harveians in more than just their concern for the implications of the circulation. Just as vigorously, they seized upon several of the secondary themes that Harvey had developed in his later writings—the blood as the locus of heat, the denial of vital spirits, the concern with the two species of blood color, the possible role of the air, the significance of embryogenesis and fetal anatomy—and used them as starting points for speculation and experimentation. Moreover, in doing so, they developed some of the technical tricks that Harvey himself had used, such as the open-thorax experiment, as well as some that Harvey's work had only suggested, such as injection.

It is perhaps inevitable that the historical process of remaking a system of physiology to bring it into line with new factual discoveries should be one centering upon a few delimited and abstruse questions. Theories of respiration, or of the origins of heat and blood color, are in themselves quite specialized and arcane topics. Yet they must be the conceptual focus of the narrative of any historian of science because only in a few such limited areas can a diverse number of observations, experiments, and speculations be brought to bear. Theories of respiration have some degree of importance in themselves, but even more as entry points into much larger questions of the organism and its environment.

The key was heat. The work of Bathurst, Willis, and Boyle makes clear that the crucial turning point was the simultaneous rejection of two interlocking propositions of the classical theory: the heat of the body is innate, and respiration serves to cool it. Once the Oxonians had made the conceptual shift from an *innate* heat to a *generated* heat, they were free to search for new functions for respiration. Not only were they free to search, they were impelled to do so.

The answers that finally emerged were compounded almost equally of four elements: the corpuscular philosophy, practical and speculative chemistry, experimentation in pneumatics, and new techniques and experiments in vivisection. From Highmore and Boyle's first joint exposure to atomism in the late 1640s, through Boyle's mature arguments for a syncretistic corpuscular philosophy, to Mayow's use of it as the basis in matter-theory for his unified conception of nitro-aerial particles, the corpuscular philosophy informed the work of the Oxford Harveians. Some, like Hooke, might prefer a more Cartesian version, while in the work of others, such as Lower, it is perceivable only under a cover of more traditional anatomical concerns. But in the work of all, it was there nonetheless, and was the *sine qua non* of any new system of physiological thought.

Within the Oxford group, the corpuscular philosophy had two progeny: chemistry and pneumatics. Both afforded radically different new ways to approach a handful of old physiological questions. Chemistry provided a transdiction between the properties of matter as men manipulated them in the laboratory, and as they were manipulated by nature in men. Generally, it made possible a new programmatic approach to the composition and properties of all organic substances. Specifically, it supplied several key concepts, such as that of a volatile salt, which suggested both new explanations for the nature of blood color, and new experiments to test those explanations. Pneumatics provided a physical way of manipulating a form of matter that could not be manipulated chemically. In that sense, it was almost the counterpart of chemistry in dealing with gases. This was not without its dangers—witness the long-standing alternative that it was simply the *pressure* of the air that was crucial to respiration—but it did afford a marvelously exact way of testing rigorously a number of the classical theories of respiration. The study of one pneumatic property, elasticity, also supplied for Mayow the link, albeit adventitious, between the effects of combustion and respiration on the air.

But such theories as were suggested by the corpuscular philosophy, chemistry, and even pneumatics, would have remained "philosophical romances" of the biological realm, without the elaboration of vivisectional techniques to corroborate the predictions that such speculations might bring forward. From the first vivisectional experiments of Highmore, through those of Boyle, Willis, Needham, and Hooke, and culminating in the unexcelled finesse of Lower, continuing experimentation by vivisection provided not only the best way of testing conjectures, but also the best link to the more traditional Harveian methods of verification.

It is also clear that such conceptual changes were intimately associated with the formation, prospering, and dissolution of a delimited scientific community. In episode after episode, crucial junctures of conceptual change are revealed as dependent upon the nature and quality of interaction among the members of the Oxford community.

The turning points were those of collaboration, of proposal and response, tied to shifting patterns of contact and communication in unique temporal and geographical loci. Harvey and Highmore worked on anatomy at Oxford. Highmore and Boyle studied atomism in rural Dorsetshire, and applied it to physiology. Petty, Willis, and Bathurst explored chemical approaches to Harveian problems at Oxford. Bathurst's "Praelectiones" circulated among Wilkins, Boyle, and Willis at Oxford. Students disputed the new physiology there. Boyle applied the corpuscular philosophy in his essay on saltpeter, and circulated the manuscript among his friends at Oxford and London. Boyle conducted pneumatic experiments at Oxford with Wilkins, Wallis, Ward, and Wren. Boyle and Hooke built their new air pump there and used it to explore both the elasticity of the air and its role in respiration. They conducted more extensive respiration trials at Chelsea. Hooke elaborated the role of the aerial niter in combustion, and discussed it with Wren. Willis developed a chemical/corpuscular concept of fermentation in the blood, and disseminated the idea in club meetings and university lectures. Wren, Boyle, and Clarke worked out injection techniques at Oxford and London. Such innovations flowered in the transfusion work of Boyle and Lower. Boyle, Lower, and Locke experimented with the color of arterial blood. Lower defended Oxford Harveianism. At London Hooke and Lower responded to Needham's theories by developing pulmonary insufflation. Lower used insufflation and injection to refute Cartesian theories of the heart, and to establish the origins of arterial blood. Willis, Mayow and Lower developed theories of muscular contraction. Willis attempted, with only partial success, to argue the identity of combustion and animal heat. Oxonians used the aerial niter in many ways. Mayow synthesized Willis's and Boyle's ideas in a general physiology of nitro-aerial particles, in the process explaining the contrary evidence found by Boyle and Hooke. In each case, the rate and, to a large degree, the nature of the innovation was dependent upon patterns of interaction.

This is not to say that the Oxford Harveians were "isolated" from outside ideas. The course of conceptual change refutes this. Just within physiology alone, Pecquet's discovery of the thoracic duct, Malpighi's findings on the structure of the lungs, and the Fracassati experiment of scraping off the florid layer—all came from the "outside," and had a great effect on the course of Oxford Harveianism. But the significance of these outside influences emerged only when integrated with, or placed in opposition to, the more general Oxford interpretation of the Harveian program. Only then did these facts become the basis of research activity that proceeded according to the normal give and take.

Nor does one assert, by finding the origins and nature of scientific ideas in social interactions, that within the group there was some kind of suspiciously totalitarian uniformity. Exactly the opposite should be obvious. There was not a man in the Oxford group who agreed precisely with any of his peers on such questions as the nature of the air, the function of respiration, the composition of the blood, or the origins of animal heat and muscular motion. Yet all agreed well enough on such subjects, within broader limits, to make fruitful communication possible.

One part of this agreement was the common perception of the "key problems." In the organized and professionalized sciences of the twentieth century, there is in each field a dominant consensus on the class of phenomena the investigation of which is most likely to produce signal advances. This consensus, which often transcends the bounds of language and geography, is established in the specialized literature of the field, and reinforced by such scientific gatherings as symposia and congresses. But the biological and chemical sciences of the mid-seventeenth century lacked such an international consensus. Rather, the choice of certain problems as "key," and their approach by certain routes, and with certain techniques, seems to have arisen more out of the idiosyncratic history of scientific concerns within a smaller temporal and spatial arena, than from some objective determination of research strategies. In carrying forward the work of Harvey, Oxford physiologists seized upon the idea of niter as a source of energy, organismic heat as a generated process, and respiration as the function that linked them. Every one of these commitments to specific "problem areas" has a genealogy traceable through the life of the Oxford scientific community. At Leyden during the 1650s and 1660s, a coterie of men organized around Franciscus Sylvius chose as "key" the problem of acid and alkali interactions, the process of digestion, and the structure and function of secretory glands. I suspect than an informed and critical treatment of the Leyden group would show patterns similar to those observable in England—objective results arising from shared perceptions of what are the "interesting" problems. A man's voluntary and close cohorts are, to a large degree, the guardians of his perceptions, and this is no less true in the perception of the relative *importance* of scientific problems, than in the determination of political opinions. There may be no such thing as a true sociology of scientific ideas, but there is most certainly one of the *valuations* placed upon their importance and intrinsic interest.

Perceptions find expression in language. It was no accident that the Oxford community not only infused words like "air," "fermentation" and "heat" with rather delimited local meanings, but also invented words like "nitro-aerial spirit." Concrete objects like muscles, tendons, nerves, saltpeter, or flame, naturally have names whose perceptual content was relatively uniform throughout Europe. Words like "vital spirit," "innate heat," or "volatile salt" are half-way between such concrete names and such fully abstract concepts as "nature" or "essence." As such, their concrete referent tends to deceive one into thinking that they have an easily specifiable and widely accepted meaning. Such was not always the case. One man's "fermentation" was often very different from another man's. Middle-level terms such as these tended to be fully intelligible only insofar as they were infused by specific language communities with agreed-upon meanings. This seems to be exactly what occurred within the Oxford community. Close interaction over many years not only evolved a set of shared concepts, it also defined a lexicon that described those concepts, even including a few new coinages. Such a pattern of discourse facilitated an easy exchange of ideas, and may well even have been, half-consciously, a statement of allegiance. One notes, for example, how many of the minor followers of Willis tended to use both his concepts and his specific shades of meaning on some words. Advantageous as such a community of discourse may be, it also has the limiting effect of tying the fate of the communal concepts to that of the community itself. The nitro-aerial particles were never quite the same after the breakup of the Oxford group.

Close scientific communities not only have an informal consensus on the problems to be addressed, and a common lexicon of words to make up an explanation, they also share a set of expectations about the overall nature and character of those explanations. For the Oxford group, the only acceptable processes were those that involved the interaction of particles having chemical characteristics. They rejected a Helmontian view of processes directed by metaphysical *archei,* and tended to shy away from the extreme corpuscular position that qualities and processes were due solely to the size, shape and motion of ultimate particles. They expected, moreover, that any explanatory scheme would be provisional. Thus they frequently used the word "hypothesis" to indicate the interim character of proposed explanations of experimental observations. Boyle, probably the man most sensitive to these issues, even went so far as to describe on several occasions the elements of a good hypothesis based on experiment. This attitude might well account for certain aspects of the famous disagreement between Hooke and Newton on the nature of light; Hooke persisted in referring to Newton's explanations as "hypotheses," which Newton understood in the Cartesian sense of unverified deductions from first principles, and was insulted. He did not understand that "hypothesis" had a perfectly respectable meaning in the experimentation of the Oxford community.

More concretely, close communities like the Oxford Harveians share a common reservoir of techniques and instruments. This may mean many things. It may signify the simple fact that a manipulative technique, such as injection or pulmonary insufflation, is taught by one member of the group to another. It may mean the equally simple fact that members have easy access within their midst to a special piece of apparatus, such as an air pump. It may also mean that the members of a community, because they have had first-hand experience with such techniques or apparatus, tend to have greater faith in the truth of the experimental results thereby obtained. Hobbes felt that ratiocination could counter Boyle's air pump experiments; Hooke, Wallis, Wren, Willis, and Mayow, among others, who had seen such experiments performed, accepted the outcome without hesitation.

Such common experience in techniques and apparatus is especially important in the medical sciences of physiology and chemistry. Experimental results in these areas may be reported in books or articles, but will never have the same impact until they have been duplicated by the reader for himself. This is in turn facilitated by learning to use a technique or instrument under the guidance of someone already experienced in it, who can demonstrate the little manipulative tricks, the shortcuts, and possible pitfalls. This aspect of the medical sciences is implicitly recognized in the great effects historians of medicine attribute to the introduction of laboratory and practical instruction into medical education. Lacking, for the most part, university facilities for learning the practical parts of anatomy and chemistry, the Oxford group worked out informal and cooperative ways to acquire skill in these sciences. When such cooperation brought forward new techniques or instruments, members of the group were quite easily and naturally socialized into their use. This was especially the case when, as in the case of injection and of the air pump, the technical advances were applicable to problems already perceived as important. Further use of the technique or apparatus only increased concentration on the "key problems" they illuminated.

A local scientific community serves yet another important function in the experimental medical sciences: that of providing a ready-made clutch of witnesses, whose presence may be adduced as evidence for a particular result. A mathematician's methods of proof are open for all to duplicate; one needs only pen and paper. Certain experiments in physics may require the construction of special apparatus, but they too are usually amenable to easy confirmation. But both in early chemistry, and in the anatomical and physiological sciences, in which the materials under scrutiny may vary so greatly, the course of the experiment itself affected by so many adventitious variables, and the necessary manipulative skills and techniques so difficult to communicate in writing, replication of experimental findings by other men may be difficult to accomplish. Having a group of one's friends participate in an experiment provides not only needed sets of extra hands and eyes, but also mouths willing to testify that the trial was carried out when and as the author describes, and with the reported results. Little wonder that Lower, once he had worked out his technique for transfusion, called in Wallis and Millington as both helpmates and witnesses.

Moreover, experimentation in the medical sciences in the post-Harveian period revolved around problems that could not be solved by the traditional Harveian methods. Ways of thinking and experimenting had to be imported into physiology from chemistry and from the physical sciences. In the Oxford group this was accomplished, not by having physicians become physicists, or physicists physicians, but by the social device of having them participate in each other's work. Not only did this provide each profession with insights into the concerns, the ways of thinking and experimenting of the other, it also established the easy and familiar intercourse by which physicians became comfortable with the idea of casting their explanations in terms dictated by physicists and chemists. It also allowed both groups to feel that they were engaged jointly in a process of intellectual innovation whose flaunting of traditional natural philosophy might have given one group alone cause to doubt the appropriateness of its audacity.

Just as differing scientific interests contributed to the success of Oxford physiology, so did differing scientific personalities. Mayow was as bold as Boyle was cautious. Willis was as speculative as Lower was factual. Some were skilled with their hands, others were less so. Some, like Willis and Boyle, were prolific writers. Others of great influence, like Wren and Millington, published nothing on their own. Some, like Oldenburg and Wilkins, were largely communicators and organizers. Others were the direct beneficiaries of such organization. Still others, many others, were interested bystanders—absorbing, applauding, and disseminating, but not producing. Such a cast of dozens may not have been necessary to assemble the array of ideas and discoveries that constituted the conceptual legacy of the Oxford physiologists. This might have been worked out by one, two, or three men, perhaps even working independently of each other. But I doubt it. The very complexity of the task of extending the Harveian innovations into the rest of physiology, to say nothing of the intricacies of the explanatory scheme that finally emerged, seems to me to have demanded much the kind of cast and script that was eventually played out.

Institutions played an important, if not determining, part in assembling and maintaining this community and thereby making possible such a local proliferation of new ideas. College headships and fellowships supported Oxonians in the leisure necessary for reading, meeting, and experimenting. Ties of college and school were the lines along which new members were recruited into the community. The case of Westminster School is particularly striking, since Busby's students comprised such a small percentage of the university as a whole, and such a disproportionately large part of the core Oxford group. Not only were Westminster boys exceptionally talented, they were also predisposed to the sciences by their training, and tended to bring like-minded schoolmates into what they regarded as a fascinating activity. Oxford academic institutions likewise provided modes of expression for the new philosophy—in lectures, in disputations, and through books brought out by Oxford booksellers. Most of all, simple propinquity conspired with facilities, talent, and leisure to make possible the close interaction that characterized the Oxford community. Its members were almost all bachelors, living within ten minutes' walk of each other, meeting each other almost daily in alehouses, coffeehouses, bookshops, and dining halls, largely undistracted by the demands of family or career. Little wonder that, given forceful intellectual leaders, and a time in which numerous scientific problems seemed to cry out for answers, they could respond so brilliantly.

Conversely, when times had changed, and men had aged, and problems seemed to have found some solutions, the cooperative ventures of youth lost their appeal. It seems no accident that each of the major figures in the Oxford tradition—Highmore, Bathurst, Willis, Wren, Boyle, Hooke, Lower and Mayow—reached his individual apogee of creativity at some time between his late twenties and early forties. As each had his own individual cycle of creativity, so did the community of which he was a part. The Oxford institutions that had helped nurture and support this community were not, of themselves, capable of perpetuating it. The Oxford physiologists who had taken Harvey's discoveries, and made them the basis of a new physiology, were able to take those discoveries no further. The ideas that had been so enthusiastically discussed in meetings, correspondence, and books, perpetuated an existence only in the printed word. The science of the Oxford physiologists had ceased to grow, and thereby ceased to be true science.

REFERENCES AND DATES

I have made no attempt to cite in the footnotes all of the primary and secondary sources examined in the preparation of this book. Nor is this necessary, since full references to books and articles on many of the figures treated here may be found in the *Isis cumulative bibliography* (London: Mansell, 1971 and 1976). However, since that work indexes secondary sources published only up until 1965, I have noted, where appropriate, more recent scholarship.

The dates used in the book are Old Style, i.e., according to the Julian calendar, which was ten days behind the Gregorian calendar in use on the Continent (New Style, or N.S.). The British Isles also generally followed the Church Year, according to which the new year began on 25 March. Therefore all dates between 1 January and 24 March of, for example, 1666, were given as 1665/6, or sometimes simply as 1665. In the text I have silently corrected all such dates to bring them into line with modern usage.

Titles of modern scholarly journals have been abbreviated according to the usual practices.

OTHER ABBREVIATIONS:

"Annals" — MS. Annals of the Royal College of Physicians, London

Aubrey, *Brief lives* — *"Brief lives," chiefly of contemporaries, set down by John Aubrey, between the years 1669 & 1696*, ed. Andrew Clark. 2 vols. (Oxford: Clarendon Press, 1898).

Birch, *History* — Thomas Birch, *The history of the Royal Society of London for improving of natural knowledge, from its first rise*. 4 vols. (London, 1756–1757).

CSPD   *Calendar of State Papers Domestic*

Correspondence   *The correspondence of Henry Oldenburg*, ed.
                                    and trans. A. Rupert Hall and Marie Boas
                                    Hall. 11 vols. to date (Madison: University
                                    of Wisconsin Press, 1965–).

DSB   *The dictionary of scientific biography.* 14
                                    vols. (New York: Charles Scribner's Sons,
                                    1970–1976).

Evelyn, *Diary*   *The diary of John Evelyn*, ed. E. S. de Beer.
                                    5 vols. (Oxford: Clarendon Press, 1955).

Foster, *Alumni*   Joseph Foster, ed., *Alumni Oxoniensis: The
                                    members of the University of Oxford, 1500–
                                    1714.* 4 vols. (Oxford: Parker, 1891–1892).

Hartlib Papers   The papers of Samuel Hartlib, Sheffield
                                    University Library, quoted by kind permis-
                                    sion of their owner, Lord Delamere.

Hartlib, "Ephemerides"   Samuel Hartlib, diary-like commonplace
                                    books (1639–1660), Hartlib Papers, Bundles
                                    XXVIII, XXIX, and XXX. Quoted from the
                                    typed transcription made by G. H. Turnbull,
                                    ibid.

Harvey, *De motu cordis*   William Harvey, *Exercitatio anatomica de
                                    motu cordis et sanguinis in animalibus*
                                    (Frankfurt, 1628).

——, *De circulatione*   William Harvey, *Exercitatio anatomica de
                                    circulatione sanguinis* (Cambridge, 1649).

——, *De generatione*   William Harvey, *Exercitationes de genera-
                                    tione animalium* (London, 1651).

——, *Generation*   William Harvey, *Anatomical exercitations
                                    concerning the generation of living creatures*
                                    (London, 1653).

——, *Works*   William Harvey, *Works*, trans. Robert Willis
                                    (London: Sydenham Society, 1847).

——, *Motion*   William Harvey, *Movement of the heart and
                                    blood in animals: An anatomical essay*, trans.
                                    Kenneth J. Franklin (London: Dent, 1963),
                                    pp. 1–111.

——, *Circulation*   William Harvey, *The circulation of the
                                    blood*, ibid., pp. 121–179.

Keynes, *Harvey*

Geoffrey Keynes, *The life of William Harvey* (Oxford: Clarendon Press, 1966).

Partington, *History*

J. R. Partington, *A history of chemistry.* 4 vols. (London: Macmillan, 1961–1964).

Phil. Trans.

The Philosophical Transactions, London.

Stationers' register

A transcript of the registers of the Worshipful Company of Stationers: from 1640–1708 A.D. 3 vols. (London: Privately Printed, 1913).

Ward, "Diary"

John Ward, diary-like commonplace books, V.a. 284–299, 16 vols. Folger Shakespeare Library, Washington D.C.

Wood, *Athenae*

Anthony Wood, *Athenae Oxoniensis: An exact history of all the writers and bishops who have had their education in the University of Oxford*, new ed., Philip Bliss. 4 vols. (London: Rivington, 1813–1820).

————, *Fasti*

Anthony Wood, *Fasti Oxoniensis*, 2 vols., bound ibid.

————, *History and antiquities*

Anthony Wood, *The history and antiquities of the colleges and halls in the University of Oxford* (Oxford: Clarendon Press, 1786).

————, *Life and times*

The life and times of Anthony Wood, antiquary, of Oxford, 1632–1695, described by himself, ed. Andrew Clark. 5 vols. (Oxford: Clarendon Press, 1891–1900).

Works

The works of the Honourable Robert Boyle, ed. Thomas Birch, 2d ed. 6 vols. (London: Rivington, 1772).

NOTES

PREFACE

1. James B. Conant, *Science and common sense* (New Haven: Yale University Press, 1951), pp. 24–27.

2. Genesis, ii 7, ix 4.

3. Michael Foster, *Lectures on the history of physiology during the sixteenth, seventeenth, and eighteenth centuries* (Cambridge: University Press, 1901), pp. 172–197; Francis Gotch, *Two Oxford physiologists: Richard Lower 1631 to 1691, John Mayow 1643 to 1679* (Oxford: Clarendon Press, 1908).

4. T. S. Patterson, "John Mayow in contemporary setting. A contribution to the history of respiration and combustion," *Isis 15* (1931): 47–96, 504–546; K. J. Franklin, intro. and trans. of Richard Lower, *Tractatus de corde* (1669) in *Early science in Oxford* (Oxford: Printed for the Subscribers, 1932); John F. Fulton, *A bibliography of two Oxford physiologists, Richard Lower 1631–1691, John Mayow 1643–1679* (Oxford: University Press, 1935); Ebbe C. Hoff and Phebe M. Hoff, "The life and times of Richard Lower, physiologist and physician (1631–1691)," *Bull. Hist. Med. 4* (1936): 517–535; Douglas McKie, "Fire and the flamma vitalis: Boyle, Hooke and Mayow," in E. A. Underwood, ed., *Science, medicine and history* (London: Oxford University Press, 1953), I, 469–488; J. R. Partington, "The life and work of John Mayow (1641–1679)," *Isis 47* (1956): 217–230, 405–417.

5. Leonard G. Wilson, "The transformation of ancient concepts of respiration in the seventeenth century," *Isis 51* (1960): 161–172; G. J. Goodfield, *The growth of scientific physiology: Physiological method and the mechanist-vitalist controversy, illustrated by the problems of respiration and animal heat* (London: Hutchinson, 1960); Everett Mendelsohn, *Heat and life: The development of the theory of animal heat* (Cambridge, Mass.: Harvard University Press, 1964).

6. Thomas Birch, *The history of the Royal Society of London for improving of natural knowledge, from its first rise*, 4 vols. (London, 1756–1757).

7. Robert K. Merton, *Science, technology and society in seventeenth century England*, 2d ed. (New York: Fertig, 1970); Charles Webster, *The great instauration: Science, medicine and reform 1626–1660* (New York: Holmes and Meier, 1976).

8. R. E. W. Maddison, "A tentative index of the correspondence of the Honourable Robert Boyle, F.R.S.," *Notes Rec. R. Soc. Lond. 13* (1958): 128–201; *The correspondence of Henry Oldenburg*, ed. and trans. A. Rupert Hall and Marie Boas Hall, 11 vols. to date (Madison: University of Wisconsin Press, 1965–).

1: THE CHALLENGE: HARVEY'S CIRCULATION AND ITS UNSOLVED PROBLEMS

1. The standard source on Harvey's career is the magisterial biography by Geoffrey Keynes, *The life of William Harvey* (Oxford: Clarendon Press, 1966).

2. British Library, MS. Sloane 230, "Prelectiones anatomie universalis." This MS. was reproduced in facsimile with a transcription as *Prelectiones anatomiae universalis* (London: J. & A. Churchill, 1886). Gweneth Whitteridge, *The anatomical lectures of William Harvey* (Edinburgh: Livingstone, 1964) gives a new transcription and translation of the lectures, arranged on facing pages; it is hereafter cited as *Prelectiones*.

3. On the issues involved in the discovery of the pulmonary transit, as well as references to the earlier literature, see: Edward D. Coppola, "The discovery of the pulmonary circulation: a new approach," *Bull. Hist. Med.* 31 (1957):44−77; Walter Pagel, *William Harvey's biological ideas: Selected aspects and historical background* (Basel/New York: Karger, 1967), especially pp. 127−209; Jerome J. Bylebyl, "Cardiovascular physiology in the sixteenth and early seventeenth centuries," unpublished doctoral dissertation, Yale University, 1969; Gweneth Whitteridge, *William Harvey and the circulation of the blood* (London: Macdonald, 1971), esp. pp. 41−77.

4. The literature on Harvey and his discovery of the circulation is vast. The best introductions to this corpus are: Pagel, *Harvey's biological ideas*; Whitteridge, *Harvey and the circulation*; and Walter Pagel, *New light on William Harvey* (Basel/New York: Karger, 1976).

5. The best summary of these changes, especially as they apply to the heart and lungs, is Bylebyl, "Cardiovascular physiology." On the Galenic medical system, see Owsei Temkin, *Galenism: Rise and decline of a medical philosophy* (Ithaca/London: Cornell University Press, 1973).

6. André du Laurens, *Historia anatomica humani corporis* (Frankfurt, 1600), pp. 7−8; Caspar Bauhin, *Theatrum anatomicum* (Frankfurt, 1605), pp. 12−16.

7. Du Laurens, *Historia*, pp. 225−258; Bauhin, *Theatrum*, pp. 86−168, 270−315.

8. Du Laurens, *Historia*, pp. 343−378; Bauhin, *Theatrum*, pp. 360−469.

9. Bauhin reflects the common practice of discussing the veins and the arteries in different parts of the same book: *Theatrum*, pp. 383−399, 432−438, 444−450.

10. Du Laurens, *Historia*, pp. 383−409; Bauhin, *Theatrum*, pp. 489−700.

11. Du Laurens, *Historia*, pp. 348−358; Bauhin, *Theatrum*, pp. 408−432, especially 417−419. In this and the following paragraphs I have attempted to summarize cardiovascular and respiratory physiology as it appeared to Harvey's contemporaries. I have made no attempt to assess in what ways these ideas were precisely indebted to Galen; on this see A. Rupert Hall, "Studies on the history of the cardiovascular system," *Bull. Hist. Med.* 34 (1960): 391−413, and especially the exhaustive work by C. R. S. Harris, *The heart and the vascular system in ancient Greek medicine, from Alcmaeon to Galen* (Oxford: Clarendon Press, 1973), pp. 267−431.

12. Du Laurens, *Historia*, pp. 358−361; Bauhin, *Theatrum*, pp. 418−422, 429−432, 458.

13. Du Laurens, *Historia*, pp. 369−372; Bauhin, *Theatrum*, pp. 428−432, 460−461.

14. Du Laurens, *Historia*, pp. 369−372; Bauhin, *Theatrum*, pp. 418, 429−432, 458, 460−461.

15. Du Laurens, *Historia*, pp. 26, 357−358; Bauhin, *Theatrum*, pp. 432−438.

16. Du Laurens, *Historia*, pp. 3, 23; Bauhin, *Theatrum*, pp. 408, 414−415, 421 and passim.

17. Du Laurens, *Historia*, pp. 241−242; Bauhin, *Theatrum*, pp. 409, 414−415, 418−419, 422, 429−432.

18. Du Laurens, *Historia*, p. 18; Bauhin, *Theatrum*, pp. 415, 420–421, 436–438.

19. Du Laurens, *Historia*, pp. 86–106; Bauhin, *Theatrum*, pp. 135–143, 383–399, 432–438, 444–450.

20. Du Laurens, *Historia*, pp. 86–106 passim; Bauhin, *Theatrum*, pp. 420–421, 432–438.

21. Du Laurens, *Historia*, pp. 314–327; Bauhin, *Theatrum*, pp. 438–444, 428, 456.

22. Harvey, *Prelectiones*, pp. 265–269.

23. Ibid., pp. 269–273.

24. Du Laurens, *Historia*, pp. 352–358.

25. Realdo Colombo, *De re anatomica* (Venice, 1559), pp. 175–180.

26. Harvey, *De motu cordis*, pp. 10–19; *Motion*, pp. 9–21; *Works*, pp. 9–19.

27. Harvey, *De motu cordis*, pp. 20–40; *Motion*, pp. 23–56; *Works*, pp. 19–45.

28. Jerome J. Bylebyl, "The growth of Harvey's *De motu cordis*," *Bull. Hist. Med.* 47 (1973): 427–470.

29. Gweneth Whitteridge, "*De motu cordis*: Written in two stages? Comment," ibid., 51 (1977): 130–139; Jerome J. Bylebyl, "*De motu cordis*: Written in two stages? Response," ibid., 51 (1977): 140–150.

30. Harvey, *De motu cordis*, pp. 41–64; *Motion*, pp. 57–97; *Works*, pp. 45–75.

31. Harvey, *Prelectiones*, pp. 126–127.

32. Ibid., pp. 162–163.

33. Ibid., pp. 14–15, 30–31.

34. Ibid., pp. 246–247, 310–311.

35. Ibid., pp. 32–33, 248–251, 256–257.

36. Ibid., pp. 126–127, 250–251.

37. Harvey, *De motu cordis*, p. 42; *Motion*, p. 59; *Works*, pp. 46–47.

38. Harvey, *De motu cordis*, pp. 49, 51; *Motion*, pp. 71, 74; *Works*, pp. 55, 58.

39. Harvey, *De motu cordis*, p. 59; *Motion*, pp. 88–89; *Works*, pp. 68–69.

40. Harvey, *Prelectiones*, pp. 272–273.

41. Ibid., pp. 214–217, 276–277, 282–283.

42. Ibid., pp. 274–275.

43. Ibid., pp. 276–277, 282–283.

44. Ibid., pp. 278–279.

45. Ibid., pp. 294–295, 298–299.

46. Ibid., pp. 274–275.

47. Ibid., pp. 290–291, 294–297.

48. Ibid., pp. 228–229, 292–293.

49. Ibid., pp. 294–295.

50. Harvey, *De motu cordis*, pp. 33, 40, 65; *Motion*, pp. 44–45, 56, 100; *Works*, pp. 35–36, 44–45, 77.

51. Harvey, *De motu cordis*, pp. 33–36, 65–66; *Motion*, pp. 45–49, 100–101; *Works*, pp. 36–39, 77–78.

52. Harvey, *De motu cordis*, p. 36; *Motion*, p. 50; *Works*, pp. 39–40.

53. Harvey, *De motu cordis*, p. 72; *Motion*, p. 111; *Works*, p. 85.

54. Harvey, *De motu cordis*, pp. 33, 36; *Motion*, pp. 45, 50; *Works*, pp. 35–36, 40.

55. Harvey, *De motu cordis*, pp. 37–40; *Motion*, pp. 51–56; *Works*, pp. 42–44.

56. Harvey, *De motu cordis*, pp. 38, 45–46; *Motion*, pp. 52, 65–66; *Works*, pp. 41, 51–52.

57. Harvey, *De motu cordis*, pp. 40, 65; *Motion*, pp. 56, 100; *Works*, pp. 45, 77.

58. Harvey, *Prelectiones*, pp. 142–143, 292–293, 318–319.

59. Ibid., pp. 292–293.

60. Ibid., pp. 258–259.

61. Ibid., pp. 256–257, 296–299.

62. Harvey, *De motu cordis*, p. 13; *Motion*, p. 12; *Works*, p. 12.

63. Harvey, *De motu cordis*, pp. 17–18; *Motion*, pp. 16–18; *Works*, pp. 15–17.

64. Harvey, *De motu cordis*, p. 18; *Motion*, p. 18; *Works*, p. 16.

65. Harvey, *Prelectiones*, pp. 292–293.

66. Ibid., pp. 296–297.

67. Ibid., pp. 4–5.

68. Harvey, *De motu cordis*, p. 8; *Motion*, p. 7; *Works*, p. 7.

69. Harvey, *De generatione*, sig. alr; *Generation*, sig. A3r; *Works*, p. 146.

70. Harvey, *Prelectiones*, pp. 22–23.

71. Harvey, *De generatione*, sig. alr-v; *Generation*, sig. A5r; *Works*, p. 146.

72. See Michael T. Ghiselin, ''William Harvey's methodology in *De motu cordis* from the standpoint of comparative anatomy,'' *Bull. Hist. Med.* 40 (1966): 314–327.

73. Harvey, *De motu cordis*, p. 20; *Motion*, p. 23; *Works*, p. 19.

74. Harvey, *De circulatione*, p. 44; *Circulation*, p. 141; *Works*, p. 109.

75. Harvey, *Prelectiones*, pp. 280–281.

76. Harvey, *De motu cordis*, pp. 18, 27–29, 33–37, 46, 62–63, 65–66, 69; *Motion*, pp. 19, 36–38, 45–50, 65, 94–95, 100, 106–107, *Works*, pp. 17, 28–30, 36–40, 51–52, 73–74, 77, 81–82.

77. Harvey, *De circulatione*, pp. 16, 36–38, 101; *Circulation*, pp. 128, 138, 168; *Works*, pp. 95, 103, 132.

78. Harvey, *Prelectiones*, pp. 268–269.

79. Harvey, *De motu cordis*, pp. 13–14, 47, 48–54, 56–58; *Motion*, pp. 13–14, 68, 70–80, 84–86; *Works*, pp. 12–13, 53, 54–62, 65–66.

80. Harvey, *De circulatione*, pp. 11–13, 46–49, 109–110; *Circulation*, pp. 126, 142–143, 172–173; *Works*, pp. 93, 110–111, 135–136.

81. Harvey, *Prelectiones*, pp. 108–109.

82. Harvey, *De motu cordis*, p. 17; *Motion*, p. 18; *Works*, p. 16.

83. Harvey, *De generatione*, pp. 4–5, 12, 16, 24, 222; *Generation*, pp. 6–8, 20–21, 29, 42, 406; *Works*, pp. 173–174, 184, 190, 200, 473.

84. Harvey, *Prelectiones*, pp. 108–109.

85. Harvey to Schlegel, 26 March 1651, in K. J. Franklin, ed., *The circulation of the blood and other writings* (London: Dent, 1963), p. 184.

86. For assessments of Harvey's quantitative methodology, see Frederick G. Kilgour, ''William Harvey's use of the quantitative method,'' *Yale J. Biol. Med.* 26 (1954): 410–421; F. R. Jevons, ''Harvey's quantitative method,'' *Bull. Hist. Med.* 36 (1962): 462–467; Pagel, *Harvey's biological ideas*, pp. 73–81.

87. Harvey, *De motu cordis*, pp. 41–42, 43–46; *Motion*, pp. 57–58, 61–66; *Works*, pp. 45–46, 48–52.

88. On Harvey's Aristotelianism see Erna Lesky, ''Harvey und Aristoteles,'' *Sudhoffs Arch.* 41 (1957): 289–316, 349–378, and the writings of Walter Pagel, especially *Harvey's biological ideas*, pp. 28–47, 83–86, 233–247, 331–335, and *New light*, pp. 13–33.

89. Harvey, *Prelectiones*, passim, esp. pp. 16–21, 30–31, 164–165, 244–245.

90. For a general discussion of Harvey's vitalism, see Pagel, *Harvey's biological ideas*, especially pp. 233–277.

91. Harvey, *De generatione*, sig. B2r–v, pp. 167–173; *Generation*, sig. ¶6r–7r, pp. 384–389; *Works*, pp. 164, 397–406.

92. Harvey, *De generatione*, pp. 29, 175, 254; *Generation*, pp. 51–52, 320, 467; *Works*, pp. 207, 409, 516.

93. Harvey, *De generatione*, pp. 112, 259; *Generation*, pp. 205, 475; *Works*, pp. 321, 524.

94. Harvey, *Prelectiones*, pp. 78–79, 100–101, 312–313.

95. Harvey, *De generatione*, pp. 255–256; *Generation*, pp. 468–469; *Works*, pp. 517–518.

96. Harvey, *De generatione*, pp. 150–151; *Generation*, p. 276, *Works*, p. 375.

97. Harvey, *De circulatione*, pp. 114–115, 128–129, 136–137; *Circulation*, pp. 169, 174, 178; *Works*, pp. 132, 137, 140.

98. Aubrey, *Brief lives*, I, 302.

2: HARVEY'S LATER WORK AND ENGLISH DISCIPLES

1. Harvey, *De motu cordis*, p. 63; *Motion*, p. 96; *Works*, p. 74.

2. See "The reception of Harvey's doctrine during his lifetime, 1628–1657," in Keynes, *Harvey*, pp. 447–455.

3. Harvey, *De motu cordis*, p. 6; *Motion*, p. 5; *Works*, p. 5.

4. Aubrey, *Brief lives*, I, 300.

5. Keynes, *Harvey*, pp. 229–263.

6. On Ent see Charles Webster, "George Ent," *DSB*, IV, 376–377, and Theodore M. Brown, "Physiology and the mechanical philosophy in mid-seventeenth century England," *Bull. Hist. Med.* 51 (1977): 25–54, esp. pp. 27–30.

7. See *Biographia Britannica* (London, 1747–1766), IV, 2267–2279, and the life Thomas Birch prefixed to John Greaves, *Miscellaneous works* (London, 1737), I, i-lxi.

8. *Laureae Apollinari* (Padua, 1636).

9. Harvey, Ent, and Greaves dined together on 5 October 1636 at the English Jesuit College in Rome, whose refectory was known for its hospitality to British travelers: Henry Foley, ed., "The Pilgrim-Book of the English Hospital of the Most Holy Trinity and St. Thomas of Canterbury, Rome," in *Records of the English Province of the Society of Jesus*, 7 vols. in 8 (London: 1877–1883), VI, 614.

10. See the evidence cited in Keynes, *Harvey*, pp. 258–259. This agrees with the biographical details on Greaves in *Biographia Britannica* and *Miscellaneous works*.

11. George Ent, *Apologia pro circulatione sanguinis* (London, 1641), sig. A4v.

12. Parigiano (latinized as Aemylius Parisanus) published an attack on Harvey in 1635 that is rare and bibliographically complicated; see Geoffrey Keynes, *A bibliography of the writings of Dr. William Harvey*, 2d ed. (Cambridge: University Press, 1953), p. 8.

13. Ent, *Apologia*, sigs. A2r–A4r.

14. On Glisson see R. Milnes Walker, "Francis Glisson and his capsule," *Annals of the Royal College of Surgeons of England* 38 (1966): 71–91. Glisson's manuscripts are in the British Library, MSS. Sloane 574B, 681, 1116, 2251, 2326, 3258, and 3306–3315.

15. British Library, MS. Sloane 3313, f. 1r, 2r.

16. Ibid., respectively MSS. Sloane 3310, ff. 327, 393–396; 3307, ff. 74–79; 3309, ff. 18–20; 3309, ff. 141–144; 3312, ff. 2–16, 106–128.

17. Ibid., respectively MSS. Sloane 3308, f. 296; 3308, f. 167; 3312, ff. 42–47; 3309, ff. 139–140.

18. Ibid., MS. Sloane 1116.

19. Ibid., MS. Sloane 3312, ff. 265–268.

20. Ibid., MS. Sloane 3310, ff. 249–255.

21. Ibid., MS. Sloane 3310, f. 301.

22. Ibid., MS. Sloane 3308, ff. 278–281.

23. F. J. Cole, "Henry Power on the circulation of the blood," *J. Hist. Med.* 12 (1957): 291–324. See esp. p. 294.

24. Wallis gave two major accounts of these meetings, in his *Defence of the Royal Society* (London, 1678), pp. 7–9, and in his autobiography of 1697, which comprises pp. cxl–clxx in Thomas Hearne, ed., *Peter Langtoft's chronicle* (Oxford, 1725), I, esp. pp. clxi–clxiv. These differ in some matters of detail.

25. In the passage written in 1697 (p. clxii) Wallis included "the Lymphatick Vessels" as a fourth specific topic of discussion. Since discovery of the lymphatics was not announced in print until 1653, this is most likely simply an anachronism. There is a slight possibility that the group knew of George Joyliffe's discovery of the lymphatics at Oxford during the mid-1640s. See below on Joyliffe.

26. Ibid., p. cl.

27. *Biographia Britannica*, IV, 2216.

28. Cf. Theodore Haak to Marin Mersenne, 24 May/3 June 1647, in Harcourt Brown, *Scientific organizations in seventeenth-century France, 1620–1680* (Baltimore: Williams & Wilkins, 1934), p. 57.

29. For details of this work, see Charles Webster, "The College of Physicians: 'Solomon's House' in Commonwealth England," *Bull. Hist. Med.* 41 (1967): 393–412, and Robert G. Frank, Jr., "The physician as virtuoso in seventeenth-century England" in Robert G. Frank, Jr. and Barbara Shapino, *English scientific virtuosi in the sixteenth and seventeenth centuries* (Los Angeles: William Andrews Clark Memorial Library, 1979), pp. 57–114.

30. The Gulstonian Lecturers, insofar as their names have been recorded, were: William Rand (1639), Francis Glisson (1640), Thomas Sheaf (1641), George Ent (1642), John Micklethwaite (1644), Assuerus Regimorter (1645), Robert Wright (1647), Jonathan Goddard (1648), Edward Emily (1649), Edmund Trench (1650), and Christopher Merrett (1654).

31. Francis Glisson, *De rachitide, sive morbo puerili, qui vulgo The Rickets dicitur, tractatus* (London, 1650), sig. A5v. Glisson, George Bate, and Assuerus Regimorter made up the committee, while Goddard, Thomas Sheaf, Nathan Paget and Edmund Trench were among those contributing clinical observations.

32. See the references to the work of such committees in Walter Charleton, *The immortality of the human soul, demonstrated by the light of nature* (London, 1657), pp. 41–44; [Christopher Merrett], *A letter concerning the present state of physick* (London, 1665), pp. 15–19; [Timothy Clarke], *Some papers writ in the year 1664. In answer to a letter, concerning the practice of physick in England* (London, [1670]), pp. 23–30.

33. See the references to Glisson's work with Wharton in British Library, MS. Sloane 3315, ff. 23–24.

34. Walter Charleton, *Immortality*, pp. 34–35.

35. Wood, *Athenae*, III, 351.

36. George Thomson, *Misochymias* (London, 1671), p. 60. For an excellent account of splenectomies in the 1650s, see Charles Webster, "The Helmontian George Thomson and William Harvey: The revival and application of splenectomy to physiological research," *Med. Hist.* 15 (1971): 154–167.

37. See Charleton, *Immortality*, p. 37; Charles Goodall, *The Royal College of Physicians of London founded and established by law* (London, 1684), sig. Uu4v; Charleton's conclusions on muscular motion are in his *Natural history of nutrition, life, and voluntary motion* (London, 1659), pp. 182–210.

38. See Keynes, *Harvey*, pp. 397–406.

39. These were catalogued by Christopher Merrett, *Catalogus librorum, instrumentorum chirurgicorum, rerum curiosarum, exoticarumque Coll. Med. Lond. quae habentur in Musaeo Harveano* (London, 1660).

40. Keynes, *Harvey*, p. 404.

41. Charleton, *Immortality*, p. 34.

42. Keynes, *Harvey*, pp. 277–291.

43. Keynes, *Harvey*, pp. 291–314; A. M. Cooke, "William Harvey at Oxford," *J. Roy. Coll. Phys. Lond.* 9 (1975): 181–188; A. H. T. Robb-Smith, "Harvey at Oxford," *Oxford Medical School Gazette* 9 (1957): 70–76.

44. *Academia Monspeliensis a Jacobo Primirosio Monspeliensi & Oxoniensi doctore descripta* (Oxford, 1631), pp. 1–2.

45. Cited in Christopher Hill, "William Harvey and the idea of monarchy," *Past and Present*, No. 27 (April, 1964), p. 59.

46. Keynes, *Harvey*, pp. 292–293, 295, 303.

47. For a detailed account of Harvey's election and tenure at Merton, see ibid., pp. 298–304.

48. Aubrey, *Brief lives*, I, 37–38, 300.

49. For biographical details on Highmore, see J. Elise Gordon, "Nathaniel Highmore, physician and anatomist, 1614–1685," *Practitioner* 196 (1966): 851–858; idem, "Haydocke and Highmore—two 17th century physicians. Part 2: Nathaniel Highmore 1614–1685," *Midwife and Health Visitor* 5 (1969): 364–369; idem, "Highmore, Nathaniel," *DSB*, VI, 386–388.

50. Nathaniel Highmore, *History of generation* (London, 1651), pp. 100–101.

51. British Library, MS. Add. 29586, f. 21r–v.

52. Harvey, *De generatione*, pp. 38, 71; *Generation*, pp. 69, 129; *Works*, pp. 220, 266.

53. Now in the National Library of Medicine, Bethesda, Maryland. Unfortunately, the copy contains no annotations by Greaves.

54. John Greaves, "The manner of hatching chickens at Cairo," *Phil. Trans.* 12 (1678): 923–925.

55. For biographical details on Scarburgh, see: Keynes, *Harvey*, pp. 304–307; Burton Chance, "Charles Scarborough, an English educator and physician to three kings: A medical retrospective into the times of the Stuarts," *Bull. Hist. Med.* 12 (1942): 274–303; J. J. Keevil, "Sir Charles Scarburgh," *Ann. Sci.* 8 (1952): 113–121.

56. Aubrey, *Brief lives*, I, 299.

57. Wood, *Fasti*, II, 97.

58. See, for example, the autopsies and dissections mentioned in his *Corporis humani disquisitio anatomica* (The Hague, 1651), pp. 60, 178–179.

59. British Library, MS. Sloane 577, (Colombo) ff. 42r, 44r; (Fallopio) ff. 38r, 40r, 41r, 43r, 47r, 51r, 52v.

60. Ibid., ff. 22r, 36r, 44r, 48r, 52r, 58r, 59r.

61. Ibid., passim and ff. 61r, 62r.

62. Ibid., ff. 64r-81r.

63. Ibid., (reverse codex), ff. 99v–86v, especially 96v, 94v, 92r, 91v.

64. Cf. Gordon, "Haydocke and Highmore," pp. 367–368.

65. Highmore, *Disquisitio*, sig. A1r.

66. This conversation was recorded by Greaves in his *Pyramidographia: Or a description of the pyramids in Aegypt* (London, 1646), pp. 101–102.

67. Edward Greaves, *Morbus epidemicus* (Oxford, 1643); *Oratio habita in edibus Collegii Medicorum Londinensium, 25 Jul. 1661. Die Harvaei memoriae dicato* (London, 1667).

68. Daniel Whistler, *Disputatio medica inauguralis, de morbo puerili Anglorum, quem patrio idiomate indigenae vocant The Rickets* (Leyden, 1645), especially pp. 5, 7, 9, 15.

69. Robert Plot, *The natural history of Oxford-shire* (Oxford, 1677), p. 300.

70. Christopher Merrett, *Pinax rerum naturalium Britannicarum* (London, 1666), pp. 78, 109, 124, 183.

71. On Charleton see the excellent article by Lindsay Sharp, "Walter Charleton's early life, 1620–1659, and relationship to natural philosophy in mid-seventeenth century England," *Ann. Sci.* 30 (1973): 311–340.

72. Thomas Willis, "De febribus," pp. 171–175, in *Diatribae duae medico-philosophicae* (London, 1659).

73. These works are now in the Radcliffe Science Library, Oxford.

74. On Cowley at Oxford, see Arthur H. Nethercot, *Abraham Cowley: The muse's Hannibal* (London: Oxford University Press, 1931), pp. 80–92.

75. See the poems by all three in William Cartwright, *Comedies, Tragi-Comedies, with other Poems* (London, 1651).

76. Ibid., sigs *8r–[11r], [*16r–17v], pp. 284–286. These items were noticed and explicated by Richard A. Hunter and Ida Macalpine, "Sir John Berkenhead's lines on William Harvey, 1651," *J. Hist. Med.* 17 (1962): 403–405, and by C. A. R. Boyd and S. A. C. Boyd, "An early reference in English poetry to the circulation of the blood," *J. Hist. Med.* 25 (1970): 212–214.

77. M[artin] Ll[ewellyn], "To the Incomparable Dr. Harvey, On his Books of the *Motion* of the *Heart* and *Blood*, And of the *Generation* of *Animals*," in Harvey, *Generation*, sig. a1v-2r.

78. Keynes, *Harvey*, pp. 299, 309, 410, 412.

79. Birch, *History*, IV, 537.

80. Sharp, "Walter Charleton's early life," pp. 317, 337–339.

81. L. M. Payne, "Sir Charles Scarburgh's Harveian Oration, 1662," *J. Hist. Med.* 12 (1957): 158–164, esp. p. 163.

82. Wood, *Fasti*, I, 515.

83. Harvey, in a letter to Johann Daniel Horst, 13 July 1655, said that he had not seen Highmore for over seven years: Harvey, *Circulation*, p. 207.

84. See the letters from Joyliffe to John Evelyn, 1648–1649, describing his attempts to obtain a skeleton, and his work on botany: Evelyn Correspondence #950–953, Evelyn MSS., Library of Christ Church, Oxford.

85. Hartlib, "Ephemerides," 1648, f. N-01.

86. Ibid., 1650, f. F-G3.

87. Keynes, *Harvey*, pp. 314–317.

88. Harvey, *De motu cordis*, pp. 36, 68; *Motion*, pp. 50, 104; *Works*, pp. 40, 80.

89. Harvey, *De generatione*, pp. 4–5; *Generation*, pp. 6–7, 8–9; *Works*, pp. 173–174.

90. Harvey, *De generatione*, p. 26; *Generation*, p. 47; *Works*, p. 204.

91. Harvey, *De motu cordis*, p. 63; *Motion*, p. 96; *Works*, p. 74.

92. Payne, "Scarburgh's Harveian Oration," p. 162.

93. Harvey, *De circulatione*, p. 85; *Circulation*, p. 156; *Works*, p. 122.

94. British Library, MS. Sloane 486; Gweneth Whitteridge, ed. and trans., *De motu locali animalium 1627* (Cambridge: University Press, 1959).

95. Harvey, *De generatione*, p. 241; *Generation*, p. 442; *Works*, p. 498.

96. Harvey, *De generatione*, p. 256; *Generation*, p. 470; *Works*, p. 518.

97. Harvey, *De generatione*, pp. 229, 301; *Generation*, pp. 418, 555; *Works*, pp. 482, 585.

98. Harvey, *De generatione*, p. 19; *Generation*, p. 34; *Works*, p. 195.

99. Hartlib, "Ephemerides," 1652, f. Y-Y5.

100. Harvey, *De circulatione*, pp. 2−3; *Circulation*, pp. 121−122; *Works*, pp. 89−90.

101. Harvey, *De generatione*, pp. 63, 275; *Generation*, pp. 114, 507−508; *Works*, pp. 254, 546.

102. Payne, "Scarburgh's Harveian Oration," p. 163.

103. See the map of old Whitehall in *Wren Society*, 1930, vol. 7, plate VI.

104. *A Deep Sigh Breath'd Through the Lodgings at White-Hall, Deploring the Absence of the Court, and the Miseries of the Pallace* (London, 1642), sigs. A2r−v. Thomason's copy is dated 4 October 1642: George Fortescue, *Catalogue of the pamphlets . . . collected by George Thomason, 1640−1661* (London: British Museum Trustees, 1908), I, 177.

105. George S. Dugdale, *Whitehall through the centuries* (London: Phoenix House, 1950), pp. 53, 173.

106. Harvey, *De generatione*, p. 229; *Generation*, p. 418; *Works*, pp. 481−482.

107. Aubrey, *Brief lives*, I, 303.

108. Hartlib, "Ephemerides," 1652, f. Y−Y5.

109. Thomas Fuller, *The History of the Worthies of England* (London, 1662), under "Kent," p. 79.

110. Payne, "Scarburgh's Harveian Oration," p. 163.

111. Abraham Cowley, *Verses, Written upon Several Occasions* (London, 1663), p. 21.

112. Goodall, *The Royal College of Physicians*, sigs. Sslv-2r.

113. On Riolan and his criticisms of Harvey, see Nikolaus Mani, "Jean Riolan II (1580−1657) and medical research," *Bull. Hist. Med.* 42 (1968): 121−144, especially pp. 122, 128−132; also Gweneth Whitteridge, *William Harvey and the circulation of the blood* (London: Macdonald, 1971), pp. 175−182.

114. Harvey, Essay I, *De circulatione*, pp. 1−41; *Circulation*, pp. 121−140; *Works*, pp. 89−105.

115. Harvey, Essay II, *De circulatione*, pp. 43−124; *Circulation*, pp. 141−179; *Works*, pp. 109−141.

116. The miscellaneous and independent nature of Essay II has been noted by many scholars, although to my knowledge none has attempted to date it. See Keynes, *Harvey*, p. 327, and Whitteridge, *Harvey*, pp. 187−188.

117. Harvey, *De circulatione*, pp. 88−89; *Circulation*, p. 162; *Works*, p. 127. Sir Robert Darcy, of Dartford, Kent, died in 1618: British Record Society, *Index of wills proved in the Prerogative Court of Canterbury* (London: British Record Society, 1912), V (1605−1619), p. 113.

118. Harvey, *De circulatione*, p. 85; *Circulation*, p. 161; *Works*, p. 126. The experiment, on the internal jugular vein of a doe, was almost certainly done at the same time in the mid-1630s as the observations on reproduction in deer described in Essays 64−70 of *De generatione*, pp. 217−244; *Generation*, pp. 396−446; *Works*, pp. 466−501.

119. Harvey, *De circulatione*, p. 108; *Circulation*, p. 172; *Works*, p. 135. See Charles

Webster, "William Harvey's conception of the heart as a pump," *Bull. Hist. Med.* 39 (1965): 508–517.

120. Harvey, *De circulatione*, pp. 112–122; *Circulation*, pp. 174–178; *Works*, pp. 137–140.

121. First noted by F. M. G. de Feyfer, "Bemerkungen zu Harvey's *Exercitatio tertia* [sic] *de circulatione sanguinis* ad Joannem Riolanum Filium," *Janus* 14 (1909): 335–346. For a full treatment of van der Wale and his experiments, see J. Schouten, *Johannes Walaeus: Zijn betekenis voor de verbreiding van de leer van de bloedsomloop* (Assen: Van Gorcum, 1972); idem, "Johannes Walaeus (1604–1649) and his experiments on the circulation of the blood," *J. Hist. Med.* 29 (1974): 259–279; and Walter Pagel, *New light on William Harvey* (Basel/New York: Karger, 1976), pp. 113–135.

122. Harvey, *De circulatione*, pp. 68, 80, 94; *Circulation*, pp. 153, 159, 165; *Works*, pp. 119, 124, 129–130.

123. The argument that *De generatione* occupies a medial rather than terminal place among Harvey's works is brilliantly developed in Charles Webster, "Harvey's *De generatione*: its origins and relevance to the theory of circulation," *Brit. J. Hist. Sci.* 3 (1967): 262–274. However, as noted below, I would not accept his contention (p. 267) that the work was completed by 1638.

124. Harvey, *De generatione*, Dedicatory Epistle; *Generation*, sig. A3r; *Works*, p. 145.

125. Thomas Smith to Samuel Hartlib, 30 October 1648: Webster, "Harvey's *De generatione*," pp. 266–267.

126. Keynes, *Harvey*, p. 333.

127. Hartlib, "Ephemerides," 1649, f. G-H6.

128. The work, in Latin and English, was registered at the Company of Stationers on 20 March 1651: *Stationers' register*, I, 363.

129. Harvey to Paul Marquand Schlegel, 26 March 1651, and Harvey to Giovanni Nardi, 15 July 1651, in K. J. Franklin, ed., *The Circulation of the Blood and Other Writings* (London: Dent, 1963), pp. 183, 192.

130. Harvey, *De generatione*, p. 15; *Generation*, p. 26; *Works*, p. 188.

131. Harvey, *De generatione*, p. 54; *Generation*, p. 97; *Works*, p. 241.

132. Harvey, *De generatione*, p. 30; *Generation*, p. 53; *Works*, p. 208. For a description of the trip, see Keynes, *Harvey*, pp. 196–201.

133. Harvey, *De generatione*, p. 48; *Generation*, p. 88; *Works*, p. 234.

134. Harvey, *De generatione*, p. 32; *Generation*, p. 57, *Works*, p. 211.

135. Harvey, *De generatione*, pp. 4–5, 20, 26, 63; *Generation*, pp. 7–8, 35, 47, 114; *Works*, pp. 173–174, 195, 204, 254.

136. Harvey, *De generatione*, p. 229; *Generation*, p. 419; *Works*, p. 482.

137. Harvey, *De generatione*, pp. 241, 256; *Generation*, pp. 442, 470; *Works*, pp. 498, 518.

138. Harvey, *De generatione*, p. 229; *Generation*, p. 418; *Works*, pp. 481–482. The translation in *Generation* calls Charles I our "late" king, but this is not in the Latin.

139. Harvey, *De generatione*, pp. 260, 263, 267, 271, 275–276; *Generation*, pp. 477, 484, 492, 499, 508–509; *Works*, pp. 526, 530, 536, 540, 546–547.

140. Harvey, *De generatione*, pp. 291, 301; *Generation*, pp. 536–537, 566; *Works*, pp. 570–571, 586.

141. Webster, "Harvey's *De generatione*," p. 266.

142. Kenelm Digby, *Two treatises, in one of which, the nature of bodies; in the other, the nature of mans soule, is looked into* (Paris, 1644), p. 221.

143. On Digby's movements from 1636 to 1641, see R. T. Petersson, *Sir Kenelm Digby: The ornament of England, 1603–1665* (London: Jonathan Cape, 1956), pp. 120–155.

144. Harvey, *De generatione*, p. 154; *Generation*, p. 283; *Works*, p. 380.

145. Harvey, *De generatione*, p. 186; *Generation*, p. 339; *Works*, p. 423.

146. Harvey, *De circulatione*, p. 17; *Circulation*, p. 129; *Works*, p. 95.

147. Harvey, *De generatione*, pp. 156−157; *Generation*, pp. 285−287; *Works*, pp. 382−384.

148. Harvey, *De generatione*, p. 157; *Generation*, p. 287; *Works*, p. 384.

149. Harvey said that young Montgomery was the eldest son of Viscount Montgomery, that he was between eighteen and nineteen years of age, and that he came to London after he traveled through France and Italy. This would be Hugh, eldest son of the second Viscount Montgomery of the Great Ardes. He was called home from the Continent after the outbreak of the Irish rebellion in Ulster on 23 October 1641, and probably after his father was given a commission on 19 November 1641. The younger Hugh became the third Viscount Montgomery on his father's death on 15 November 1642. See William Montgomery, *The Montgomery manuscripts*, ed. George Hill (Belfast: Cleeland and Dargan, 1869), pp. 151−153. Since the king left Whitehall in January 1642, Harvey must have seen young Hugh sometime between November 1641 and January 1642.

150. Harvey, *De generatione*, Essays 16−19 (misnumbered 15−18 in Latin), pp. 48, 49−55, 59, 61−64; *Generation*, pp. 87, 89−100, 108, 111−116; *Works*, pp. 233, 235−242, 249, 252−255.

151. Harvey, *De generatione*, Essay 18 (misnumbered 17), p. 59; *Generation*, p. 108; *Works*, p. 249.

152. Harvey, *Prelectiones*, pp. 126−127.

153. Harvey, *De motu cordis*, pp. 28−29; *Motion*, pp. 36−37, 38; *Works*, pp. 28−31.

154. Harvey, *De generatione*, pp. 244−251; *Generation*, pp. 447−462; *Works*, pp. 501−512.

155. Harvey, *De generatione*, p. 245; *Generation*, p. 449; *Works*, p. 503.

156. Harvey, *De generatione*, p. 249; *Generation*, p. 458; *Works*, p. 510.

157. Harvey, *De generatione*, pp. 149−153; *Generation*, pp. 274−280; *Works*, pp. 374−378.

158. Harvey, *De circulatione*, pp. 102−103, 110−115; *Circulation*, pp. 169, 173−175; *Works*, pp. 133, 136−138.

159. Harvey, *De circulatione*, pp. 59−68; *Circulation*, pp. 149−153; *Works*, pp. 115−120.

160. Harvey, *De circulatione*, pp. 23−24, 66−67; *Circulation*, pp. 132, 152; *Works*, pp. 98, 118−119.

161. Harvey, *De circulatione*, pp. 101−103, 110; *Circulation*, pp. 169, 173; *Works*, pp. 133, 136.

162. Harvey, *De generatione*, pp. 262−263; *Generation*, pp. 482−484; *Works*, pp. 529−530.

163. Harvey, *De generatione*, p. 263; *Generation*, p. 483; *Works*, p. 530.

164. Harvey, *De circulatione*, pp. 54, 58, 111; *Circulation*, pp. 146−148, 173; *Works*, pp. 113−115, 136.

165. Harvey, *De generatione*, pp. 159−160; *Generation*, p. 292; *Works*, pp. 387−388.

3: THE SCIENTIFIC COMMUNITY IN COMMONWEALTH AND RESTORATION OXFORD

1. For an introduction to this topic, see Mark H. Curtis, *Oxford and Cambridge in transition, 1558−1642* (Oxford: Clarendon Press, 1959); Kenneth Charlton, *Education in renaissance England* (London: Routledge & Kegan Paul, 1965); Hugh Kearney, *Scholars and gentlemen: Universities and society in preindustrial Britain, 1500−1700*

(Ithaca, New York: Cornell University Press, 1970); Lawrence Stone, ed., *The university in society* (Princeton, New Jersey: Princeton University Press, 1974).

2.   For a summary of admissions, B.A.s, and M.A.s at Oxford, see my "Science, medicine, and the universities of early modern England: Background and sources," *Hist. Sci.* 11 (1973): 194–216, 239–269, esp. p. 213.

3.   *Victoria history of the county of Oxford*, ed. H. E. Salter and M. D. Lobel, vol. III (London: Institute of Historical Research, 1954), pp. 96, 102, 115–116, 127–128, 169–170, 216–217, 240–242, 272–274, 283–285, 289.

4.   See Wilbur K. Jordan, *Philanthropy in England, 1480–1660* (London: Allen & Unwin, 1959).

5.   On scientific benefactions see my "Science, medicine, and the universities," pp. 239–242, 264–265.

6.   *The Bodleian Library in the seventeenth century* (Oxford: Bodleian Library, 1951), pp. 7–13, 21.

7.   Thomas Hyde, *Catalogus impressorum librorum bibliothecae bodlejanae in academia Oxoniensi* (Oxford, 1674), esp. I, 58, 143, 279.

8.   Derived from a computer manipulation of biographical information found in R. B. McKerrow et al., *A dictionary of printers and booksellers in England, Scotland, and Ireland, and of foreign printers of English books, 1557–1640* (London: Bibliographical Society, 1910).

9.   The auction catalogues of his entire stock are *Catalogus in quaevis lengua & facultate insignium tam antiquorum quam recentium librorum Richardi Davis bibliopolae*, three parts (London, 1686–1688).

10.   Frank, "Science, medicine, and the universities," p. 213.

11.   The following conclusions are drawn largely from information in Montagu Burrows, ed., *The register of the Visitors of the University of Oxford, from A.D. 1647 to A.D. 1658* (London: Camden Society, 1881). See also the treatment of the Visitors in Charles Webster, *The great instauration: Science, medicine and reform, 1626–1660* (New York: Holmes & Meier, 1976), pp. 131–133.

12.   Anthony Wood, *The history and antiquities of the colleges and halls in the University of Oxford* (Oxford, 1786), p. 596.

13.   Walter Pope, *The life of the right reverend father in God Seth, Lord Bishop of Salisbury* (London, 1697), pp. 18–21.

14.   Burrows, *Register*, pp. xcvi, cii, cix, cxii, cxxi.

15.   On the medical course at Oxford, see my "Science, medicine, and the universities," pp. 207–211, and the sources cited therein.

16.   [Seth Ward], *Vindiciae academiarum* (Oxford, 1654), pp. 2, 29–30, 46. Pages 1–7 are an introductory letter by John Wilkins.

17.   The Laudian statutes (1636) required the Sedleian Professor to lecture twice a week in term on any or all of Aristotle's scientific works, including *Physica, De caelo, De meteoris, Parva naturalia, De anima*, and *De generatione et corruptione*: John Griffith, ed., *Statutes of the University of Oxford codified in the year 1636 under the authority of Archbishop Laud, Chancellor of the University* (Oxford: Clarendon Press, 1888), pp. 36–37. Some of Willis's very un-Aristotelian lectures, in the hand of Richard Lower, are in the Royal Society, Boyle Papers, XIX, ff. 1–33. See Chapter VII.

18.   Wood, *History and antiquities*, II, 884.

19.   See the discussion of these disputations in Chapter IV.

20.   Ward's manuscripts are now in the Folger Shakespeare Library, Washington, D.C. They and their contents are described in my article, "The John Ward diaries: Mirror of seventeenth century science and medicine," *J. Hist. Med.* 29 (1974): 147–179, and are

hereafter cited as "Ward, Diary." The Locke manuscripts are described in P. Long, *A summary catalogue of the Lovelace Collection of the papers of John Locke in the Bodleian Library* (Oxford: University Press, 1959), and again in "The Mellon donation of additional manuscripts of John Locke from the Lovelace Collection," *Bodleian Library Record* 7 (1964): 185–193.

21. Bodleian Library, MS. Locke e. 4 [c. 1652–1659], pp. (Helmont) 43–46, 57, 125; (Borel) 43, 45, 57, 107, 130–131; (Glisson) 73; (Ent) 35; (Rudbeck) 152. For a detailed treatment of this manuscript see Kenneth Dewhurst, "An Oxford medical student's notebook," *Oxford Medical School Gazette* 11 (1959): 141–145.

22. Bodleian Library, MS. Locke e. 4, pp. 31, 86–95, 99–106.

23. Bodleian Library, MS. Locke f. 20 [c. 1654–1658], passim; Ward, "Diary," III [c. 1652–1658], ff. 12r–18v, 26r–38v, 44r–49v, 52r–61v, 87r–89r.

24. Bodleian Library, MSS. Locke f. 18, d. 9 and 11, f. 19; British Library MS. Add. 32,554.

25. Ward, "Diary," VII, f. 30v, 55r, 88r; VIII, f. 5v, 31v, and 52r–104v passim.

26. Bodleian Library, MSS. Locke c. 41, b. 7, and f. 49. For a description, see J. W. Gough, "John Locke's herbarium," *Bodleian Library Record* 7 (1962): 42–46.

27. Ward, "Diary," IX, f. 76v.

28. Samuel de Sorbière, *Relation d'un voyage en Angleterre* (Paris, 1664), pp. 105–106.

29. [Richard Watkins], *Newes from the dead. Or a true and exact narrative of the miraculous deliverance of Anne Greene*, 2d ed., (Oxford, 1651), pp. 1–8, quote from p. 8.

30. Petty to Hartlib, 16 December 1650: Hartlib Papers, Bundle VIII (23).

31. Oxford University Archives, "Registrum Convocationis, 1647–1659," pp. 124–125.

32. Wood, *Life and times* (4 May 1658), I, 250–251; Ward, "Diary," VII, ff. 65r–66r; X, f. 54r.

33. Ward, "Diary," VII, f. 52r.

34. Lower to Boyle, 4 June 1663 and 24 June 1664: *Works*, VI, 467, 470–471.

35. Thomas Willis, *Pathologiae cerebri, et nervosi generis specimen* (Oxford, 1667), pp. 50, 61, 84, 106–108, 121–123, 161–165, 200, 219.

36. Ward, "Diary," VIII, ff. 40r–41r, 43r; X, f. 56v; IV, ff. 25v, 35v.

37. John Lydall to John Aubrey, 23 January 1648/9, 20 February 1648/9, and 5 April 1649, in Bodleian Library, MS. Aubrey 12 ff. 294r, 296r, 302r. See also my article, "John Aubrey, F.R.S., John Lydall, and science at Commonwealth Oxford," *Notes Rec. R. Soc. Lond.* 27 (1973): 193–217.

38. Wood, *Athenae*, IV, 477.

39. Aubrey, *Brief lives*, I, 410–411.

40. Ward, "Diary," VII, f. 32r passim; VIII, f. 72v passim.

41. Wood, *Life and times*, I, 290, 472–473. See also G. H. Turnbull, "Peter Stahl, the first public teacher of chemistry at Oxford," *Ann. Sci.* 9 (1953): 265–270.

42. Ward, "Diary," VIII, ff. 1v, 10r, 71r, 72r, 87v, 90r, 98r–v.

43. Ward, "Diary," XIII.

44. Bodleian Library, MS. Locke f. 25. See mention of this in Locke to Boyle, 24 February 1666/7: *Works*, VI, 537.

45. Frank, "John Ward Diaries," pp. 152–154, and sources cited there.

46. John Wallis, *Opera mathematica* (Oxford, 1699), I, 359.

47. John Wallis, *A treatise of algebra, both historical and practical* (London, 1685), p. 293.

48. John Wallis, *A fourth letter, concerning the Sacred Trinity* (London, 1681), p. 18.

49. Wallis, *Opera*, pp. 479, 483.

50. Hartlib, "Ephemerides," 1655, f. 34—34-3.

51. See Albert Van Helden, "Christopher Wren's *De corpore Saturni*," *Notes Rec. R. Soc. Lond.* 23 (1968): 213—229.

52. John Wallis, *Exercitationes tres* (London, 1678), p. 1.

53. Evelyn, *Diary*, III, 384—385.

54. Robert Boyle, *New experiments physico-mechanical touching the spring of the air* (Oxford, 1660), pp. 145—147.

55. Hartlib, "Ephemerides," 1657, f. 54—54-4.

56. Boyle, *New experiments*, pp. 111, 132, 303—304.

57. Robert Boyle, *A continuation of new experiments physico-mechanical, touching the spring and weight of the air, and their effects. The I. part* (Oxford, 1669), pp. 3, 5, 10, 48. Boyle, *Tracts . . . of a discovery of the admirable rarefaction of the air* (London, 1671), pp. 5, 17, 21.

58. The emphasis on Wadham began with Thomas Sprat, *The history of the Royal-Society of London, for the improving of natural knowledge* (London, 1667), pp. 53—58, and has recently, if somewhat overenthusiastically, been defended in Margery Purver, *The Royal Society: Concept and creation* (Cambridge, Mass.: M. I. T. Press, 1967).

59. Birch, *History*, I, 1—3, first noticed Wallis's statements about the "1645 Group" in London, and suggested it might be the same as the "Invisible College" mentioned by Boyle. Many subsequent historians have taken this conjecture as fact. Most of the older histories of the Royal Society place more emphasis on the continuity at London, and treat Oxford in the 1650s as a temporary efflorescence: see, for example, C. R. Weld, *A history of the Royal Society* (London: John W. Parker, 1858), I, 38—54; Sir Henry Lyons, *The Royal Society, 1660—1940: A history of its administration under its charters* (Cambridge: University Press, 1944), pp. 1—18; and Dorothy Stimson, *Scientists and amateurs: A history of the Royal Society* (New York: Schuman, 1948), pp. 1—69. Christopher Hill, in his *Intellectual origins of the English revolution* (Oxford: Clarendon Press, 1965), pp. 14—84, emphasized the development of science and medicine at London in the period before 1640. Purver's immoderate advocacy of Oxford origins, as well as P. M. Rattansi, "The intellectual origins of the Royal Society," *Notes Rec. R. Soc. Lond.* 23 (1968): 129—143, prompted a response from Hill, "The intellectual origins of the Royal Society—London or Oxford?" ibid., pp. 144—156. An interpretation recognizing both groups was argued by A. Rupert Hall and Marie Boas Hall, "The intellectual origins of the Royal Society—Oxford and London," ibid., pp. 157—168. See also R. H. Syfret, "The origins of the Royal Society," ibid., 5 (1948): 75—137, and a judicious summary by Douglas McKie, "The origins and foundation of the Royal Society of London," ibid., 15 (1960): 1—37. Charles Webster vigorously criticized many aspects of Purver's book in "The origins of the Royal Society," *Hist. Sci.* 6 (1967): 106—128.

60. Webster, *Great instauration*, pp. 153—178.

61. Locations of meetings are given in John Wallis, *A defence of the Royal Society* (London, 1678), pp. 5, 8; Anthony Wood, *History and antiquities*, II, 633.

62. Lt. Col. Kelsey, the deputy governor of the Oxford garrison, recommended Petty to Convocation on 27 September 1649, and he was given leave to proceed D.M. on 29 October 1649: Oxford University Archives, "Registrum Convocationis, 1647—1659," pp. 73, 75. The location at Buckley (or Bulkeley) Hall was worked out by Purver, *Royal Society*, pp. 119—120. On the building itself, which dates from the fourteenth century, see W. A. P., "Tackley's Inn," *The Oriel Record*, June, 1941, pp. 139—155.

63. Petty to Boyle, 17 February 1657[/8]: *Works*, VI, 139.

64. Sharrock to Boyle, 13 July 1664: Royal Society, Boyle Letters, V, f. 98.

65. Lord Edmund Fitzmaurice, *The life of Sir William Petty* (London: John Murray, 1895), pp. 5—13; Charles Webster, "New light on the Invisible College: The social relations of

English science in the mid-seventeenth century,'' *Trans. Roy. Hist. Soc.*, 5th series, 24 (1974): 19−42.

66. Wallis, *Defence*, p. 8.

67. C. L. Shadwell and H. F. Salter eds., *Oriel College records* (Oxford: Clarendon Press, 1926), p. 215; Hartlib, ''Ephemerides,'' 1653, f. HH−HH1.

68. On Wilkins see Barbara Shapiro, *John Wilkins, 1614−1672: An intellectual biography* (Berkeley and Los Angeles: University of California Press, 1969). John Evelyn saw Wilkins's collection on his visit to Wadham on 13 July 1654: Evelyn, *Diary*, III, 110.

69. William Coles, *The art of simpling* (London, 1657), p. 12; *Adam in Eden: Or, natures paradise* (London, 1657), sig. a2r.

70. Ward to Justinian Isham, 27 February 1651/2, in H. W. Robinson, ''An unpublished letter of Dr. Seth Ward relating to the early meetings of the Oxford Philosophical Society,'' *Notes Rec. R. Soc. Lond.* 7 (1949): 69; Wallis, *Defence*, p. 8.

71. Hartlib, ''Ephemerides,'' 1651, f. A−B2.

72. See Purver, *Royal Society*, pp. 121−126. The three lists of participants are preserved in the Bodleian Library, MS. Wood donat. 1, pp. 1−3.

73. Wallis, *Defence*, p. 8.

74. Boyle to Hartlib (draft), 14 September 1655: Royal Society, Boyle Papers, XXXVII (unfoliated).

75. Robert Hooke, autobiographical fragment in *The posthumous works* (London, 1705), p. iii.

76. Bodleian Library, MS. Ashmole 1810; transcribed in Douglas McKie, ''Origins and foundation,'' pp. 25−26. The attribution of these rules to Langbaine was first made by Purver, *Royal Society*, pp. 111−112.

77. Petty to Hartlib, 16 December 1650: Hartlib Papers, Bundle VIII (23).

78. Royal Society, Boyle Papers, XIX, f. 7r; see also Chapter 7.

79. Meetings of 31 July and 7 August 1661: Birch, *History*, I, 36.

80. Meetings of 25 February and 4 March 1663: ibid., I, 204, 206; John Wallis, *Mechanica: sive, de motu, tractatus geometricus, pars tertia* (London, 1671), pp. 759−767.

81. Robinson, ''Unpublished letter,'' p. 69; Hartlib, ''Ephemerides,'' 1655, f. 32−32-8; 1656, f. 44−44-4.

82. Robinson, ''Unpublished letter,'' pp. 69−70.

83. Willis, MS. 799[A], verso of final leaf and recto of inside back cover, in the Wellcome Library, London.

84. [Seth Ward], *Vindiciae academiarum* (Oxford, 1654), p. 35.

85. Boyle to Oldenburg: *Correspondence*, II, 453−652, passim.

86. John Wallis to Thomas Smith, 29 January 1696/7, in Thomas Hearne, *Peter Langtoft's Chronicle* (London, 1725), I, clxii.

87. Worsley to Hartlib, 12/22 June 1648: Hartlib Papers, Bundle VIII (27).

88. John Wilkins, *A discourse concerning the beauty of providence* (London, 1649), p. 49.

89. Hartlib, ''Ephemerides,'' 1655, ff. 34−34-3 to 6.

90. James Harrington, *The prerogative of popular government* (London, 1658), sig. A2v, responding to Matthew Wren, *Considerations on Mr. Harrington's Common-Wealth of Oceana* (London, 1657), which had been dedicated (sigs. A2r−7v) to Wilkins.

91. Matthew Wren, *Monarchy asserted* (London, 1659), sigs. A7v−8r.

92. John Britton, *Memoir of John Aubrey* (London, 1845), p. 93.

93. Hartlib, ''Ephemerides,'' 1650, ff. A−B8, G−H6, F−G8, H−J3; 1651, ff. A−B2,

A−B8; 1652, f. CC−CC5; 1653, ff. GG−GG6 & 7, LL−LL1 & 2, MM−MM1, NN−
NN8; 1654, f. ZZ−ZZ4; 1655, ff. 23−23-4, 34−34-3 to 6; 1657, ff. 50−50-4, 51−51-3,
53−53-5, 53−53-8, 54−54-4; 1658, f. 56−56-4.

94.   Evelyn, *Diary* (13 July 1654), III, 110.

95.   Walter Charleton, *The immortality of the human soul, demonstrated by the light of nature*
(London, 1657), pp. 43−48, quote on p. 48.

96.   See the biographical notes in *Correspondence*, I, xvii−xl.

97.   Oldenburg to Adam Boreel, Edward Lawrence, and Thomas Shirley, [April, 1656]:
Correspondence, I, 89/90−91, 92−93/94, 96/98.

98.   Oldenburg to Thomas Coxe, 24 January 1656/7: ibid., I, 112/113.

99.   Dedication to Seth Ward, *Phil. Trans.*, 1669, Preface to vol. 4.

100.   Statistics computed from the Poll Tax of 1667 printed in H. E. Salter, ed., *Surveys and
tokens* (Oxford: Clarendon Press, 1923), p. 216.

101.   Joseph Wells, *Wadham College* (London: F. E. Robinson, 1898), pp. 5−6, 15, 20−21,
23−24; Bernard W. Henderson, *Merton College* (London: F. E. Robinson, 1899),
pp. 40−41, 79, 109−110; Charles Grant Robinson, *All Souls College* (London: F. E.
Robinson, 1899), pp. 17−19, 74−76, 96−100; Henry L. Thompson, *Christ Church*
(London: F. E. Robinson, 1900), pp. 12, 35−36.

102.   Wood, *Athenae*, IV, 97; Edward Millington, *Bibliotheca Oweniana* (London, 1684)
contains a listing of almost 3,000 works, see especially pp. 23−31.

103.   Wood, *Life and times*, I, 306.

104.   See John Sargeaunt, *Annals of Westminster School* (London: Methuen, 1898), pp. 1−10,
21−24, 28−31, 81−93; G. F. Russell Barker, *Memoir of Richard Busby D.D. (1606−
1695), with some account of Westminster School in the seventeenth century* (London:
Lawrence and Bullen, 1895), pp. 4−9.

105.   Hartlib, "Ephemerides," 1653, f. FF−FF-5. Hooke discussed Pell's papers with Busby
on 31 October 1689: R. T. Gunther, *Early science in Oxford* (Oxford: For the author,
1935), X, 161.

106.   Aubrey, *Brief lives*, I, 410. Hooke continued to see Busby into the 1690s, primarily to
supervise construction at the school.

107.   British Library, MS. Add. 4292, ff. 84−87.

108.   Wood, *History and antiquities*, IV, 436−437. There is some doubt about whether this
lectureship was endowed by Busby, or founded in his name, since the bequest is not
mentioned in the probated form of Busby's will: Barker, *Memoir*, pp. 131−147.

109.   Joseph Welch, *The list of Queen's Scholars of St. Peter's College, Westminster*, ed.
Charles Bagot Phillimore (London: G. W. Giner, 1852), pp. 122−139. See also G. F.
Russell Barker and Alan H. Stenning, *The record of old Westminsters* (London: Chiswick
Press, 1928). 2 vols.

110.   Oxford University Archives, "Registrum examinatorum et candidatorum, 1638−1669.

111.   Oxford University Archives, "Admissiones ad regendum, 1634−1852."

112.   Oxford University Archives, "Registrum Congregationis, 1648−1659," f. 154r.

4: OXONIANS AND NEW APPROACHES TO PHYSIOLOGY

1.   An excellent assessment of almost every aspect of Paracelsianism is now available in
Allen G. Debus, *The chemical philosophy: Paracelsian science and medicine in the
sixteenth and seventeenth centuries* (New York: Science History Press, 1977).

2. The literature on the history of atomism is voluminous. For the period before the Renaissance, see Cyril Bailey, *The Greek atomists and Epicurus* (Oxford: Clarendon Press, 1928) and Andrew van Melson, *From atomos to atom* (Pittsburgh: Duquesne University Press, 1952).

3. For an entrée into the problems and sources of early modern atomism, see Kurd Lasswitz, *Geschichte der Atomistik vom Mittelalter bis Newton* (Hamburg: Leopold Voss, 1890); Marie Boas, "The establishment of the mechanical philosophy," *Osiris* 10 (1952): 412–541; Robert Hugh Kargon, *Atomism in England from Hariot to Newton* (Oxford: Clarendon Press, 1966).

4. Henrik de Roy, *Fundamenta physices* (Amsterdam, 1646) and *Fundamenta medica* (Utrecht, 1647); Cornelis Hooghelande, *Cogitationes, quibus Dei existentia et animae spiritualitas, et possibilis cum corpore unio, demonstrantur: necnon brevis historia oeconomiae corporis animalis proponitur, atque mechanice explicatur* (Amsterdam, 1646).

5. Marin Mersenne, *Cogitata physico-mathematica in quibus tam naturae quam artis effectus admirandi certissimis demonstrationibus explicantur* (Paris, 1644). On Mersenne see the excellent study by Robert Lenoble, *Mersenne ou la naissance du mécanisme* (Paris: Vrin, 1944).

6. On Gassendi's atomism see Bernard Rochot, *Les travaux de Gassendi sur Epicure et sur l'atomisme, 1619–1658* (Paris: Vrin, 1944), and Lasswitz, II, 126–188.

7. Pierre Gassendi, *Animadversiones in decimum librum Diogenis Laertii* (Lyon, 1649), I, 201–202.

8. Ibid., I, 202–222.

9. Ibid., I, 221–222.

10. Ibid., I, 222–236.

11. Ibid., I, 236–362.

12. Ibid., I, 317–331.

13. Petty to More, [March 1649]: Hartlib Papers, Bundle VII (123); the letter is quoted in full and explicated in Charles Webster, "Henry More and Descartes: Some new sources," *Brit. J. Hist. Sci.* 4 (1969): 359–377.

14. Cavendish to Petty, 7/17 April 1648: Hartlib Papers, Bundle VIII (29).

15. Boyle to Hartlib, 19 March and 8 May 1647: *Works*, I, xxxvii, xli.

16. Hartlib to Boyle, 9 May 1648: *Works*, VI, 77–78.

17. Petty to Boyle, 15 April 1653: *Works*, VI, 138.

18. Wallis to Oldenburg, 2 August 1666: *Correspondence*, III, 203.

19. Lindsay Sharp, "Walter Charleton's early life, 1620–1659, and relationship to natural philosophy in mid-seventeenth century England," *Ann. Sci.* 30 (1973): 311–340, especially 323–327.

20. Walter Charleton, *The darknes of atheism dispelled by the light of nature* (London, 1652), sigs. b3r, c2r, pp. 105, 274, 276, 314, 329.

21. Walter Charleton, *Physiologia Epicuro-Gassendo-Charltoniana: Or a fabrick of science natural, upon the hypothesis of atoms* (London, 1654).

22. Sharp, "Charleton's early life," p. 327.

23. Cf. Boyle to unknown correspondent, [early March 1647]: *Works*, VI, 39–40; Boyle to Katherine Jones, 6 March 1647: *Works*, I, xxxvi.

24. "An Invitation to a free and generous Communication of Secrets and Receits in Physick." Boyle promised this to Hartlib in his letter of 8 May 1647 (*Works*, I, xli), and it was published by Hartlib in his *Chymical, medicinal, and chyrurgical addresses* (London, 1655), pp. 113–150. See also Margaret Rowbottom, "The earliest published writing of Robert Boyle," *Ann. Sci.* 6 (1950): 376–389.

25. Robert Boyle, *Some motives and incentives to the love of God* (London, 1659), p. 26 [the work was completed by 6 August 1648]; *Occasional reflections upon several subiects* (London, 1665), Section II, pp. 188, 194 [written 1649].

26. Boyle, *Motives,* p. 56.

27. See Boyle's own list of his treatises, dated 25 January 1650, in the Royal Society, Boyle Papers XXXVI.

28. Royal Society, Boyle Papers XXVI, ff. 162−175. A small portion of this essay was printed by R. S. Westfall, "Unpublished Boyle papers relating to scientific method," *Ann. Sci.* 12 (1956): 63−73, 103−117. However, since the folio sheets are misbound in the MS. volume, Westfall thought the essay broke off after only two leaves, whereas it continues for twelve more. The proper sequence is ff. 162−163, 168−175, 166−167, 164−165.

29. Ibid., f. 162r.

30. Ibid., ff. 162r−163r.

31. Ibid., ff. 163r, 168r.

32. Ibid., f. 169r. The reference to Aristotle is to the *Historia animalium,* 5.2.539^{b}17-540^{a}26.

33. Robert Boyle, *Some considerations touching the usefulnesse of experimental naturall philosophy* (Oxford, 1663), First Part, Essay II [from internal evidence, written c. 1649−1651], pp. 39−40.

34. Royal Society, Boyle Papers XXVI ff. 170r−175r, 166r−167r, 164r−165r.

35. Hartlib, "Ephemerides," 1651, f. A−4.

36. Boyle, *Usefulnesse,* First Part, Essay I [written c. 1648−1650], p. 5.

37. Ibid., p. 19.

38. Petty to Boyle, 15 April 1653: *Works,* VI, 138.

39. Boyle to Frederick Clodius, [April 1654]: *Works,* VI, 54−55.

40. Boyle, *Usefulnesse,* First Part, Essay II [c. 1649−1651], pp. 37−38; Essay V [c. 1653−1654], pp. 94−96.

41. Ibid., Essay III [c. 1651−1653], pp. 54−55.

42. Ibid., Essay II [c. 1649−1651], pp. 35−36.

43. Ibid., Essay V [c. 1653−1654], p. 113.

44. Ibid., Essay IV, pp. 65−68.

45. Ibid., Essay IV, pp. 68−69.

46. Ibid., Essay IV, p. 70.

47. Ibid., Essay IV, p. 71.

48. Ibid., Essay IV, pp. 72−74.

49. Ibid., Essay IV, pp. 75−78.

50. Ibid., Essay IV, pp. 79−81.

51. Ibid., Essay IV, p. 86.

52. Nathaniel Highmore, *Corporis humani disquisitio anatomica* (The Hague, 1651), sig. A1r−v.

53. Ibid., p. 144.

54. Ibid., p. 145.

55. Ibid., p. 146.

56. Ibid., pp. 157−160.

57. Ibid., pp. 168−171.

58. Ibid., p. 171.

59. Andreas Vesalius, *De humani corporis fabrica* (Basel, 1543), pp. 660–661.

60. Harvey, *De motu cordis,* p. 17; *Motion,* p. 18; *Works,* p. 16.

61. Highmore, *Disquisitio,* pp. 171–172.

62. Ibid., pp. 180–182.

63. Ibid., pp. 189–190.

64. Ibid., pp. 178–179.

65. Nathaniel Highmore, *The history of generation* (London, 1651), p. 115. I have used Highmore's own copy, now in the British Library. It has a number of interesting corrections and additions in Highmore's hand.

66. Ibid., pp. 116–124.

67. Ibid., pp. 125–141.

68. Kenelm Digby, *Two treatises* (Paris, 1644), pp. 203–204, 221–225, 233–239.

69. Highmore, *History,* pp. 5–24, 40.

70. Ibid., pp. 40–46.

71. Ibid., pp. 46–66.

72. Ibid., pp. 67–83.

73. Ibid., pp. 96–105.

74. Ibid., Dedicatory Epistle, sig. ¶ 3r.

75. In Vol. III of the Petty Papers, labelled "Sir William Petty's Medical Studies and Other Papers." The volume is divided into numbered items, each of which is paginated or foliated separately. The Petty Papers are now in the possession of his descendant, the Marquis of Lansdowne, and are kept at Bowood, Calne, Wiltshire. These MSS. were used with the kind permission of Lord Lansdowne, and with the assistance of his agent, Mr. J. R. Hickish.

76. Petty Papers, III, (1) and (7). Petty referred to these in a catalogue of his own writings as "Six Physico-Medical Lectures read at Oxford 1649[/50]": see Lord Edmund Fitzmaurice, *The life of William Petty, 1623–1687* (London: John Murray, 1895), p. 317.

77. Petty Papers, III, (27), "Oratio inauguralis anatomica scholis publicis publico habenda. 4to Martij, 1651/50 Oxonii," 12 pp.

78. Petty Papers, III, (4), 57 ff. The notes are divided into five sections, I–V, and deal with strictly anatomical matters. They are in a rough, hasty hand and have been heavily corrected by Petty himself.

79. Petty Papers, III, (26), "Lectio anatomica proemialis habenda in actu publico, Oxonii Annj 1651," 13 pp.

80. Petty Papers, III, (28) and (30). These seem to correspond to the second and third of the "Three Ostiological Lectures 1651" mentioned in his catalogue of writings: see Fitzmaurice, *Life,* p. 317.

81. Petty Papers, III, (10), 38 pp. These seem to correspond to the item "Seven Months Practice in a chemical Laboratory 1645" in Petty's catalogue of his own writings.

82. Ibid., pp. 2, 9, 29–30.

83. Ibid., p. 16.

84. Ibid., pp. 4–5, 7–8, 21–23, 35.

85. Petty Papers, III, (6).

86. Wood, *Fasti,* II, 189.

87. The *Syntagma* had appeared in three editions by the 1650s: Frankfurt (1641), Padua (1647) and Padua (1651). The two Padua editions are identical in text; I have used them

because they are the ones most likely to have been in use at Oxford in the 1650s.

88. Johannes Vesling, *Syntagma anatomicum* (Padua, 1651), sig. +4v.

89. Ibid., pp. 117–121.

90. Ibid., pp. 121–123.

91. Ibid., pp. 38–40.

92. Ibid., p. 122.

93. Oxford University Archives, "Registrum Congregationis, 1634–1647," f. 179v.

94. Ibid.

95. Ibid., f. 181r.

96. Ibid., f. 180v.

97. "Registrum Congregationis, 1648–1659," f. 154r; (1659–1669), f. 176v.

98. "Registrum Congregationis, 1659–1669," f. 176v.

99. Ibid. (1648–1659), f. 150v; (1659–1669), f. 175r.

100. Ibid., (1648–1659), f. 151v.

101. Ibid., ff. 150v, 152v.

102. Ibid., f. 150v.

103. Ibid. (1659–1669), f. 176v.

104. Ibid. (1648–1659), f. 152v.

105. Ibid. (1659–1669), f. 174r.

106. Ibid., f. 175v.

107. Ibid., f. 176v.

108. Ibid. (1648–1659), f. 150r.

109. Ibid., f. 152r, 153v; (1659–1669), f. 176v.

110. Ibid. (1648–1659), f. 154r.

111. Ibid., f. 151v.

112. Ibid. (1659–1669), f. 175r; (1669–1680), f. 203r.

113. Ibid. (1648–1659), ff. 152v, 153v.

114. Ibid., f. 152v.

115. Ibid., f. 153v.

116. Ibid. (1659–1669), f. 175v.

117. Ibid. (1648–1659), f. 150v.

118. Ibid., f. 151v.

119. Ibid., ff. 153v–154r.

120. Ibid. (1659–1669), f. 174r.

121. Ibid. (1659–1669), f. 176v; (1669–1680), f. 203v.

122. Thomas Warton, *The life and literary remains of Ralph Bathurst, M. D.* (London, 1761), pp. 204–205, quoting an autobiographical fragment by Bathurst.

123. See Robert G. Frank, Jr., "John Aubrey, F. R. S., John Lydall, and science at Commonwealth Oxford," *Notes Rec. R. Soc. Lond.* 27 (1973): 193–217.

124. Warton, *Life,* p. 39; both Warton p. vi, and the *Biographia Britannica* (London, 1747–1766), VI, 9, mention two volumes of MS. letters from Bathurst to Willis among the Ashmolean manuscripts, but these seem not to have survived into the twentieth century.

125. [Richard Watkins], *Newes from the dead* (Oxford, 1651), p. 3. According to Warton, *Life,* p. xi, Bathurst also wrote (for friends, above whose names they appeared) five of the commendatory poems that were published in this pamphlet.

126. Oxford University Archives, "Registrum Convocationis, 1647–1659," pp. 199, 201.

127. See the letters by Whistler of 25 August 1653 and 14 October 1653, praising Bathurst's diligence: CSPD (1653–1654), pp. 104, 507.

128. Montagu Burrows, ed., *The register of the Visitors of the University of Oxford, from A.D. 1647 to A.D. 1658* (London: Camden Society, 1881), pp. 389–390.

129. W[illiam] D[urham], *The life and death of that judicious divine and accomplished preacher, Robert Harris, D. D., late President of Trinity Colledge in Oxon.* (London, 1660), p. 50.

130. Thomas Hobbes, *Humane nature* (Oxford, 1650), sig. A12r–v; Bathurst to Hobbes, 27 May 1651: Warton, *Life,* p. 49.

131. "Registrum Congregationis, 1648–1659," f. 150v. These were printed in Warton, *Life,* pp. 211–238, who identified them as being *pro gradu M.B.,* and assigned them to 1654. This is clearly wrong; they were defended in 1651. Bathurst argued a different set of questions in 1654.

132. Ibid., pp. 211–221. Bathurst drew his arguments from *exercitationes* 58, 69, 70, and "On the uterine membranes and humours": Harvey, *De generatione,* pp. 194–199, 229–244; *Generation,* pp. 352–359, 420–446; *Works,* pp. 434–442, 482–501.

133. Warton, *Life,* pp. 222–228.

134. Ibid., pp. 229–238.

135. Warton, *Life,* pp. 127–210. In his will (Oxford University Archives, Will of Ralph Bathurst, D. M., August 29, 1698, p. 6) Bathurst left all his papers to his nephew, and executor, Richard Healy of Wells, Somersetshire. Healy passed the MSS. on to his grandson, [Thomas?] Payne, Prebendary of Wells. Payne, in turn, communicated them to Warton in the 1750s. I have been unable to locate any subsequent references to the existence or location of the MSS.

   The "Praelectiones" existed then in two copies, one holograph, and a second in another hand, with corrections by Bathurst. Throughout the MS. as printed are a number of cryptic references to "a" followed by a number and sometimes further by a second Greek letter. Warton notes (p. ix) that these symbols probably refer to Bathurst's *Loci commones medicinae,* books of observations and experimental results, which Bathurst quoted on one occasion in the "foul" MS.

136. Ibid., p. 127.

137. Ibid., pp. 128–132.

138. Ibid., pp. 132–147. The Harvey reference is to the *De generatione,* the Highmore to the *Disquisitio* (1651), and the Ent to the *Apologia pro circulatione sanguinis* (London, 1641).

139. Bathurst, "Praelectiones," pp. 148–155.

140. Ibid., pp. 153–155.

141. Ibid., p. 152.

142. Ibid., pp. 155–157.

143. Ibid., pp. 160–173.

144. Ibid., p. 162.

145. Petty to Hartlib, 16 December 1650: Hartlib Papers, Bundle VIII (23); John Wallis, *Mechanica: sive, de motu, tractatus geometricus, pars tertia* (London, 1671), pp. 759–767.

146. Bathurst, "Praelectiones," pp. 174–178.

147. Ibid., pp. 178–183.

148. Ibid., pp. 184–185.

149. Ibid., pp. 185–186.

150. Ibid., pp. 187−193.

151. Ibid., pp. 193−203.

152. Ibid., pp. 204−210.

153. Ibid., pp. 187−188, 191−192.

154. Ent, *Apologia,* p. 18.

155. Ibid., pp. 96−98.

156. Warton, *Life,* p. 159.

157. Bathurst, ''Praelectiones,'' p. 192.

158. Ibid., p. 197.

159. Ibid., p. 195.

160. Partington, *History,* II, 574, cites Warton's statement that the lectures had remained in Bathurst's hands ''for some years, received his repeated corrections, and were highly approved by Mr. Boyle.'' (Warton, *Life,* p. 60) He concludes from this that Bathurst's lectures, as printed, may include materials added up until his death in 1704, and that we do not know which and when. This is a most unlikely interpretation that vitiates a valuable piece of evidence. Warton noted two recensions of the lectures among the Bathurst MSS., one a corrected copy, and the other from which the text was printed. It seems likely that corrections were made up until 1656, when Bathurst decided not to publish. Internal evidence bears this out; of the works cited, none was published after 1654.

161. Boyle is often incorrectly said to have moved to Oxford in 1654. For an accurate précis of Boyle's movements in the years before he came to Oxford, see R. E. W. Maddison, ''Studies in the life of Robert Boyle, F.R.S. Part VI, the Stalbridge period, 1645−1655, and the Invisible College,'' *Notes Rec. R. Soc. Lond.* 18 (1963): 104−124. Boyle's sister, Katherine Jones, Viscountess Ranelagh, inspected Boyle's prospective lodgings, and noted that he was ''here much desired'': Katherine Jones to Boyle, 12 October [1655]: *Works,* VI, 523−524.

162. Boyle to Bathurst, 14 April 1656: Warton, *Life,* pp. 163−165.

163. Hartlib, ''Ephemerides,'' 1656, f. 40−40-5.

164. British Library, MS. Sloane 548, ff. 14r−17v.

5: ROBERT BOYLE ON NITER AND THE PHYSICAL PROPERTIES OF THE AIR

1. Wilhelm Adolf Scribonius, *Naturall philosophy* (London, 1621), pp. 14−16, 19; Christoph Scheibler, *Philosophia compendiosa* (Oxford, 1639), pp. 55, 57, 60−62; Frans Burgersdijck, *Idea philosophiae naturalis* (Oxford, 1637), pp. 46−47; Eustace, *Summa philosophiae quadripartita* (Cambridge, 1640), pp. 226−228; Daniel Sennert, *Epitome naturalis scientiae* (Oxford, 1632), pp. 189−190; and especially Johann Magirus, *Physiologiae peripateticae libri sex* (Cambridge, 1642), pp. 100−104. For an overview of science textbooks in this period, see Patricia Reif, ''The textbook tradition in natural philosophy, 1600−1650,'' *J. Hist. Ideas* 30 (1969): 17−32.

2. For Aristotle's ideas on air, see: *De caelo,* $311^b4−11$; *Meteorologica,* $339^a36−^b17$, $340^a22−26$, $341^a17−20$, $341^b1−351^a18$, $359^b26−371^b17$.

3. For Aristotle's arguments against the void, see *Physica,* $213^a11 − 217^b28$.

4. The best scholarly book on the events of the 1640s, set in the context of classical and medieval arguments on the void, is Cornelis de Waard, *L'expérience barométrique: ses antécédents et ses explications* (Thouars: J. Gamon, 1936). See also W. E. Knowles Middleton, ''The place of Torricelli in the history of the barometer,'' *Isis* 45 (1963):

11–28, and his *History of the barometer* (Baltimore: Johns Hopkins Press, 1964), pp. 3–32.

5. On the interpretation of the Torricellian experiment, see Middleton, *Barometer*, pp. 33–54, and the excellent treatment by Charles Webster, "The discovery of Boyle's Law, and the concept of the elasticity of air in the seventeenth century," *Arch. Hist. Exact Sci.* 2 (1965): 441–502.

6. John Wallis, *Defence of the Royal Society* (London, 1678), p. 8.

7. Haak to Mersenne, 24 March/3 April 1648: Bibliothèque Nationale, Nouvelles Acquisitions Françaises 6206, f. 91, quoted from Harcourt Brown, *Scientific organizations in seventeenth century France, 1620–1680* (Baltimore: Williams & Wilkins, 1934), p. 58.

8. Haak to Mersenne, 3/13 July 1648: N.A.F. 6206, f. 64, quoted from Brown, *Scientific organizations*, p. 271.

9. Cavendish to Petty, 7/17 April 1648: Hartlib Papers, Bundle VIII (29), also printed in Webster, "Discovery," p. 456. Hartlib to Boyle, 9 May 1648: *Works*, VI, 77–78.

10. Pierre Gassendi, *Animadversiones in decimum librum Diogenis Laertii* (Lyons, 1649), I, 425–427, 434–440.

11. Walter Charleton, *Physiologia Epicuro-Gassendo-Charltoniana: or a fabrick of science natural, upon the hypothesis of atoms* (London, 1654), pp. 35–61. See especially the discussion of this in Webster, "Discovery," pp. 457–458.

12. Jean Pecquet, *Experimenta nova anatomica* (Harderwijk, 1651), pp. 91–109; see also Webster, "Discovery," pp. 451–454.

13. Webster, "Discovery," pp. 459–464.

14. [Seth Ward], *Vindiciae academiarum* (Oxford, 1654), pp. 2, 11, 28, 33, 36, 53.

15. Seth Ward, *In Thomae Hobbii philosophiam exercitatio epistolica* (Oxford, 1656), pp. 93, 120–124, 132–142, 157, 161, 167–168, 170–173, 189–194.

16. Henry Guerlac, "John Mayow and the aerial nitre," *Actes du VIIe Congrès international d'Histoire des Sciences* (1953), 332–349; and his "The poets' nitre," *Isis* 45 (1954): 243–255; Partington, *History*, II passim; Allen G. Debus, "The Paracelsian aerial niter," *Isis* 55 (1964): 43–61.

17. Francis Bacon, *Sylva sylvarum or a naturall history in ten centuries* (London, 1627), pp. 23–24, 27, 97, 249; Guerlac, "Poets' nitre," p. 248.

18. Bacon, *Sylva sylvarum*, pp. 117, 150.

19. Debus, "Paracelsian aerial niter," pp. 46–47.

20. Ibid., pp. 47–48.

21. Ibid., pp. 52–54.

22. Ibid., pp. 49–50.

23. Guerlac, "Mayow and the aerial nitre," pp. 339–341.

24. Guerlac, "Poets' nitre," p. 251.

25. Daniel Sennert, *Hypomnemata physica* (1636) in *Opera omnia* (Paris, 1641), I, 427. Guerlac, "Poets' nitre," pp. 252–253, notes that Sennert's explanation was adopted, *sans* atomism, by Jan Amos Comenius in 1643 and by Thomas Browne in 1646.

26. Guerlac ("Mayow and the aerial nitre," p. 342) argues that Seton and Sendivogius were possibly Ent's sources.

27. Especially Bodleian Library MS. Locke f. 18 passim.

28. Bodleian Library MS. Locke f. 20, p. 174.

29. It continued as a tutorial favorite through the Commonwealth and Restoration periods; further editions were published at Oxford in 1653 and 1664.

30. Royal Society, Boyle Papers, XXVIII, pp. 309–311.

31. Ibid., VIII, f. 140r.

32. Ibid., VIII, f. 140v.

33. Ibid., XXV, p. 157.

34. Boyle to Worsley, [c. 1647]: *Works*, VI, 40.

35. Hartlib Papers, Bundles XXXIX (1), LXVI (15), and LXXI (11). For an excellent treatment of these saltpeter projects, see Charles Webster, *The great instauration: science, medicine and reform 1626–1660* (New York: Holmes & Meier, 1976), pp. 59–61, 377–380, 386, 457.

36. Hartlib, "Ephemerides," 1654, f. RR–RR1. Cf. also ff. SS–SS5, WW–WW2.

37. Royal Society, Boyle Papers, VIII, f. 142r; Hartlib, "Ephemerides" 1655, f. 24–24-3.

38. Royal Society, Boyle Papers, VIII, f. 142v.

39. Ibid., f. 147r.

40. Ibid., f. 147v.

41. Hartlib to Boyle, 28 February 1654: *Works*, VI, 79. Hartlib passed on information about saltpeter to Boyle in this letter (p. 80), and in his subsequent one of 8 May 1654 (p. 85).

42. Marie Boas, "An early version of Boyle's Sceptical chymist," *Isis* 45 (1954): 153–168.

43. Robert Boyle, *Certain physiological essays, written at distant times, and on several occasions* (London, 1661). The work consists of five essays: (1) "A Proemial Essay," pp. 1–35; (2) "Of the Unsuccessfulness of Experiments," pp. 37–66; (3) "Of the Unsucceeding Experiments," pp. 67–105; (4) "A Physico-Chymical Essay, Containing an Experiment with some Considerations touching the differing Parts and Redintegration of Salt-Petre," pp. 107–135; (5) "The History of Fluidity and Firmnesse," pp. 137–249. Boyle noted in the prefatory "Advertisement," sig. A2r, that (1) was written about four years before, i.e., 1657, and that (2), (3) and (4) were written before it, i.e., before 1657. In a "Preface" intercalated between p. 105 and p. 107, Boyle noted, sig. P5r, that (5) was written "but the last year save one," i.e., 1659.

44. Ibid., "Preface" sig. P3v, pp. 6–7.

45. Ibid., "Preface," sig. P4r.

46. Ibid., pp. 18–19; "Preface," sig. P4r–v.

47. Ibid., pp. 107–135.

48. The terminal date may be deduced from the closing comment of the essay, pp. 134–135, which refers to Johann Rudolf Glauber's "small Treatises freshly publish'd." These seem to be the tracts published by Glauber simultaneously in Latin and German, *Des Teutschlandts Wohlfart*, 7 vols. (Amsterdam, 1656–1661), and *Prosperitatis Germaniae*, 7 vols. (Amsterdam, 1656–1661). About September 1656 Hartlib recorded Boyle's comments that in Glauber's treatise "the annexed discourse of salpeeter De Nitro is the most substantial rational et real piece, wherein many secrets are discovered which himself had before": Hartlib, "Ephemerides" 1656, f. 45–45-8.

49. Boyle, *Certain physiological essays*, "Preface," [leaf between sigs. P4 and P5]r–v.

50. Henry Stubbe, *Legends no histories* (London, 1670), p. 64.

51. Boyle, *Certain physiological essays*, pp. 107–110.

52. These five propositions are an ordered summary of Boyle's rambling remarks, ibid., pp. 115–132.

53. Ibid., pp. 125–127.

54. Ibid., p. 127.

55. Ibid., p. 125.

56. Ibid., pp. 137–249.

57. Ibid., pp. 158–164.

58. Ibid., p. 190.

59. Royal Society, Boyle Papers, X, f. 82–83; XXII, pp. 201–246; XXVII, pp. 253–285.

60. Robert Boyle, *The origine of formes and qualities, (according to the corpuscular philosophy,) illustrated by considerations and experiments, (written formerly by way of notes upon an essay about nitre)* (Oxford, 1666).

61. Robert Boyle, *New experiments and observations touching cold, or an experimental history of cold, begun* (London, 1665), p. 412.

62. Ibid., pp. 412–447.

63. Ibid., p. 448.

64. Ibid., pp. 449–450.

65. Ibid., pp. 453–460, 591–596.

66. Ibid., pp. 458–459, 594–595.

67. Ibid., pp. 451, 457.

68. Ibid., pp. 461–462.

69. Ibid., pp. 453, 457.

70. See the letters in Oldenburg's *Correspondence*: Oldenburg to Hartlib, 18 July 1658 (I, 171); Oldenburg to Boyle, 10 September 1658 (I, 177); Oldenburg to Hartlib, 30 April 1659 N.S. (I, 220); Oldenburg to Saporta, 6 May 1659 N.S. (I, 226); Oldenburg to Freiherr von Friesen, 6 May 1659 N.S. (I, 234); Oldenburg to Boyle, 7 May 1659 N.S. (I, 245–246); Oldenburg to Tollé, 9 May 1659 N.S. (I, 248); John Beale to Oldenburg, 30 September 1659 (I, 318).

71. Stubbe, *Legends*, pp. 35–36, 48, and 51–87 passim.

72. For Sharrock's work with Stahl, and his residence at John Crosse's, see Sharrock to Boyle, 16 and 29 December 1660: Royal Society, Boyle Letters, V, ff. 92–95.

73. Sharrock to Boyle, 16 February 1661, enclosing Boyle's "piece about Nitre": ibid., f. 83.

74. See Evelyn, *Diary*, II & III passim; Hartlib "Ephemerides," passim, esp. 1649–1650, and 1652.

75. 14 August 1661: Birch, *History*, I, 41. The English edition of *Certain physiological essays* was registered at the Company of Stationers on 29 October 1660, and the Latin on 13 June 1661: *Stationers' register*, II, 281, 298.

76. Thomas Sprat, *The history of the Royal-Society of London, for the improving of natural knowledge* (London, 1667), pp. 260–283; Stubbe, *Legends*, pp. 35–116.

77. Oldenburg to Giuseppe Francesco Borri, 7 September 1661: *Correspondence*, I, 417–418.

78. See the letters exchanged between Oldenburg and Spinoza, August 1661 to August 1663: *Correspondence*, I, 415, 432, 439, 458–466, 472–473; II, 40–42, 92–96, 99–100, 103–104.

79. Digby gave his lecture at Gresham College on 23 January 1661 and presented the Society with printed copies on 14 August 1661: Birch, *History*, I, 13, 41.

80. Kenelm Digby, *A discourse concerning the vegetation of plants* (London, 1661), pp. 60–63.

81. Ibid., pp. 64–67.

82. Ibid., pp. 68–70.

83. Ibid., p. 71.

84. Ibid., pp. 83–84.

85. Boyle to Hartlib, 19 March 1647: *Works*, I, xxxviii.

86. Hartlib to Boyle, 9 May 1648: _Works_, VI, 77−78.

87. Robert Boyle, _Some considerations touching the usefulnesse of experimental naturall philosophy_ (Oxford, 1663), Part I, pp. 68−69.

88. Robert Boyle, _New experiments physico-mechanicall, touching the spring of the air_ (Oxford, 1660), pp. 129−133. Boyle does not give the year of these experiments, but they had to have been done before the autumn of 1659, and neither Boyle nor Wren were in Oxford during the appropriate months in 1658.

89. Ibid., p. 8, where Boyle noted in late 1659 that he had had these special glasses blown several years ago.

90. Fritz Krafft, "Otto von Guericke," _DSB_, V, 574−576.

91. Hartlib to Boyle, 7 January 1658: _Works_, VI, 98−99.

92. Boyle, _New experiments_, p. 5. Dungarvan had matriculated at Christ Church on 25 November 1656 (Foster, _Alumni_, I, 163). On 11 November 1658, he and his brother Richard Boyle were granted a pass to France; they had still not departed on 3 March 1659, when another pass was issued, and an order in Council given to recommend to the Admiralty Commission to find a fit vessel to transport them to France (_CSPD_, 1658−1659, pp. 295, 581−582).

93. Boyle, _New experiments_, p. 6.

94. Ibid., p. 7.

95. Richard Waller, "The Life of Dr. Robert Hooke," in _The posthumous works of Dr. Robert Hooke_ (London, 1705), p. iii. There is some doubt about when Hooke arrived in Oxford, for Anthony Wood, who knew Hooke, said he came up to Christ Church in 1650 as a chorister (Wood, _Athenae_, IV, 628). As was common for members not on the foundation, Hooke did not matriculate until much later, on 31 July 1658 (Foster, _Alumni_, II, 740).

96. Aubrey, _Brief lives_, I, 410.

97. Waller, "Life," pp. iii−iv, quoting an autobiographical fragment by Hooke.

98. Aubrey, _Brief lives_, I, 410−411.

99. Robert Hooke, _Micrographia_ (London, 1665), p. 44, in which he records reading Descartes "8 years since"; the book was completed by 1664.

100. Aubrey, _Brief lives_, I, 411.

101. Waller, "Life," p. iii, quoting an autobiographical fragment by Hooke.

102. The most complete description of Hooke's pump is Boyle's own, _New experiments_, pp. 8−17.

103. Waller, "Life," p. iv.

104. Boyle, _New experiments_, p. 242. Hooke, in an autobiographical fragment, said that he contrived and perfected the air pump for Boyle "in 1658, or 9" (Waller, "Life," p. iii).

105. Boyle, _New experiments_, p. 1.

106. Pope to Boyle, 10 September 1659 [N.S.?]: _Works_, VI, 636. Boyle may even have asked Pope to contact some of the Frenchmen interested in pneumatics, since Pope wrote: "I am not yet acquainted with any whom you enquire after, except monsieur _Petit_"; Pierre Petit (1589−1677) had collaborated with Pascal in 1646 on the first Torricellian experiments done in France.

107. Oldenburg was in Paris with Jones from April 1659 to May 1660. Some of his letters to Boyle and Hartlib, describing scientific affairs in the French capital are in Correspondence, I, 218−370 passim. Other letters from Oldenburg and Jones to Hartlib have survived in the extracts Hartlib sent on to Evelyn; see British Library, MS. Add. 15,948, ff. 66−100 passim.

108. Boyle, _New experiments_, p. 2.

109. Boyle to Hartlib, 3 November 1659; letter not extant, but quoted in Hartlib to John Worthington, 7 November 1659: James Crossley, ed., *The diary and correspondence of Dr. John Worthington* (Manchester: Chetham Society, 1847), I, 161−162.

110. Hartlib to Boyle, 15 November 1659: *Works*, VI, 131.

111. Boyle, *New experiments*, pp. 394−395, 399.

112. See Sharrock to Boyle, 26 January and 9 April 1660: *Works*, VI, 319−320.

113. Wood to Hartlib, 5 December 1660: Hartlib Papers, Bundle XXXIII (1). Wood was in London from approximately November 1659 to September 1660 when he returned once again to Dublin.

114. Oldenburg to Boyle, 20 March 1660 N.S.: *Correspondence*, I, 364.

115. Evelyn, *Diary*, III, 255.

116. Hartlib to Worthington, 4 June 1660: Crossley, *Correspondence*, I, 199. Thomason received his copy in August 1660: G.K. Fortescue, *Catalogue of the pamphlets . . . collected by George Thomason, 1640−1661* (London: Trustees of the British Museum, 1908), II, 334.

117. Boyle, *New experiments*, p. 398.

118. Ibid., p. 111.

119. Ibid., pp. 303−304.

120. Ibid., sig. A5r.

121. Ibid., pp. 1−73, 233−258.

122. Ibid., pp. 20−44.

123. Ibid., pp. 45−52.

124. Ibid., pp. 53−61.

125. Ibid., pp. 66−73.

126. Ibid., pp. 236−242. Boyle's results work out to an atmospheric pressure of 15.9 lb/in^2.

127. Ibid., p. 22.

128. Ibid., pp. 23−27.

129. Boyle, *Certain physiological essays*, pp. 20−21.

130. Ibid., pp. 154−155, 187−189, 193.

131. Boyle, *New experiments*, pp. 28−36.

132. Ibid., pp. 286−301.

133. Ibid., pp. 32−33.

134. Ibid., pp. 106−124.

135. Ibid., pp. 140−143.

136. Ibid., pp. 105, 202−213.

137. Ibid., pp. 319−325.

138. Ibid., pp. 217−228.

139. Ibid., pp. 74−83.

140. Ibid., p. 80.

141. Ibid., pp. 81−82, 88−89, 100−101.

142. Ibid., pp. 147−162.

143. Ibid., p. 161.

144. Ibid., pp. 162−171.

145. Ibid., pp. 167, 176−179.

146. Birch, *History*, I, 123−124.

147.   John Evelyn recommended Pope as George Evelyn's tutor on 4 April 1663 (Evelyn, *Diary*, III, 354), and Pope's last recorded attendance at the Royal Society was on 10 June (Birch, *History*, I, 255).

148.   Hooke to Boyle, 6 October 1664: *Works*, VI, 492.

149.   See John Ward's (obit. 1758) notes on Gresham College, British Library, MS. Add. 6195, f. 70v.

150.   Robert Hooke, *An attempt for the explication of the phaenomena, observable in an experiment published by the Honourable Robert Boyle, Esq; in the XXXV. experiment of his epistolical discourse touching the aire* (London, 1661), pp. 5–8, 23–25, 46–50.

151.   Ibid., pp. 28–29.

152.   Robert Boyle, *A defence of the doctrine touching the spring and weight of the air* (London, 1662). *An examen of Mr. T. Hobbes his Dialogus physicus de natura aeris* (London, 1662). Both of these works, although paginated separately, were appended to the second edition of *New experiments* published in 1662.

153.   Franciscus Linus [Hall], *Tractatus de corporum inseparabilitate; in quo experimenta de vacuo, tam Torricelliana, quam Magdeburgica, & Boyliana, examinantur, veraque eorum causa detecta, ostenditur, vacuum naturaliter dare non posse* (London, 1661).

154.   Thomas Hobbes, *Dialogus physicus, sive de natura aeris conjectura sumpta ab experimentis nuper Londini habitis in Collegio Greshamensi. Item de duplicatione cubi* (London, 1661).

155.   Boyle, *Defence*, pp. 69–92; *Examen*, pp. 42–86.

156.   Boyle, *Defence*, pp. 19–32; the quote is from p. 20.

157.   Boyle, *Examen*, pp. 3–4.

158.   John Wallis, *Hobbius heuton-timorumenos. Or a consideration of Mr Hobbes his dialogues. In an epistolary discourse, addressed, to the Honourable Robert Boyle, Esq.* (Oxford, 1662), pp. 148, 151–152.

159.   Boyle, *Examen*, pp. 16–20.

160.   Ibid., pp. 26–30; Wallis, *Heuton-timorumenos*, pp. 154–158.

161.   Boyle, *Examen*, p. 5.

162.   Ibid., pp. 13–15.

163.   Wallis, *Heuton-timorumenos*, pp. 153–154.

164.   Boyle, *Defence*, pp. 3–15, 48–54; *Examen*, pp. 9–13, 20–25.

165.   Boyle, *Defence*, pp. 94–98.

166.   Ibid., pp. 57–68.

167.   Ibid., p. 64; Hooke, *Micrographia*, p. 225.

168.   Webster, "Discovery," pp. 470–484; also see his "Richard Towneley and Boyle's Law," *Nature* 197 (1963): 226–228; I. B. Cohen, "Newton, Hooke, and 'Boyle's Law," *Nature* 204 (1964): 618–621.

169.   See the correspondence between Power and Croone in the British Library, MS. Sloane 1326, especially Croone to Power, 20 July 1661 (f. 26v); Power to Croone, 15 August 1661 (f. 27r); and Croone to Power, 14 September 1661 (f. 25r–v).

170.   Hooke, *Micrographia*, p. 225.

171.   Meetings of 11 and 18 September 1661: Birch, *History*, I, 45. The table of results demonstrating reciprocality, taken from the MS. Register Book of the Royal Society, is reproduced in Douglas McKie, "Boyle's Law," *Endeavour* 7 (1948): 148–151.

172.   Boyle, *Examen*, p. 4.

173.   Meetings of 13 February, 27 March, and 15 May 1661, and 29 January 1661/2: Birch, *History*, I, 16, 19, 23, 75.

174. Wren had promised to do a book of microscopical observations along the line of some drawings he had already done and presented to the King. Wren then begged off and suggested Hooke. Moray wrote to Wren on 13 August 1661 that they had "persuaded Mr. Hook" to take up the task "of drawing the Figures of small Insects by the Help of the Microscope": Stephen Wren, *Parentalia* (London, 1750), p. 211.

175. Birch, *History*, I, 442.

176. Hooke, *Micrographia*, pp. 13–14.

177. Ibid., p. 103.

178. Ibid., p. 104.

179. See, for example, the discussion of these theories in J. D. Lysaght, "Hooke's theory of combustion," *Ambix* 1 (1937): 93–108; Douglas McKie, "Fire and the flamma vitalis: Boyle, Hooke, and Mayow," in E. Ashworth Underwood, ed., *Science, medicine, and history* (London: Oxford University Press, 1953), I, 469–488, esp. pp. 471–476; H. D. Turner, "Robert Hooke and theories of combustion," *Centaurus* 4 (1956): 297–310; Richard S. Westfall, "Robert Hooke," *DSB*, VI, 481–488, esp. pp. 484–485; and Partington, *History*, II, 550–566.

180. Ibid., p. 105.

181. Ibid., pp. 103–105.

182. *Phil. Trans.* 1 (3 April 1665): 27–32.

183. Ibid., p. 29.

184. Meeting of 4 January 1665: Birch, *History*, II, 2.

185. Meetings of 11, 18, and 25 January 1665: ibid., II, 4, 7–8, 10.

186. Meetings of 18 and 25 January 1665: ibid., II, 8, 10.

187. Meetings of 15 and 22 February, 1 and 8 March 1665: ibid., II, 15, 17, 19, 20.

188. Meeting of 28 June 1665: ibid., II, 60.

6: NEW EXPERIMENTS ON RESPIRATION, 1659–1665

1. Robert Boyle, *Certain physiological essays* (London, 1661), pp. 85–88 [written c. 1655–1656].

2. Robert Boyle, *Some considerations touching the usefulnesse of experimental naturall philosophy* (Oxford, 1663), Part II [written 1656–1657], pp. 9, 11, 46–47.

3. Ibid., pp. 10, 46.

4. Ibid., pp. 12–13.

5. Boyle, *Certain physiological essays*, p. 87.

6. Boyle, *Usefulnesse*, Part II, pp. 18–19.

7. Ibid., pp. 21–26.

8. Ibid., pp. 19–21.

9. Ibid., pp. 13–14.

10. Ibid., pp. 14–18.

11. Ibid., pp. 11, 13.

12. Ibid., pp. 68, 85, 86, 105, 113, 210, 223.

13. Ibid., pp. 120, 160.

14. Ibid., pp. 72–73, 231; Hartlib to Boyle, 30 June 1657: *Works*, VI, 92.

15. Boyle, *Usefulnesse*, Part II, pp. 10–11. On splenectomy experiments in the late 1650s, see Charles Webster, "The Helmontian George Thomson and William Harvey: The revival and application of splenectomy to physiological research," *Med. Hist.* 15 (1971): 154–167.

16. Robert Boyle, *New experiments physico-mechanicall, touching the spring of the air* (Oxford, 1660), pp. 328–332.

17. Ibid., pp. 331–332.

18. Ibid., pp. 333–334.

19. Ibid., pp. 335–383.

20. Ibid., pp. 336–343.

21. Ibid., pp. 346–349.

22. Ibid., pp. 349–350.

23. Ibid., pp. 350–358.

24. Ibid., pp. 359–361.

25. Ibid., p. 362.

26. Ibid., pp. 362–365. On Drebbell and his chemistry, see Partington, *History*, II, 321–324.

27. Boyle, *New experiments*, pp. 365–368.

28. Ibid., pp. 368–370, quote on p. 368.

29. Ibid., pp. 373–375, quote on p. 374.

30. Ibid., pp. 370–373.

31. Ibid., p. 383.

32. Robert Boyle, *An examen of Mr. T. Hobbes his Dialogus physicus de natura aeris* (London, 1662), pp. 49–51.

33. Robert Boyle, *A defence of the doctrine touching the spring and weight of the air* (London, 1662), pp. 88–90.

34. Evelyn, *Diary*, III, 284, where he reports such a meeting on 25 April 1661. This is not recorded in Birch, *History*, I, 21, which notes only that the Society stood adjourned between its meetings of 10 April and 1 May. As Evelyn's date was a Thursday, at a time in the Society's history when it normally met on Wednesdays, this may simply have been an informal gathering at Gresham to try out the air pump that was presented to the Society by Boyle three weeks later.

35. Evelyn, *Diary*, (19 July 1661), III, 292. The diving engine was discussed at the meetings of 1 May and 24 July: Birch, *History*, I, 21, 35.

36. Meetings of 30 April and 14 May 1662: ibid., I, 82, 83.

37. Robert Boyle, "New pneumatical experiments about respiration," *Phil. Trans.* 5 (1670): 2011–2031, 2035–2056.

38. Ibid., pp. 2011–2013, 2026–2027, 2031, 2036–2037, 2039–2040.

39. Ibid., pp. 2013–2017, 2037.

40. Ibid., pp. 2049–2052.

41. Ibid., pp. 2052–2054.

42. Ibid., pp. 2054–2056.

43. Ibid., pp. 2028–2031.

44. Ibid., pp. 2036–2039.

45. Ibid., pp. 2045–2046.

46. Ibid., pp. 2017–2019.

47. Ibid., pp. 2041–2043.

48. Ibid., pp. 2046–2048.

49. Ibid., pp. 2024–2026.

50. Ibid., pp. 2043–2044.

51. Ibid., pp. 2027–2028.

52. Ibid., pp. 2048–2049.

53. Birch, *History*, I, 106.

54. Meeting of 1 October 1662: ibid., I, 114.

55. Meetings of 10 September and 8 October 1662: ibid., I, 110, 115.

56. Ibid., I, 118.

57. Meetings of 12, 19 and 26 November 1662: ibid., I, 125, 126, 131. Boyle's attendance is recorded at only one meeting between 8 October 1662 and 21 January 1663, and not again regularly until April 1663.

58. Ibid., I, 179–180.

59. Ibid., I, 180.

60. Meeting of 11 February 1663: ibid., I, 194.

61. Meeting of 28 January 1663: ibid., I, 180.

62. Meetings of 25 March and 1 April: ibid., I, 214, 215.

63. Meetings of 1 and 8 April 1663: ibid., I, 214, 216.

64. Meeting of 20 May 1663: ibid., I, 244.

65. Meeting of 8 April 1663: ibid., I, 216.

66. Meetings of 25 March, 1 and 22 April, 10 and 17 June, 1 and 16 July 1663: ibid., I, 212, 214, 220, 254–255, 260, 268, 275.

67. Meetings of 31 December 1662, 7 and 14 January 1663: ibid., I, 167, 172, 179.

68. Meeting of 15 April 1663: ibid., I, 219.

69. Meeting of 6 May 1663: ibid., I, 235–236.

70. Meeting of 10 June 1663: ibid., I, 255.

71. Meetings of 3 December 1662 and 14 January 1663: ibid., I, 133–136, 177–178.

72. Balthasar de Monconys, *Journal des voyages* (Lyons, 1665–1666), II, 27–29, 38, 43–46.

73. Ibid., II, 33, 40–42.

74. Wren to Brouncker, 30 July 1663: British Library, MS. Sloane 2903, f. 105r. There are copies of this letter in the Royal Society, Early Letters W.3.3, and Letter Book I, 97. It was printed in a slightly changed and abbreviated version in Stephen Wren, *Parentalia* (London, 1750), pp. 224–226, and misdated as 1661. On Wren's biological work in this period, see also J. A. Bennett, "A note on theories of respiration and muscular action in England *c*. 1660," *Med. Hist.* 20 (1976): 59–69.

75. Harvey, *De motu cordis*, pp. 48, 51; *Motion*, pp. 70, 75; *Works*, pp. 54–55, 58. For commentary on the point, see Yehuda Elkana and June Goodfield, "Harvey and the problem of the 'capillaries,' " *Isis* 59 (1968): 61–73.

76. Bodleian Library, MS. Locke e. 4, p. [111].

77. James Young, "Malpighi's 'De Pulmonibus,' " *Proc. R. Soc. Med.* 23 (1929): 1–11.

78. John F. Fulton and Leonard G. Wilson, eds., *Selected readings in the history of physiology*, 2d ed. (Springfield, Ill.: Charles C. Thomas, 1966), pp. 71–72 (footnote).

79. Thomas Bartholin, *De pulmonum substantia* (Copenhagen, 1663), attacked Malpighi's views. James Allestry, one of the Royal Society's two printers, had a copy of this work in late summer 1663: Allestry to Power, August 1663, British Library, MS. Sloane 1326, f. 39r–v.

80. Nathaniel Highmore to [John Wilkins?], [c. May 1663] (rough draft): British Library, MS. Sloane 548, ff. 14v–20r. This item is listed in the British Library's manuscript catalogue as Highmore to Boyle, but this is clearly wrong, since Boyle is referred to in the third person. I suggest John Wilkins as the addressee because he and Oldenburg were Secretaries of the newly chartered Royal Society, because Highmore and Wilkins had many common friends at Oxford (Bathurst, Willis, Boyle, etc.), and because Oldenburg was quite careful about keeping his correspondence in order, whereas little has survived relating to Wilkins's tenure as Secretary. The experiments on fish referred to in the letter were done at Gresham between late March and late May 1663.

81. Ibid., ff. 14v–17r.

82. Nathaniel Henshaw, *Aero-chalinos: or a register for the air* (Dublin, 1664). The following citations are to the London, 1677 edition, which is an identical, though not page-by-page, reprint of the Dublin edition.

83. Ibid., pp. 33–54.

84. Ibid., p. 89.

85. Ibid., pp. 84–86.

86. Ibid., pp. 66–73.

87. Ibid., pp. 60–62.

88. Ibid., pp. 73–75.

89. Ibid., pp. 110–166.

90. Meeting of 2 March 1664: Birch, *History*, I, 389.

91. Meetings of 9, 16, 23, and 30 March, 6 and 27 April, 11, 18, and 25 May 1664: ibid., I, 392, 395, 401, 404, 408, 418, 423, 427, 428.

92. Meeting of 1 June 1664: ibid., I, 433.

93. Boyle to Oldenburg, 29 August 1664: *Works*, VI, 63.

94. Birch, *History*, I, 489.

95. Hooke to Boyle, 8 September 1664: *Works*, VI, 490.

96. Meeting of 30 March 1664: Birch, *History*, I, 404.

97. Ibid., I, 474.

98. Oldenburg to Boyle, 6 October 1664: *Correspondence*, II, 248.

99. Birch, *History*, I, 475.

100. Meeting of 2 November 1664: ibid., I, 482; Oldenburg to Boyle, 22 and 27 October, 3 November 1664: *Correspondence*, II, 269, 273, 280–281.

101. Birch, *History*, I, 482.

102. Meetings of 26 October and 2 November 1664: ibid., I, 478–479, 480–482; Hooke to Boyle, 29 October 1664: *Works*, VI, 496–497.

103. Boyle mentioned Needham in a letter to Hooke sometime in late October or early November. See Hooke to Boyle, 24 November 1664: *Works*, VI, 499.

104. Hooke's report to the Royal Society at its meeting of 9 November 1664: Birch, *History*, I, 485–486.

105. Oldenburg to Boyle, 10 November 1664: *Correspondence*, II, 296–297.

106. Oldenburg to Boyle, 17 November 1664: ibid., II, 310.

107. Hooke to Boyle, 10 November 1664: *Works*, VI, 498.

108. Hooke to Boyle, 24 November 1664: *Works*, VI, 499.

109. Meeting of 9 November 1664: Birch, *History*, I, 486.

110. Meetings of 16 and 23 November 1664: ibid., I, 488, 496–497.

111. Meetings of 16 November and 14 December 1664: ibid., I, 488, 504.

112. Meeting of 4 January 1665: ibid., II, 2.

113. Meetings of 25 January and 8 February 1665: ibid., II, 10, 12.

114. Ibid., II, 20−21.

115. Meeting of 15 March 1665: ibid., II, 22−23.

116. Ibid., II, 25−26.

117. *The Diary of Samuel Pepys*, ed. Robert Latham and William Matthews (Berkeley, Los Angeles, London, University of California Press, 1970−1976), VI, 64.

118. Meetings of 29 March, 12 and 19 April 1665: Birch, *History*, II, 27, 30, 31.

119. Meetings of 12 and 19 April 1665: ibid., II, 29, 32−40.

120. Meeting of 12 April 1665: ibid., II, 31.

121. Meetings of 3, 10, 17, 24, and 31 May, 7 June 1664: ibid., II, 45, 46, 49, 51, 53, 54.

122. Wren to Hooke, 20 April [1665]: Royal Society, Early Letters W.3.6.

123. Thomas Sprat, *History of the Royal-Society of London, for the improving of natural knowledge* (London, 1667), p. 232.

7: OXONIANS ON ANIMAL HEAT AND THE NATURE OF THE BLOOD, 1656−1666

1. The only full-length historical treatment of Willis is Hansruedi Isler, *Thomas Willis, 1621−1675, doctor and scientist* (New York: Hafner, 1968). Audrey B. Davis, *Circulation physiology and medical chemistry in England, 1650−1680* (Lawrence, Kansas: Coronado Press, 1973), pp. 13−30, surveys Willis's life at Oxford. I have given a synoptic view of his life and ideas in the *DSB*, XIV, 404−409. The fundamental study by Alfred Meyer and Raymond Hierons, "On Thomas Willis's concepts of neurophysiology," *Med. Hist.* 9 (1965): 1−15, 142−155, has an extensive bibliography that provides the best entree into the literature on specialized topics of Willis's life and work.

2. Aubrey, *Brief lives*, II, 303.

3. John Lydall to John Aubrey, 23 January, 20 February, 13 March and 5 April 1649: Bodleian Library, MS. Aubrey 12, ff. 294r, 298r, 302r. See also Robert G. Frank, Jr., "John Aubrey, F. R. S., John Lydall, and science at Commonwealth Oxford," *Notes Rec. R. Soc. Lond.* 27 (1973): 193−217, especially pp. 196−198, 213.

4. Seth Ward to Sir Justinian Isham, 27 February 1652, in H. W. Robinson, "An unpublished letter of Dr. Seth Ward relating to the early meetings of the Oxford Philosophical Society," *Notes Rec. R. Soc. Lond.* 7 (1949): 69−70; MS 799[A], Wellcome Institute of the History of Medicine, London, verso of final leaf and recto of inside back cover.

5. Hartlib, "Ephemerides" 1654, ff. WW−WW 7−8.

6. Wood to Hartlib, 9 February 1659: Hartlib Papers, Bundle XXXIII (1).

7. Hartlib, "Ephemerides" 1656, f. 48−48-3.

8. Thomas Willis, "De fermentatione," in *Diatribae duae medico-philosophicae* (London, 1659); the tract is separately paginated. It was published by James Allestry, the brother of Willis's close friend and fellow member of Christ Church, Richard Allestree.

9. On Willis's doctrine of fermentation, especially as it compared to that of Charleton and Sylvius, see Davis, *Circulation physiology*, pp. 65−92.

10. Willis, "De fermentatione," pp. 1−16.

11. Ibid., p. 10.

12. Ibid., p. 17.

13. Ibid., pp. 20−25.

14. Ibid., pp. 25−32.

15. Ibid., pp. 33−48.

16. Ibid., pp. 49−65.

17. Ibid., pp. 81−97.

18. Such ideas of fermentations in the animal body are summarized in Partington, *History*, II, 234−237, 280, 285−287, and in Davis, *Circulation physiology*, pp. 46−63.

19. Willis, "De fermentatione," pp. 25−26.

20. Ibid., pp. 7−8, 66−72.

21. Ibid., pp. 76−78.

22. Ibid., pp. 78−81.

23. Ibid., p. 20.

24. Thomas Willis, "De febribus," pp. 37−39, 201−209, 216−239, in *Diatribae duae* (paginated separately).

25. It was registered at the Company of Stationers on 26 November 1658: *Stationers' register*, II, 207.

26. Hartlib to Boyle, 16 December 1658: *Works*, VI, 115; Wood to Hartlib, 9 February and 2 March 1659: Hartlib Papers, Bundle XXXIII (1).

27. Willis, "De fermentatione," p. 31.

28. Willis, "De febribus," sigs. H2* 1v−2r.

29. Ibid., pp. 1−14.

30. Ibid., pp. 6−7, 18−19.

31. Ibid., pp. 20−22.

32. Ibid., pp. 143−145.

33. Ibid., pp. 25−239.

34. The best introduction to the general subject of injection and transfusion in this period is Heinrich Buess, *Die historischen Grundlagen der intravenösen Injektion* (Aarau: Sauerländer, 1946), which cites much of the earlier literature. See also Horace Manchester Brown, "The beginnings of intravenous medication," *Ann. Med. Hist.* 1 (1917): 177−197.

35. Wallis to Oldenburg, 1 November 1675: *Correspondence*, (in press).

36. For details on Potter's attempts at transfusion, see Charles Webster, "The origins of blood transfusion: A reassessment," *Med. Hist.* 15 (1971): 387−392.

37. William Harvey, *De motu cordis*, pp. 60−61; *Motion*, p. 92; *Works*, p. 71.

38. In the earliest extant holograph listing of works that Boyle had either written, or intended to write, he recorded a treatise "Of Turning Poysons into excellent Remedyes": Royal Society, Boyle Papers, XXXVI, unfoliated. Hartlib noted in the summer of 1655 that Boyle and Clodius had experimented using a hot iron to draw out the poison of adders ("Ephemerides," 1655, f. 32−32-3).

39. Boyle quotes from this unpublished piece in *Some considerations touching the usefulnesse of experimental naturall philosophy* (Oxford, 1663), Part II, pp. 57−60.

40. Boyle's account of the origin of injection experiments is contained in a "Postscript" added to *Usefulnesse*, Part II, pp. 62−65. Boyle gives no dates for these first trials, but Wren wrote to Robert Wood of their success in June, 1656, and the previous March was the last time Boyle, Wilkins, and Wren were all in Oxford together. For observations on these experiments, and on the other medical interests of Wren, see: William Carleton Gibson, "The bio-medical pursuits of Christopher Wren," *Med. Hist.* 14 (1970): 331−341; Tibor Doby, "Sir Christopher Wren and medicine," *Episteme* 7 (1973): 83−106.

41. Wren to Petty, [June, 1656]: see J. A. Bennett, "A study of *Parentalia*, with two unpublished letters of Sir Christopher Wren," *Ann. Sci.* 30 (1973): 129–147, esp. pp. 146–147. Wren says in the letter that it is being transmitted by Wood; we know from the "Ephemerides" that Wood visited Hartlib on 3 July 1656, before his departure to Ireland.

42. Neile to Oldenburg, 15 December 1667: *Correspondence*, IV, 54–57.

43. Because of defective records at Balliol, nothing is known of Clarke's career prior to his appearance before the Visitors in 1648 as a member of that college. Although he refused to submit until the legality of Visitation was clarified (Burrows, *Register*, p. 101), Clarke presumably stayed on at Oxford, proceeding D. M. in 1652. He seems to have moved to London in late 1654, and was examined for candidacy at the Royal College of Physicians in early 1655 ("Annals," IV, f. 51a). Clarke knew John Evelyn (possibly from Balliol in the early 1640s) and George Joyliffe.

44. Clarke to Oldenburg, April/May 1668, in *Phil. Trans.* 3 (18 May 1668): 672–678; reprinted in *Correspondence*, IV, 350–369, quote on 355, 365.

45. Oxford University Archives, "Registrum Congregationis, 1648–1659," f. 151v.

46. T[imothy] C[larke], *Some papers writ in the year 1664* (London, 1670), p. 14.

47. *Correspondence*, IV, 355, 365.

48. Ibid., 357, 366; Boyle, *Usefulnesse*, Part II, p. 64.

49. Boyle, *Usefulnesse*, Part II, p. 64.

50. Petty to Boyle, 17 February 1657/8: *Works*, VI, 139.

51. Boyle, *Usefulnesse*, Part II, p. 64.

52. Henry Stubbe, *The plus ultra reduced to a non plus* [London, 1670], p. 117.

53. F[rancis] V[ernon], *Detur pulchriori: Or, a poem in the praise of the University of Oxford* (Oxford, 1658), p. 3. This poem is sometimes attributed to Francis Vaux, but both the knowledge of science and the similarity to Vernon's later poems suggests him as a more likely author.

54. Henry Stubbe, *Otium literatum* (Oxford, [1658]), p. 10.

55. Hartlib, "Ephemerides" 1658, f. 55–55-6.

56. *Correspondence*, IV, 354, 362–363.

57. Meetings of 28 May and 5 June 1661: Birch, *History*, I, 25.

58. Meeting of 9 April 1662: ibid., I, 80.

59. Entry for 16 May 1664: *The diary of Samuel Pepys*, ed. Robert Latham and William Matthews (Berkeley, Los Angeles, London: University of California Press, 1970–1976), V, 151.

60. Meeting of 16 September 1663: Birch, *History*, I, 303.

61. Johann Daniel Major, *Prodromus inventae a se chirurgiae infusoria* (Leipzig, 1664). See Major to Oldenburg, 13 December 1664 N. S. and Oldenburg to Major, 11 March 1665: *Correspondence*, II, 334–337, 379–380.

62. Meetings of 11 January and 12 April 1665: Birch, *History*, II, 6, 30.

63. Meeting of 17 May 1665: ibid., II, 48.

64. Ibid., II, 50.

65. Ward, "Diary," VII, f. 29v.

66. Ibid., VIII, f. 40r.

67. The anatomical use of injections to demonstrate the vascularity of tissues had a history that was to some degree independent of physiological injections; see F. J. Cole, "The history of anatomical injections," in Charles Singer, ed., *Studies in the history and method of science* (Oxford: Clarendon Press, 1921), II, 285–343.

68. Lower to Boyle, 18 January 1662: *Works*, VI, 464.

69. Lower to Boyle, 4 June 1663: ibid., VI, 467.

70. Lower to Boyle, 24 June 1664: ibid., VI, 473.

71. Thomas Willis, *Cerebri anatome* (London, 1664), pp. 13, 56−57, 83, 104, 106, 147−148, 413. These and the following citations are to the quarto edition.

72. Ibid., pp. 60−61, 94−97.

73. Lower to Boyle, 4 June 1663: *Works*, VI, 467.

74. Lower to Boyle, 24 June 1664: ibid., VI, 474.

75. Lower to Boyle, 18 January 1662: ibid., VI, 464.

76. Lower to Boyle, 8 June 1664: ibid., VI, 478.

77. Ibid.

78. See Oldenburg to Boyle, 4 July, and Boyle to Oldenburg, 8 and 23 July 1665: *Correspondence*, II, 430, 437, 444. Oldenburg printed an account in the *Phil. Trans.* 1 (6 November 1665): 100, 117−118.

79. Boyle to Oldenburg, 23 July 1665: *Correspondence*, II, 444; published in *Phil. Trans.* 1 (6 November 1665): 100−101.

80. Boyle to Oldenburg, 23 July 1665: *Correspondence*, II, 444−445.

81. Boyle to Oldenburg, 8 and 12 August 1665: ibid., II, 454, 475−476.

82. Lower to Boyle, 3 September 1666: *Works*, VI, 480.

83. Boyle to Lower, 26 June 1666, in Richard Lower, *Tractatus de corde* (London, 1669), pp. 177−179.

84. Wood, *Life and times*, II, 42, 44.

85. Boyle to Oldenburg, 27 August 1665: *Correspondence*, II, 484.

86. Wood, *Life and times*, II, 43.

87. See the letters from John Evelyn to Richard Browne of 10 and 15 November, 7, 9 and 16 December 1665, and 25 January 1666, passing messages to Timothy Clarke at Oxford: Evelyn Correspondence # 1484, 1485, 1489, 1490, 1492, and 1499 in the Evelyn Papers, Christ Church, Oxford.

88. Boyle to Oldenburg, 22 October 1665: *Correspondence*, II, 577. Johann Sigismund Elsholtz, *Clysmatica Nova, Oder Newe Clystier-Kunst* (Berlin, 1665).

89. Oldenburg to Boyle, 24 October; Boyle to Oldenburg, 28 October; Oldenburg to Boyle, 31 October 1665: *Correspondence*, II, 579, 580, 585.

90. "An account of the rise and attempts, of a way to conveigh liquors immediately into the mass of blood," *Phil. Trans.* 1 (4 December 1665): 128−130; see also Moray to Oldenburg, 27 November 1665: *Correspondence*, II, 625.

91. Anthony Wood received a letter from Lower on 3 February, and the two were drinking together at the Blue Boar in Oxford on 24 February 1666: *Life and times*, II, 72.

92. Pepys met Clarke at Whitehall on 18 February 1666: *Diary*, VII, 48.

93. Lower, *De corde*, pp. 172−176.

94. Hooke to Boyle, 21 March 1666: *Works*, VI, 505.

95. Beale to Boyle, 31 March 1666: ibid., VI, 397.

96. Birch, *History*, II, 67−68.

97. Ward, "Diary," XI, f. 32r.

98. Birch, *History*, II, 98.

99. Lower to Boyle, 6 [June] 1666, in *De corde*, pp. 180−184. In the text the letter is dated 6 July 1666, but this is corrected on p. 188.

100. Birch, *History*, II, 98.

101. Boyle to Lower, 26 June 1666, in *De corde*, pp. 177−179.

102. Lower to Boyle, 3 September 1666: *Works*, VI, 480–481.

103. Oldenburg to Boyle, 25 September 1666: *Correspondence*, III, 233–235.

104. Meeting of 26 September 1666: Birch, *History*, II, 115.

105. Ibid., II, 123.

106. Pepys, *Diary*, VII, 370–371.

107. Ibid., VII, 373.

108. See Harcourt Brown, "Jean Denis and transfusion of blood, Paris, 1667–1668," *Isis* 39 (1948): 15–28; N. S. R. Maluf, "History of blood transfusion," *J. Hist. Med.* 9 (1954): 59–107; Hebbel E. Hoff and Roger Guillemin, "The first experiments on transfusion in France," ibid., 18 (1963): 103–124; Joseph Schiller, "La transfusion sanguine et les débuts de l'Académie des Sciences," *Clio Medica* 1 (1965): 33–40; A.C. Crombie, "Bluttransfusion im 17. Jahrhundert," *Bild Wiss.* 5 (1968): 237–246; Davis, *Circulation physiology*, pp. 173–205; Jean-Jacques Peumery, *Les origines de la transfusion sanguine* (Amsterdam: B. M. Israel, 1975).

109. Gideon Harvey, *The conclave of physicians* (London, 1683), p. 197. Harvey was a student at Exeter College in 1655–1657.

110. The register of the Restoration Visitors is printed in F.J. Varley, "The Restoration Visitation of the University of Oxford and its colleges," *Camden Miscellany*, vol. 18 (London: Royal Historical Society, 1948); on Crosse see pp. 7, 9, 10.

111. The documents on Willis's election are in the Oxford University Archives, "Registrum Convocationis, 1659–1671," reverse codex f. 4r–v.

112. Oxford University Archives, "Registrum Congregationis, 1659–1669," f. 174r.

113. Liber 11 (1660–1667), f. 50r–v, "Dr. Willis his Lease 1662" 15 October 1662, Archives, Corpus Christi College, Oxford. Willis to Aubrey, 17 and 21 February 1662/3: Bodleian Library, MS. Aubrey 13, ff. 254r, 255r.

114. Marjorie Hope Nicolson, *Conway letters: The correspondence of Anne, Viscountess Conway, Henry More, and their friends, 1642–84* (New Haven: Yale University Press, 1930), pp. 113–114, 116, 407.

115. Cf. Thomas Smith to Daniel Fleming, 19 May 1667, telling of Willis leaving a "throng of Patients" to consult with them as they passed through Oxford: John Richard Magrath, ed., *The Flemings in Oxford* (Oxford: Oxford Historical Society, 1904), I, 170.

116. Kenneth Dewhurst, "Willis in Oxford: Some new MSS," *Proc. R. Soc. Med.* 57 (1964): 682–687.

117. Balthasar de Monconys, *Journal des voyages* (Lyon, 1665–1666), II, 54; Samuel Sorbière, *Relation d'un voyage en Angleterre* (Paris, 1664), p. 94, other references to Willis pp. 67–68, 82.

118. Thomas Sprat, *Observations on Monsieur de Sorbier's voyage into England* (London, 1665), pp. 218–223.

119. Chapterbook (1649–1688), pp. 77, 88, 104, Chapter Archives, Christ Church, Oxford.

120. Lower to Boyle, 24 June 1664: *Works*, VI, 474.

121. Wood, *Life and times*, I, 428; II, 10.

122. Richard Griffith, *A-la-mode phlebotomy no good fashion* (London, 1681), p. 171.

123. Bodleian Library, MS. Locke f. 11, f. 24r.

124. Wood, *Life and times*, II, 12.

125. Ward, "Diary," VII, f. 83r.

126. Wood, *Life and times*, II, 4.

127. Walter Needham, *Disquisitio anatomica de formato foetu* (London, 1667), sig. A3r.

128. Needham to Busby, 8 February 1654[/5]: British Library, MS. Add. 4292, ff. 84r–85v.

129. Needham to Busby, 2 May 1655: ibid., f. 87r.

130. Needham to Busby, [undated but clearly c. 1655]: British Library, MS. Add. 4293, f. 85r. Vopiscus Fortunatus Plemp (1601–1671) was Professor of Medicine at Louvain 1633–1671. On his denial of the circulation in 1638 and subsequent affirmation of it in 1644, see Davis, *Circulation physiology*, pp. 102–105, 111–112.

131. Hartlib, "Ephemerides" 1658, f. 55–55-7.

132. Needham's name was dropped from the list of Fellows of Queens' College in mid-1660; see treasurer's Account Book, 1648–1660, Queens' College, Cambridge.

133. Needham, *De formato foetu*, sig. A3r-v.

134. Needham to Busby, 8 February 1654[/5]: British Library, MS. Add. 4292, f. 84r.

135. Ward, "Diary," VII, f. 30v.

136. See J. A. Bennett, "A note on theories of respiration and muscular action in England c. 1660," *Med. Hist.* 20 (1976): 59–69.

137. Cf. Jeamson to Wren, 19 March 1667[/8]: Royal Society Early Letters I.1.156.

138. Samuel Morris, *Disputatio medica inauguralis, de catarrho* (Leyden, 1668), sig. A3r.

139. Wood, *Life and times*, I, 290, 472–475; Sharrock to Boyle, 26 January, 6 and 29 December 1660, 21 February, 3 and 13 December 1661: *Works*, VI, 320, 321 and Royal Society, Boyle Letters, V, ff. 85r, 92v, 95r; Stahl to Williamson, 28 December 1661: *CSPD* (1661–1662), p. 193.

140. Ward, "Diary," VIII, f. 39v.

141. Lower to Boyle, 24 June 1664: *Works*, VI, 474.

142. Lower to Boyle, 26 November 1662: *Works*, VI, 465.

143. Royal Society, Boyle Papers, XIX, ff. 1–35.

144. Bodleian Library, MS. Locke, f. 19, pp. 1–68 passim.

145. Royal Society, Boyle Papers, XIX, ff. 7r, 27r.

146. Willis, *Cerebri anatome*, sig. a6r–v.

147. Lower to Boyle, 18 January 1661[/2]: *Works*, VI, 462.

148. Willis, *Cerebri anatome*, sig. A2v–A3v, pp. 4, 68–69, 75, 102, 125, 224, 357, 391.

149. Cf. Wallis to Oldenburg, 17 February 1673: *Correspondence*, IX, 466, in which Wallis shows an intimate knowledge of Willis's *Cerebri anatome*.

150. Ward, "Diary," IX, f. 126r.

151. Lower to Boyle, 18 January 1661[/2] and 26 November 1662: *Works*, VI, 462–465.

152. Lower to Boyle, 27 April and 4 June 1663: *Works*, VI, 466–468.

153. James Allestry to Henry Power, August 1663: British Library, MS. Sloane 1326, f. 39r–v.

154. John Wallis, "Letter against Mr. Maidwell, 1700," in the Oxford Historical Society *Collectanea* (Oxford: Clarendon Press, 1885), pp. 269–337, especially pp. 315–317.

155. Lower wrote to Boyle that concerning recent anatomical work at Oxford, "My letters to Mr. Hooke will inform you more fully"; Lower to Boyle, 24 June 1664: *Works*, VI, 474.

156. Ibid., pp. 468–474.

157. Wallis to Oldenburg, 12 May 1666: *Correspondence*, III, 122–125; *Phil. Trans.* 1 (4 June 1666): 222–226.

158. Birch, *History*, I, 485, 504, 509; Royal Society, Boyle Papers, XVIII, pp. 176–177.

159. Ward, "Diary," X, f. 42v, 56v; IV, f. 35v.

160. Samuel Parker, *Tentamina physico-theologica de Deo* (Oxford, 1665), especially pp. 79–98, 100–108, 116–120, 138–139.

161. Samuel Parker, "An Account of the Nature and Extent of the Divine Dominion and Good-

nesse, Especially as they refer to the Origenian Hypothesis concerning the Preexistence of Souls," pp. 194−195, in *A free and impartial censure of the Platonick Philosophie*, 2d ed., (Oxford, 1667).

162.   Robert Sharrock, *De finibus virtutis Christianae* (Oxford, 1673), pp. 114−115.

163.   Lower to Boyle, 24 June 1664: *Works*, VI, 472−473.

164.   Meeting of 21 December 1664: Birch, *History*, I, 509.

165.   Robert Boyle, *Memoirs for the natural history of humane blood* (London, 1683/4). Boyle to "J. L.," sig. A2r−A8r; quote on sig. A3v−A4r.

166.   There are no less than seven copies of these titles, in various states and hands, in the Royal Society, Boyle Papers XVIII, pp. 52, 55−56, 58, 60−61, 65, 66, and XXVI, f. 46r. The subsequent discussion refers to what seems to be the only holograph, on pp. 55−56.

167.   British Library, MS. Add. 32,554, f. 53v−54r, 100v−101r.

168.   Ibid., ff. 48r, 49r.

169.   The following passages in Locke's notebooks were first noted by Kenneth Dewhurst, *John Locke, physician and philosopher* (London: Wellcome, 1963), pp. 11−15, although his admirably concise treatment contains some errors of dating that obscure the true sequence of concepts and experiments.

170.   Bodleian Library, MS. Locke f. 19, pp. 224−225.

171.   Ibid., p. 226.

172.   Ibid., p. 158.

173.   Ibid.

174.   Ibid., pp. 212−213.

175.   Ward, "Diary" IV, f. 25v. D'Arcy Power, in "John Ward and his diary," *Trans. Med. Soc. Lond.* 40 (1917): 1−26, on p. 15 misdated this passage as 1658.

176.   Bodleian Library, MS. Locke f. 19, p. 226.

177.   Ibid., p. 227.

178.   Ibid.

179.   Edmund Meara, *Examen diatribae Thomae Willisii doctoris et professoris Oxoniensis de febribus* (London, 1665).

180.   Richard Lower, *Diatribae Thomae Willisii . . . de febribus vindicatio* (London, 1665). The first edition has an imprimatur dated 22 March 1664[/5]. The copy in the Bodleian Library, which was used for the following summary, was given by Lower and is inscribed (sig. A1v) "Bibliothecae Bodlei . . . ex dono doctissimi Authoris."

181.   Ibid., sig. A3r−A8v.

182.   Ibid., pp. 7, 23−24, 33.

183.   Ibid., pp. 7, 24, 32.

184.   Ibid., pp. 36−38.

185.   Ibid., pp. 89, 106.

186.   Ibid., pp. 73−75.

187.   Ibid., pp. 88−106.

188.   Ibid., p. 90.

189.   E.g., ibid., pp. 18−21.

190.   Ibid., pp. 57−59.

191.   Ibid., pp. 52−55.

192.   Ibid., pp. 59−64.

193.   Ibid., pp. 7−8, 23−24, 39−40.

194. Lower's idea that the process of ascension served to educe the heat from the blood rather than create it completely *de novo*, can be seen in the title of his chapter discussing ascension: "Sanguinis calor per accensionem a cordis igne, aut fermento recte exprimitur, contra Mearam" (ibid., p. 107).

195. Ibid., pp. 114–115.

196. Ibid., p. 120.

197. Ibid., pp. 118–119.

198. Ibid., p. 115.

199. Ibid., pp. 117–118.

200. Ibid., pp. 116, 128.

201. Ibid., p. 116.

202. Ibid., cf. also p. 125.

203. Ibid., pp. 125–126.

204. [Henry Oldenburg], "Account of Mr. Richard Lower's Newly Published Vindication of Doctor Willis's Diatriba de Febribus," *Phil. Trans.* 1 (5 June 1665): 75–76.

8: A DISCUSSION AMONG FRIENDS: RESPIRATION WORK AT LONDON, 1667–1669, AND RICHARD LOWER'S *TRACTATUS DE CORDE*

1. Lower to Anthony Wood, 4 September 1666: Bodleian Library, MS. Tanner 45, f. 103, telling of his intention to visit Oxford about Michaelmas (29 September).

2. "Tryals proposed by Mr. *Boyle* to Dr. Lower, to be made by him, for the improvement of transfusing blood out of one live animal into another," *Phil. Trans.* 1 (11 February 1666/7): 385–388.

3. Their chemical exercises are recorded in Bodleian Library, MS. Locke f. 25.

4. Wood, *Life and times*, II, 99; Locke to Boyle, 24 February 1666/7: *Works*, VI, 537.

5. Meeting of 20 June 1667: Birch, *History*, II, 181.

6. The essay, P. R. O. 30/42/2, ff. 35–36, is one of several medical essays by Locke among the Shaftesbury Papers in the Public Record Office, London. The Latin text, and an English translation, was published by Kenneth Dewhurst, "Locke's Essay on Respiration," *Bull. Hist. Med.* 24 (1960): 257–273, and is cited from this source.

7. George Castle, *The chymical Galenist: A treatise wherein the practise of the ancients is reconcil'd to the new discoveries in the theory of physick* (London, 1667). The dedicatory epistle is dated at Westminster, 10 March 1666[/7], sig. A8r. The book was registered with the Stationers Company on 25 April 1667: *Stationers' register*, II, 376. On sig. A3r Castle notes he showed some of the book to Millington two years previously.

8. Ibid., sig. A3r, A4r.

9. Ibid., pp. 3–5.

10. Ibid., pp. 5–7, 10.

11. Ibid., p. 4.

12. Ibid., p. 42.

13. Locke, "Respirationis usus," pp. 265, 271.

14. Ibid., pp. 264, 270.

15. Cf. Locke to Boyle, 5 May 1666: *Works*, V, 686–687.

16. Locke, "Respirationis usus," pp. 265, 271.

17.	Castle, *Chymical Galenist*, pp. 40, 42. Other statements about the lungs drawing in niter are on pp. 78 and 139.

18.	Ibid., pp. 35, 40–42, 54–55, 59–61, 78–84, 136–139.

19.	Birch, *History*, II, 172.

20.	Meeting of 9 May 1667: ibid., II, 173.

21.	Meeting of 23 May 1667: ibid., II, 176.

22.	Meeting of 27 June 1667: ibid., II, 184.

23.	Council meeting of 3 June 1667: ibid., II, 178–179.

24.	Meetings of 20 and 27 June, 4 July 1667: ibid., II, 181, 184, 185.

25.	Meetings of 11 and 18 July 1667: ibid., II, 187–188.

26.	Meetings of 27 June, 11 and 25 July 1667: ibid., II, 184, 188, 189.

27.	Hooke to Boyle, 5 September 1667: *Works*, VI, 509.

28.	The imprimatur is dated 4 August, and the book was registered on 12 August 1667: *Stationers' register*, II, 380.

29.	Walter Needham, *Disquisitio anatomica de formato foetu* (London, 1667), sigs. A2r, A5r–6r.

30.	Meetings of 3, 6 and 20 June 1667: Birch, *History*, II, 178–179, 181.

31.	Ibid., II, 194.

32.	Needham, *De formato foetu*, sigs. A3r, A5v–A6r.

33.	Ibid., sig. A6r–v.

34.	Ibid., pp. 12, 16.

35.	Ibid., p. 97.

36.	Ibid., pp. 64–83.

37.	Ibid., pp. 107–119.

38.	Ibid., pp. 120–128.

39.	Ibid., pp. 129–131, 136–137.

40.	Ibid., pp. 131–133.

41.	Ibid., pp. 133–141.

42.	Ibid., pp. 142–148. This is, incidentally, the first reference I have found in England to Malpighi's work on the lungs.

43.	Ibid., pp. 149–162.

44.	Ibid., p. 162.

45.	*Phil. Trans.* 2 (23 September 1667): 509–516.

46.	Meeting of 3 October 1667: Birch, *History*, II, 197.

47.	Ibid., II, 198. Note especially that this entry implies that Hooke and Lower merely repeated the 1664 experiment, which was most emphatically not the case. The best description is "An account of an experiment made by M. *Hook*, of preserving animals alive by blowing through their lungs with a bellows," *Phil. Trans.* 2 (21 October 1667): 539–540.

48.	Evelyn, *Diary*, III, 497.

49.	Meeting of 10 October 1667: Birch, *History*, II, 198.

50.	*The diary of Samuel Pepys*, ed. Robert Latham and William Matthews (Berkeley, Los Angeles, London: University of California Press, 1970–1976), VII, 21 (22 January 1666).

51.	Meetings of 10 and 17 October, 7 November 1667: Birch, *History*, II, 198, 200, 201, 207.

52.	Meetings of 10, 17 and 24 October, 7 and 14 November 1667: ibid., II, 200, 201, 207, 209; *Phil. Trans.* 2 (11 November 1667): 544–546.

53. Meetings of 24 and 31 October, 7, 14, and 21 November 1667, 27 February and 5 March 1668: ibid., II, 202, 203, 207, 208, 209, 215, 254, 255.

54. Meetings of 31 October and 14 November 1667: ibid., II, 203, 209−210.

55. Meetings of 23 January and 6 February 1668: ibid., II, 242, 245; *Phil. Trans.* 2 (10 February 1667/8): 613−614.

56. Oldenburg to Boyle, 12 November 1667: *Correspondence*, III, 592.

57. Meetings of 5 and 16 November 1667, 25 and 30 January 1668: Birch, *History*, II, 206, 212, 242, 243.

58. Meetings of 21 and 28 November 1667: ibid., II, 214−215, 216. Oldenburg and King both reported the outcome in letters to Boyle of 25 November 1667: *Correspondence*, III, 611−612, and *Works*, VI, 646−647.

59. Meeting of 12 December 1667: Birch, *History*, II, 225; Oldenburg to Boyle, 17 December 1667: *Correspondence*, IV, 59.

60. *Phil. Trans.* 2 (21 October 1667): 517−525; 2 (9 December 1667): 557−564; 2 (10 February 1667/8): 617−624. For letters exchanged with Wallis, Neile, Sluse, Hevelius, Beckman, and Travagino on transfusion in the period November 1667−April 1668, see *Correspondence*, III, 617; IV, 8, 82, 108, 132, 137, 212, 266, 280, 328.

61. See letters exchanged between Oldenburg and Justel and Denis over the period November 1667−March 1668 in *Correspondence*, III, 580; IV, 39, 48−54, 77, 86, 89, 129, 150, 175, 195−196, 245, 247.

62. Meetings of 7, 14 and 28 November 1667: Birch, *History*, II, 207, 209, 216.

63. Meetings of 12 December 1667; 2, 9, and 16 January 1667/8: ibid., II, 227, 234, 237, 239.

64. Ibid., II, 232−233.

65. Needham to Oldenburg, 10 March 1667/8: *Correspondence*, IV, 237.

66. Boyle to Oldenburg, 3 April 1668: *Works*, VI, 75.

67. Needham to Oldenburg, 10 March 1667/8: *Correspondence*, IV, 237−240, 242.

68. Ibid., pp. 239−243.

69. Oldenburg to Boyle, 14 April 1668: *Correspondence*, IV, 318; Birch, *History*, II, 272.

70. Oldenburg to Boyle, 17 March 1667/8: *Correspondence*, IV, 248.

71. Birch, *History*, II, 273−274. Boyle's experiments on respiration had been mentioned in Oldenburg to Boyle, 14 April 1668: *Correspondence*, IV, 318.

72. Reported at meeting of 14 May 1668: Birch, *History*, II, 282−283.

73. Meeting of 28 May 1668: ibid., II, 287.

74. See the Oldenburg *Correspondence*, III, 548−549, 575; IV, 137, 396, 581; V, 41, 244, 354.

75. Carlo Fracassati, *Dissertatio epistolica responsoria de cerebro*, pp. 410−421 in Marcello Malpighi, *Tetras anatomicarum epistolarum* (Bologna, 1665). Fracassati's injection experiments had been done at Bologna in the winter of 1665; see Giovanni Borelli to Marcello Malpighi, 7 March 1665: *The correspondence of Marcello Malpighi*, ed. Howard B. Adelmann (Ithaca: Cornell University Press, 1975), I, 248−250.

76. *Phil. Trans.* 2 (23 September 1667): 490−491.

77. Ibid., p. 493.

78. Oldenburg to Boyle, 17 September 1667: *Correspondence*, III, 476.

79. Boyle to Oldenburg, 17 October 1667: *Works*, VI, 69; this part of the letter was published by Oldenburg in the *Phil. Trans.* 2 (11 November 1667): 551−552.

80. Boyle to Oldenburg, 26 October 1667: *Works*, VI, 70−71.

81. Birch, *History*, II, 274.

82. Meetings of 28 May and 11 June 1668: ibid., II, 288, 296.

83.	Meetings of 11 and 25 June; 9, 16, 23, and 30 July; 6 August 1668: ibid., II, 296, 301, 304, 307, 309, 312.

84.	Meeting of 16 July 1668: ibid., II, 307.

85.	*Phil. Trans.* 3 (18 May 1668): [6]72−678; reprinted and translated in Oldenburg, *Correspondence*, 350−367.

86.	Pepys, *Diary*, IX, 254−255.

87.	Oldenburg to Boyle, 12 November 1667: *Correspondence*, III, 592.

88.	Richard Lower, *Tractatus de corde* (London, 1669): facsimile and translation by Kenneth J. Franklin in *Early science in Oxford*, ed. R.T. Gunther (Oxford: For the Subscribers, 1932), IX, 106−107, 115−118. On the book see John F. Fulton, *A bibliography of two Oxford physiologists, Richard Lower, 1631−1691, John Mayow, 1643−1679* (Oxford: University Press, 1935), pp. 16−23. Useful overviews of the work include: K. J. Franklin, "Some notes on Richard Lower (1631−1691), and his 'De corde,' London, 1669," *Ann. Med. Hist.*, new series, 3 (1931): 599−602; idem, "The works of Richard Lower (1631−1691)," *Proc. R. Soc. Med.* 25 (1932): 113−118; idem, "Some textual changes in successive editions of Richard Lower's *Tractatus de corde item de motu & colore sanguinis et chyli in eum transitu*," *Ann. Sci.* 4 (1939): 283−294; J. Lankhout, "R. Lower's *Tractatus de corde*," *Bijdragen tot de geschiedenis der geneeskunde* 21 (1941): 16−24; and R. Rullière, "Le *Tractatus de corde item de motu et colore sanguinis* de Richard Lower (1669)," *Histoire Sci. Méd.* 8 (1974): 85−98.

89.	Lower, *De corde*, pp. 189−190.

90.	Ibid., pp. 208−211.

91.	Ibid., pp. 165−168.

92.	Justel to Oldenburg, 15/25 April 1668: *Correspondence*, IV, 321.

93.	Oldenburg to Huet, 27 April 1668: ibid., IV, 340.

94.	Oldenburg to Sir John Finch, 4 September 1668: ibid., V, 36.

95.	Oldenburg to Malpighi, 22 December 1668: ibid., V, 280.

96.	Oldenburg to Christiaan Huygens, 8 March 1668/9: ibid., V, 437.

97.	Lower, *De corde*, p. 211.

98.	*Phil. Trans.* 3 (19 October 1668): 791−792.

99.	Lower, *De corde*, sigs. A3r−A8v.

100.	Ibid., sigs. A3r−A5r.

101.	Ibid., pp. 1−57.

102.	Ibid., pp. 17−36.

103.	Ibid., pp. 60−61.

104.	Ibid., p. 61.

105.	Ibid., pp. 61−70.

106.	Ibid., pp. 66−67.

107.	Ibid., p. 71.

108.	Ibid., p. 73.

109.	Ibid., pp. 73−74.

110.	Ibid., p. 74.

111.	Ibid., pp. 75−85.

112.	Ibid., pp. 86−87.

113.	Ibid., pp. 88−151.

114.	Ibid., pp. 153−154.

115.	Ibid., pp. 155−156.

116. Lower was using the Troy pound of twelve ounces. This gives a rate of over thirteen circulations per hour. Even using the avoirdupois pound of sixteen ounces, the rate is still about ten circulations per hour.

117. *Phil. Trans.* 4 (25 March 1669): 909–912, esp. p. 911.

118. Lower, *De corde*, pp. 158–163.

119. Ibid., p. 163 (my italics).

120. Ibid., pp. 163–164.

121. Ibid., pp. 165–166.

122. Ibid., p. 167.

123. Ibid., pp. 167–168.

124. Ibid., pp. 168–169.

125. Ibid., pp. 169–170.

126. Compare the two extracts:

> Si, per quos meatus hic transitus fiat, quaeras; ostende et tu mihi, quibus porulis alter ille spiritus nitrosus qui in nive est, per delicatulorum pocula transit, ut aestiva vina refrigeret: Quod si vitrum aut metallum spiritui huic non sint impervia; certe laxioris pulmonis vasa quanto minus?
>
> Et si fuliginibus aut seroso humori exitum per brachias non negamus, quidni per eosdem vel similes porulos, nitroso huic pabulo introitum pariter concedimus?
>
> (Bathurst, "Praelectiones," pp. 209–210.)
>
> Si per quos pulmonum meatus spiritus aeris nitrosus in sanguinem transit, eumque copiosius imbuit, a me quaeras, ostende & tu mihi quibus porulis alter ille spiritus nitrosus qui in nive est, per delicatulorum pocula transit & aestiva vina refrigerat; Quod si vitrum aut metallum spiritui huic non sint impervia, quanto facilius laxiora pulmonum vasa penetrabit? Denique si fuliginibus & seroso humori exitum non negamus, quidni per eosdem porulos vel similes nitroso huic pabulo introitum in sanguinem concedamus.
>
> (Lower, *De corde*, pp. 169–170.)

9: NITER, NITER EVERYWHERE

1. See the exhaustive treatment of early theories in E. Bastholm, *The history of muscle physiology* (Copenhagen: Munksgaard, 1950), pp. 1–124.

2. [William Croone], *De ratione motus musculorum* (London, 1664). For an analysis, see Leonard G. Wilson, "William Croone's theory of muscular contraction," *Notes Rec. R. Soc. Lond.* 16 (1961): 158–178.

3. Walter Charleton, *Natural history of nutrition, life and voluntary motion* (London, 1659), pp. 182–210.

4. For an overview of Willis's theories and their general context, see Raymond Hierons and Alfred Meyer, "Willis's place in the history of muscle physiology," *Proc. Roy. Soc. Med.* 57 (1964): 687–692.

5. Thomas Willis, *Cerebri anatome* (London, 1664), pp. 269–271, 323, 398–399.

6. Ibid., pp. 324–327.

7. Ibid., pp. 249, 382–383.

8. Thomas Willis, *Pathologiae cerebri, et nervosi generis specimen* (Oxford, 1667).

9. Oldenburg to Boyle, 12 September 1667; Oldenburg to Hevelius, 31 January 1667/8; Justel to Oldenburg, 5/15 February 1668: *Correspondence*, III, 475; IV, 136, 148. Reviewed in *Phil. Trans.* 2 (6 January 1667/8): 600–602.

10. Willis, *Pathologiae cerebri*, pp. 2−5.

11. Ibid., p. 5.

12. See especially ibid., pp. 6−19, 33−34, 95, 99, 240−243.

13. Ibid., pp. 144−169.

14. Ibid., p. 199.

15. Royal Society, Boyle Papers, XIX, f. 7r.

16. For some of the pertinent dates of Mayow's biography, see R.B.Gardiner, *The registers of Wadham College* (London: George Bell & Sons, 1889), p. 224; Douglas McKie, "The birth and descent of John Mayow: A tercentenary note," *Phil. Mag.*, 7th series, 33 (1942): 51−60. The most recent biographical overview is Theodore M. Brown, "John Mayow," *DSB*, IX, 242−247.

17. Bodleian Library, MS. D. D. All Souls College b. 90−91 (Steward's Books, 1661−1664).

18. Bodleian Library, MS. D. D. All Souls College b. 89−96 (Steward's Books, 1659−1674).

19. See especially the notebooks kept by Browne in 1663−1667: British Library, MS. Sloane 1892, ff. 7−58; 1906, ff. 1−159. See also the account of Browne's medical education in Norman Moore, *The history of the study of medicine in the British Isles* (Oxford: Clarendon Press, 1908), pp. 69−83.

20. Woodroffe first appears as Thomas Mayow's tutor in 1664: Christ Church, Oxford, Ch. Ch. MS. xii.b.107, p. 83. Boyle proposed Woodroffe for admission to the Royal Society on 30 April 1668: Birch, *History*, II, 273.

21. Locke used the rhetorical exercises of his students, including Thomas Mayow, to press his herb collection; those by Mayow are in the Bodleian Library, MS. Locke c.41, pp. 60, 766, 1106, 1366, 1418 and 1630; MS. Locke b.7, f. 75v, 208v.

22. Ward, "Diary," XIII, f. 2r.

23. Ibid., ff. 11r, 15r.

24. Ibid., ff. 12r, 17v, 23v−24r; also 33v−34r.

25. Ibid., f. 18r.

26. Ibid., passim.

27. Lower to Boyle, 3 September 1666: *Works*, VI, 480.

28. Wood, *Life and times*, II, 239.

29. Colepresse to Oldenburg, 16 August 1669: *Correspondence*, VI, 194.

30. Robert Boyle, *A continuation of new experiments physico-mechanical, touching the spring and weight of the air, and their effects* (Oxford, 1669), pp. 10, 48, 72.

31. Colepresse to Boyle, 1 February 1666/7: *Works*, VI, 547.

32. Feake was from London, born c. 1643. He matriculated at Magdalen Hall in April 1660 and at Leyden in August 1669. He graduated M. D. there in July 1670 and practiced in London, where he was L. R. C. P. in 1676. See Foster, *Alumni Oxoniensis*, II, 488; Robert W. I. Smith, *English-speaking students of medicine at Leyden* (Edinburgh & London: Oliver & Boyd, 1932), p. 83.

33. The book was completed at Oxford on 24 March 1667/8, just before Boyle returned permanently to London: *Continuation*, p. 176.

34. Ibid., pp. 3, 5, 9−10, 17−18, 48−49, 140.

35. Ibid., pp. 48−49.

36. Ward, "Diary," XIII, ff. 22r, 25r−v, 27r−28r, 31r.

37. John Mayow, *Tractatus duo. Quorum prior agit de respiratione: alter de rachitide* (Oxford, 1668). The most complete recent treatment of Mayow's works is in Partington, *History*, II, 577−613. This incorporates some, but not all, of the points Partington made in "The life and work of John Mayow (1641−1679)," *Isis* 47 (1956): 217−230, 405−417.

38. Justel to Oldenburg, c. 18 October 1668: *Correspondence*, V, 91.

39. *Phil. Trans.* 3 (16 November 1668): 833–835.

40. Thomas Browne to Edward Browne, 21 December 1668; Browne mentioned the book again in another letter two days later: Geoffrey Keynes, ed., *The works of Sir Thomas Browne* (Chicago: University of Chicago Press, 1964), IV, 37, 39.

41. Mayow, *Tractatus duo*, pp. 1–7.

42. Ibid., pp. 7–12.

43. Ibid., pp. 23–36.

44. Ibid., pp. 13–23.

45. Ibid., pp. 36–40.

46. Ibid., pp. 40–43. "... ut jam asserere non dubitemus *aerem non tam pulmonum motui inservire, sed etiam sanguini nonnihil communicare* ..." (40).

47. Ibid., pp. 43–44. "Et versimile est, *tenuiores* esse, & *nitrosas particulas*, quibus abundat aer, quae quae per pulmones communicantur" (43).

48. Ibid., pp. 45–48.

49. Ibid., pp. 48–49.

50. Ibid., pp. 49–52.

51. Ibid., p. 54: "Plane ut *praecipuus* respirationis *usus* esse videatur, ut *musculorum* & praecipue *cordis motibus* inserviat."

52. Ibid., pp. 53–54.

53. Ibid., pp. 55–56.

54. Ibid., pp. 57–60.

55. British Library MS. Sloane 2148, leaf at end, not in Millington's hand.

56. See Bathurst to the Duke of Ormond, 16 November 1675: *The life and literary remains of Ralph Bathurst*, ed. Thomas Warton (London, 1761), pp. 138–139.

57. Nicolaus Steno, *Elementorum myologiae specimen, sive musculi descriptio geometrica* (Florence, 1667). On his descriptions of muscle, see Bastholm, *History*, pp. 142–163.

58. Oldenburg reviewed it in the *Phil. Trans.* 2 (10 February 1667/8): 627–628.

59. Richard Lower, *Tractatus de corde* (London, 1669), pp. 17–23, 75–78.

60. Oldenburg to Huygens, 8 March 1669; Oldenburg to Downes, 9 March 1669: *Correspondence*, V, 435, 439.

61. Lower's book was presented at the meeting of 4 March, and Goddard made his suggestion at the meeting of 18 March 1669: Birch, *History*, II, 353, 356.

62. Ibid., II, 356.

63. Ibid., II, 412. Goddard did not bring in his account of the experiment until the meeting of 16 December 1669.

64. Meeting of 1 April 1669: ibid., II, 356.

65. Jan Swammerdam of Amsterdam, using a nerve-muscle preparation in an apparatus designed along similar lines, had demonstrated in 1667 that muscles undergo no intumescence in contraction. But his experiments were carried out privately, and lay in manuscript until they were published in his *Bijbel der Natur* (Amsterdam, 1752). See Hierons and Meyer, "Willis's place," pp. 3–4.

66. It was mentioned by Francis Glisson in his *Tractatus de ventriculo et intestinis* (London, 1677).

67. British Library, MS. Sloane 1586, ff. 140–141.

68. Nathaniel Highmore, *De hysterica et hypochondriaca passione, responsio ad Doctorem Willis* (London, 1670).

69. Thomas Willis, *Affectionum quae dicuntur hystericae & hypochondriacae pathologia spasmodica vindicata contra responsionem epistolarem Nathanael Highmore, M.D. Cui accesserunt exercitationes medico-physicae duae. 1. De sanguinis accensione. 2. De motu musculari.* (London, 1670). This work, like many of Willis's, was published by James Allestry.

70. *Phil. Trans.* 5 (25 March 1670): 1178.

71. Willis, "De motu musculari," pp. 82–93.

72. Willis, "De sanguinis accensione," pp. 48–49.

73. Ibid., p. 52: ". . . attamen nulli uspiam liquores, sive tenues, sive crassi, utut fermentescentes, aut putrescentes propterea incalescunt . . ."

74. Ibid., pp. 52–53.

75. Ibid., p. 54: ". . . ut pabulum nitrosum propter cujusvis rei incendium necessario requisitum . . ."

76. Ibid., pp. 55–57.

77. Ibid., pp. 57–58: ". . . hinc ut recta cum ratione concludere liceat, animalis vitam esse aut ignem, aut aliquid ei valde analogum . . ."

78. Ibid., pp. 58–60.

79. Ibid., pp. 62–63.

80. Ibid., pp. 64–66.

81. Some of the following writers (Clarke, Hodges, Browne, and Thruston) have been mentioned in Partington, *History*, II, 559, 571–573, 575–576.

82. See British Library, MS. Sloane 35, which is a catalogue of Stubbe's library. He had a complete collection of Descartes and Gassendi, and all the scientific work of his Oxford contemporaries: Bohun, Boyle (18 works), Castle, Charleton, Cole, Glanvill, Hodges, Hooke, Lower, Mayow, Needham, Parker, Petty, Sharrock, Sprat, Wallis, Ward, Willis (6 works), and Wren; as well as other physiological works by Glisson, Ent, Harvey, Thruston, and Wharton.

83. Henry Stubbe, *The Indian nectar, or a discourse concerning chocolata* (London, 1662), sig. A2r, dedicated the work to Willis, whom he thanked for his testimonials "When I first entered upon the practise of Physick," and whose work on fermentation and fevers Stubbe thought second in greatness only to Harvey's.

84. Ibid., sig. A2v, pp. 32–33, 76, 124–126. See also Henry Stubbe, *The miraculous conformist* (Oxford, 1666), written as a letter to Boyle, dedicated to Willis, using Willis's doctrine of fermentation: sig. A2r–v, pp. 1, 8, 10–14, 18–24, and 29.

85. Stubbe to Moray, 27 May 1667, 24 and 30 March 1668: Royal Society, Early Letters S.1.89–91. These were abbreviated and published by Oldenburg in the *Phil. Trans.* 2 (1667): 493–500, esp. 499; 3 (1668): 699–709, 717–722.

86. Thomas Henshaw, "The history of the making of salt-peter," pp. 260–276, and "The history of making gunpowder," pp. 277–283 in Thomas Sprat, *The history of the Royal-Society of London, for the improving of natural knowledge* (London, 1667).

87. Henry Stubbe, *Legends no histories: or, a specimen of some animadversions upon the History of the Royal Society* (London, 1670), pp. 36, 45–46, 51–52, 58, 64, 66, 77–78.

88. Ibid., pp. 35–116.

89. Ibid., pp. 38, 45–48, 50–52, 85–86.

90. Ibid., pp. 72–87.

91. William Clarke, *The natural history of nitre* (London, 1670), pp. 19–23.

92. Ibid., pp. 23–30.

93. Ibid., pp. 36–39.

94. Thomas Guidott, "An appendix concerning Bathe," pp. 58–59, in Edward Jorden, *A discourse of natural bathes*, 4th ed. (London, 1669).

95. See Evelyn, *Diary*, III, 416, 474, 566. Evelyn MS. 12, Christ Church, Oxford, is a catalogue of Bohun's library, including books by Boyle, Wallis, Locke, Wilkins, Sharrock, Glanvill, Trapham, Henshaw, Parker, Digby, and Willis: ff. 10r, 12v, 13r–15v, 17v–19v, 22r–26v.

96. Ralph Bohun, *A discourse concerning the origine and properties of wind* (Oxford, 1671), pp. 8, 11, 15, 21, 41, 130–135.

97. Ibid., pp. 180–185.

98. Ibid., pp. 229–230, 233–234, 248, 263–264, 299–302.

99. *Phil. Trans.* 8 (20 January 1672/3): 5147–5150. See Oldenburg to Vogel, 26 December 1671; Erasmus Bartholin to Oldenburg, 11 January 1672/3; and Oldenburg to Bartholin, 27 February 1672/3: *Correspondence*, VIII, 430; IX, 402, 499.

100. For Morton's biography, see Samuel Eliot Morrison, "Charles Morton," *Publications of the Colonial Society of Massachusetts* 33 (1940): vii–xxix.

101. See the references to their work in Morton's "Compendium physicae," ibid., pp. 11, 15, 23–24, 39, 62–69, 107, 111, 115, 133, 159, 164.

102. Daniel Cox communicated Morton's observations on Cornish sea sand for publication in the *Phil. Trans.* 10 (26 April 1675): 293–296.

103. For an overall analysis of this work, see Samuel Eliot Morrison, *Harvard College in the seventeenth century* (Cambridge, Mass.: Harvard University Press, 1936), I, 236–249.

104. Morton, "Compendium physicae," pp. 43–52.

105. Ibid., pp. 81–86.

106. Ibid., pp. 102–107.

107. Ward, "Diary," IX, f. 123r.

108. Thomas Trapham, Jr., *A discourse of the state of health in the island of Jamaica* (London, 1679), pp. 14–17, 123.

109. He and Millington had come together from Cambridge to Oxford in 1648; see mention of Hodges's visit to Cambridge in Needham to Busby, 8 February 1654/5: British Library, MS. Add. 4292, f. 84r.

110. Nathaniel Hodges, *Vindiciae medicinae & medicorum* (London, 1665), pp. 94–95, 110–138.

111. Ibid., pp. 149–150, 157.

112. See British Library, MS. Sloane 810, labelled "N. Hodges Observat: Medicae," ff. 178r, 183r, 409r–417v, 453r–v, 520v.

113. See the letters to and from Willis, ibid., f. 229r–230v, 353r–354v.

114. Ward, "Diary," IX, f. 24v, 26v; XI, f. 35v.

115. Ibid., IX, f. 24v.

116. On 12 June 1665, Hodges was one of a group appointed by the Royal College of Physicians to advise the Lord Mayor on how to control the plague: "Annals," IV, f. 88a. He and Thomas Witherly put out a broadside dated 13 July 1665 telling of their arrangements for treating the sick.

117. Walter Bell, *The Great Plague* (London: Murray, 1928), p. 355, says that Hodges wrote the book in 1667, but gives no evidence for this statement.

118. Nathaniel Hodges, *Loimologia, sive pestis nuperae apud populum Londinensem grassantis narratio historica* (London, 1671), pp. 39, 44.

119. Ibid., pp. 44–46.

120. Ibid., pp. 48–53.

121.　William Cole, *De secretione animali cogitata* (Oxford, 1674), pp. 62–63, 121, and 177.

122.　Ibid., pp. 29, 52, 61, 120, 135–136, 147–148, 162, 166, 196, 217, 271.

123.　Ibid., pp. 22, 32, 74, 96, 98, and 102.

124.　Ibid., p. 23.

125.　Thomas Browne, *Pseudodoxia epidemica: Or, enquiries into very many received tenents and commonly presumed truths*, 6th ed. (London, 1672), p. 172. This passage was first noted by Partington.

126.　Thruston was born in West Buckland, Somerset, in 1628, and admitted a Pensioner, aged 17, at Sidney Sussex College in 1645. He took his B. A. in 1649/50, his M. A. in 1653, and his M. D. (by Royal Letters) in 1665. He was a Fellow of Sidney Sussex, 1651–55, and of Gonville and Caius, 1655–1701. See John Venn, *Biographical History of Gonville and Caius College, 1349–1897* (Cambridge: University Press, 1897), I, 393. On Thruston see also Partington, *History*, II, 571–573.

127.　Malachias Thruston, *De respirationis usu primario, diatriba* (London, 1670), pp. 1–64, passim, especially pp. 21, 48, and also 157. The work was published by Martyn and Allestry, the Royal Society's publishers.

128.　Ibid., pp. 39–42.

129.　Ibid., pp. 54, 62–63.

130.　Ibid., pp. 40–41.

131.　Ibid., p. 41.

132.　Ibid., pp. 76–77.

133.　Ibid., sigs. A8r–(a)2r.

134.　Meetings of 17 and 24 May 1665: Birch, *History*, II, 48–49.

135.　During his visit to Oxford in April 1666 John Ward had recorded the news that Stahl was "now att Exciter": "Diary," XI, f. 30v.

136.　British Library, MS. Sloane 1624, "Cursus Chymicus Petri Sthalij Germani. Inceptus Mense Maio A. D. 1668. Exonij in aedib' Bampfieldianis." Thruston's name appears in the reverse codex.

137.　Meeting of 29 October 1668: Birch, *History*, II, 316–317.

138.　Thruston, *De respirationis usu* "Ad Lectorem," sig. A5v.

139.　*Phil. Trans.* 4 (17 February 1669/70): 1142. Oldenburg to Malpighi, 15 January 1669/70; Oldenburg to Gornia, 7 February 1669/70; Oldenburg to Travagino, 14 March 1669/70; Oldenburg to Winthrop, 26 March 1670: *Correspondence*, VI, 430, 467, 557, 595.

10: FIRE AND LIFE: JOHN MAYOW'S *TRACTATUS QUINQUE* (1674) AND A GENERAL PHYSIOLOGY OF ACTIVE PARTICLES

1.　Council meeting of 14 November 1670: Birch, *History*, II, 451.

2.　Needham first attended on 2 March, and was admitted at the meeting of 6 April 1671; he communicated letters from Templer at the meetings of 30 March, 8 June, 2 November, and 7 December 1671, as well as 29 February and 26 June 1672: Birch, *History*, II, 470, 476, 487, 498; III, 17, 55.

3.　*Phil. Trans.* 5 (8 August 1670): 2011–2031; 5 (12 September 1670): 2035–2056.

4.　Ibid., 5 (14 November 1670): 2093–2095; 6 (22 May 1671): 2149–2150.

5.　*Novae hypotheseos de pulmonum motu & respirationis usu specimen* (London, 1671).

6. *Phil. Trans.* 6 (17 April 1671): 2141−2142.

7. Meetings of 10, 17, and 24 February 1670: Birch, *History*, II, 423, 424.

8. Templer to Needham, 30 March 1672; published in *Phil. Trans.* 7 (19 August 1672): 5031−5033.

9. Meeting of 26 June 1672: Birch, *History*, III, 55−56.

10. Hooke proposed the air chamber at the meeting of 12 January 1671, and described the completed apparatus on 9 February: ibid., II, 463, 467−468.

11. Meetings of 23 February and 2 March 1671: ibid., II, 469, 470.

12. Ordered at the meeting of 9 March, and reported at that of 23 March 1671: ibid., II, 471, 472−473.

13. Thomas Willis, *De anima brutorum quae hominis vitalis ac sensitiva est, exercitationes duae* (Oxford, 1672). The imprimatur is dated 12 December 1671 (verso of first leaf). The book was registered at the Company of Stationers on 5 January 1672 (*Stationers' register*, II, 438), and first advertised in the term catalogue of 13 May 1672 (Edward Arber, ed., *The term catalogues, 1668−1709 A. D.* [London: Privately printed, 1903], I, 105).

14. Cf. Boyle Papers, Royal Society, XIX, ff. 6v−35r.

15. Willis, *De anima brutorum*, pp. 138−563.

16. Ibid., pp. 1−137.

17. Ibid., sig. b4v, where he thanks King and Master for their help. King consulted with Willis as early as April 1668: British Library, MS. Sloane 1588, f. 190r. References to professional and anatomical work with Willis occur throughout King's surviving notebooks: ibid., MS. Sloane 1586, ff. 20, 22, 70−71, 106−107, 154; 1588, ff. 112v, 190r, 209r, 231v, 259v.

18. Willis, *De anima brutorum*, sigs. A3r−4r.

19. *Phil. Trans.* 7 (20 May 1672): 4071−4073. Oldenburg to Sluse, 21 November, and to Vogel, 26 December 1671; Oldenburg to Cornelio, 9 February, to Erasmus Bartholin, 22 February 1671/2; Oldenburg to Mauritius, 24 April 1672: *Correspondence*, VIII, 370, 430, 529, 548; IX, 36. Willis made a rare appearance at the Royal Society on 1 May 1672 to present a copy of his book in person: Birch, *History*, III, 48.

20. Willis, *De anima brutorum*, sigs. blr−4r, pp. 1−8.

21. Ibid., pp. 9−16.

22. Ibid., pp. 34−37.

23. Ibid., pp. 38−41.

24. Ibid., pp. 41−46.

25. Ibid., pp. 46−51.

26. Ibid., pp. 51−54.

27. Ibid., pp. 54−58, 70−72.

28. Ibid., pp. 158−162.

29. Ibid., pp. 189−194, 203−206.

30. *Tracts written by the Honourable Robert Boyle, containing new experiments, touching the relation betwixt flame and air* (London, 1672), sig. A2r, pp. 1−2. This work, and the related ideas of Hooke and Mayow on fire, air and the *flamma vitalis*, have been analyzed by Douglas McKie, "Fire and the flamma vitalis: Boyle, Hooke and Mayow," in E. Ashworth Underwood, ed., *Science, medicine and history* (London: Oxford University Press, 1953), I, 469−488.

31. Oldenburg to Magalotti, 13 June 1672: *Correspondence*, IX, 108/109.

32. Oldenburg to Huygens, 5 September and 11 November 1672: ibid., IX, 235, 319.

33. The last item in the last tract of *Flame and air*, pp. 147−176, was dated 13 February 1672/3.

34. *Phil. Trans.* 8 (25 March 1673): 5197−6001; *Flame and air* was first advertised for sale in the term catalogue of 6 May 1673: *Term catalogues*, I, 135.

35. Oldenburg to Swammerdam, 10 February 1672/3; Oldenburg to Malpighi, 18 February 1672/3; Oldenburg to Erasmus Bartholin, 27 February 1672/3; Oldenburg to Gornia, 15 March 1672/3: *Correspondence*, IX, 460, 472, 499, 573.

36. Robert Boyle, *A continuation of new experiments physico-mechanical, touching the spring and weight of the air, and their effects* (Oxford, 1669), sigs. **4r−A1v.

37. Ibid., sig. **3r.

38. Boyle, *Flame and air*, pp. 67−71.

39. Ibid., pp. 21−30, 54−62.

40. Ibid., p. 57.

41. Ibid., pp. 63−66.

42. Ibid., pp. 31−34, 46−49.

43. Ibid., pp. 35−45.

44. Ibid., pp. 94−101.

45. Ibid., pp. 71−79.

46. Ibid., pp. 109−114.

47. Ibid., pp. 115−116.

48. Ibid., pp. 116−117.

49. Leonard A. Cohen, "An evaluation of the classical candle-mouse experiment," *J. Hist. Med.* 11 (1956): 127−132.

50. Ibid., p. 130.

51. Boyle, *Flame and air*, pp. 117−118.

52. Ibid., pp. 118−126.

53. Ibid., pp. 129−130.

54. Ibid., pp. 133−142.

55. Ibid., "New Experiments about Explosions," pp. 1−17.

56. Boyle, *Flame and air*, p. 17.

57. Ibid.

58. Birch, *History*, III, 61.

59. Meetings of 20 and 27 November 1672: ibid., III, 61, 63.

60. *The diary of Robert Hooke M. A., M. D., F. R. S. 1672−1680. Transcribed from the original in the possession of the Corporation of the City of London (Guildhall Library)*, ed. Henry W. Robinson and Walter Adams (London: Taylor & Francis, 1935), p. 14.

61. Birch, *History*, III, 68.

62. Meetings of 19 February and 5 March 1672/3: ibid., III, 76, 77.

63. Ibid., III, 77.

64. Hooke, *Diary* (5 March 1673), p. 32.

65. Ibid., (11 and 12 March 1673), p. 33.

66. Meeting of 12 March 1673: Birch, *History*, III, 78.

67. Hooke, *Diary* (18 March 1673), p. 34.

68. Birch, *History*, III, 78.

69. Hooke, *Diary* (19 March 1673), p. 35.

70. Birch, *History*, III, 84.

71. Meeting of 7 May 1673: ibid., III, 85.

72. Meeting of 14 May 1673: ibid., III, 89.

73. Ibid., III, 89.

74. Ibid., III, 89–90.

75. Hooke, *Diary* (28 May 1673), p. 45.

76. Meeting of 4 June 1673: Birch, *History*, III, 90–91.

77. Bodleian Library, MS. D. D. All Souls College b. 93–96 (Stewards' Books, 1667–1674).

78. Wood, *Fasti*, II, 320.

79. John Mayow, *Tractatus quinque medico-physici* (Oxford, 1674), "De motu musculari," pp. 96–100; a translation by A. Crum Brown and Leonard Dobbin, *Medico-physical works* (Edinburgh: Alembic Club, 1907), pp. 295–298. All subsequent translations are taken from this edition, modified where necessary by comparison to the original Latin.

80. Wood, *Athenae*, III, 1199.

81. Mayow, *Tractatus quinque*, pp. 246–265; *Medico-physical works*, pp. 170–182.

82. For details on editions of Mayow's published works, see John F. Fulton, *A bibliography of two Oxford physiologists* (Oxford: University Press, 1935), pp. 39–59. A detailed analysis of the *Tractatus quinque*, with special reference to Mayow's chemical opinions, has been made by Partington, *History*, II, 587–613. Aspects of Mayow's physiological ideas have been well treated by Diana Long Hall, "From Mayow to Haller: A history of respiratory physiology in the early eighteenth century," unpublished dissertation, Yale University, 1966.

83. Mayow, *Tractatus quinque*, pp. 16–17; *Medico-physical works*, pp. 12–13.

84. *Tractatus quinque*, front leaf verso.

85. *Term catalogues*, I, 183; *Phil. Trans.* 9 (20 July 1674): 101–113.

86. Mayow, *Tractatus quinque*, p. 1; *Medico-physical works*, p. 1.

87. Mayow, *Tractatus quinque*, pp. 2–8; *Medico-physical works*, pp. 2–6.

88. Mayow, *Tractatus quinque*, pp. 9–17; *Medico-physical works*, pp. 7–12.

89. Mayow, *Tractatus quinque*, p. 17; *Medico-physical works*, p. 13.

90. Mayow, *Tractatus quinque*, p. 18; *Medico-physical works*, p. 13.

91. Mayow, *Tractatus quinque*, p. 18; *Medico-physical works*, pp. 13–14.

92. Mayow, *Tractatus quinque*, p. 22; *Medico-physical works*, p. 16.

93. Mayow, *Tractatus quinque*, pp. 23–24; *Medico-physical works*, p. 17.

94. Mayow, *Tractatus quinque*, pp. 24–31; *Medico-physical works*, pp. 18–22.

95. Mayow, *Tractatus quinque*, pp. 47–58; *Medico-physical works*, pp. 34–42.

96. Mayow, *Tractatus quinque*, p. 59; *Medico-physical works*, p. 42.

97. Mayow, *Tractatus quinque*, pp. 59–65; *Medico-physical works*, pp. 42–47.

98. Mayow, *Tractatus quinque*, pp. 66–95; *Medico-physical works*, pp. 47–66.

99. Mayow, *Tractatus quinque*, pp. 195–213; *Medico-physical works*, pp. 134–147. Walter Böhm has recently argued that Mayow's system is directly descended from Descartes's, and that it therefore provides, against the positivists, an example of theory-directed experimentation. While I would certainly agree that much of Mayow's experimental work was brilliantly directed by his theories, and that he had read Descartes closely, it should by now be clear that his conceptual framework was, if anything, derived from Willis and Boyle. See Walter Böhm, "John Mayow und Descartes," *Sudhoffs Arch.* 46 (1962): 45–68; "John Mayow and his contemporaries," *Ambix* 11 (1963): 105–120; and "John Mayow und die Geschichte des Verbrennungsexperiments," *Centaurus* 11 (1967): 241–258.

100. Mayow, *Tractatus quinque*, pp. 172–181; *Medico-physical works*, pp. 118–124.

101. Mayow, *Tractatus quinque*, pp. 181–195, 214–265; *Medico-physical works*, pp. 125–134, 147–182.

102. Mayow, *Tractatus quinque*, pp. 96–103; *Medico-physical works*, pp. 66–71. Mayow's originality in this experiment is well defended by Partington, *History*, II, 593–596.

103. Mayow, *Tractatus quinque*, pp. 103–104; *Medico-physical works*, p. 72.

104. Mayow, *Tractatus quinque*, pp. 104–105; *Medico-physical works*, pp. 72–73.

105. Mayow, *Tractatus quinque*, p. 106; *Medico-physical works*, pp. 73–74.

106. Mayow, *Tractatus quinque*, pp. 106–107; *Medico-physical works*, pp. 74–75.

107. Mayow, *Tractatus quinque*, pp. 107–108; *Medico-physical works*, p. 75.

108. Mayow, *Tractatus quinque*, p. 108; *Medico-physical works*, p. 75.

109. Mayow, *Tractatus quinque*, pp. 108–114; *Medico-physical works*, pp. 75–79.

110. Mayow, *Tractatus quinque*, pp. 114–119; *Medico-physical works*, pp. 79–83.

111. Mayow, *Tractatus quinque*, p. 116; *Medico-physical works*, pp. 80–81.

112. Mayow, *Tractatus quinque*, pp. 136–140; *Medico-physical works*, pp. 94–97.

113. Mayow, *Tractatus quinque*, p. 146; *Medico-physical works*, p. 101.

114. Mayow, *Tractatus quinque*, pp. 147–149; *Medico-physical works*, pp. 101–103.

115. Mayow, *Tractatus quinque*, pp. 149–150; *Medico-physical works*, p. 103.

116. Mayow, *Tractatus quinque*, pp. 150–151; *Medico-physical works*, p. 104.

117. Boyle, *Continuation*, sig. *2v.

118. Mayow, *Tractatus quinque*, pp. 165–166, 191, 319, and 328; *Medico-physical works*, pp. 114, 131, 216, and 223.

119. Mayow, *Tractatus quinque*, pp. 152–157; *Medico-physical works*, pp. 105–108.

120. Mayow, *Tractatus quinque*, p. 157; *Medico-physical works*, p. 108.

121. Mayow, *Tractatus quinque*, pp. 267–308 passim; *Medico-physical works*, pp. 183–210 passim.

122. Mayow, *Tractatus quinque*, pp. 309–319; *Medico-physical works*, pp. 211–217.

123. Mayow, *Tractatus quinque*, pp. 319–320; *Medico-physical works*, p. 217.

124. Mayow, *Tractatus quinque*, p. 321; *Medico-physical works*, p. 218.

125. Mayow, "De motu musculari," pp. 16–52 in *Tractatus quinque; Medico-physical works*, pp. 238–264.

126. Mayow, "De motu musculari," in *Tractatus quinque*, pp. 77–84; *Medico-physical works*, pp. 282–287.

11: THE DECLINE OF THE OXFORD TRADITION

1. Oldenburg to Boyle, 10 July 1674: *Correspondence*, XI, 50.

2. *Phil. Trans.* 9 (20 July 1674): 101–113.

3. Ibid., pp. 101–110.

4. Ibid., pp. 101, 102, 104, 107, 108.

5. Ibid., pp. 110–111.

6. *The diary of Robert Hooke*, ed. H. W. Robinson and W. Adams (London: Taylor & Francis, 1935), p. 128 (29 October 1674).

7. Ibid., p. 130.

8. Ibid.

9. Meeting of 12 November 1674: Birch, *History*, III, 143.

10. Hooke, *Diary* (12 November 1674), p. 130.

11. Robert Hooke, *Lectures De potentia restitutiva, or of spring, explaining the power of springing bodies* (London, 1678), p. 1.

12. Ibid; i.e. in Robert Hooke, *A description of helioscopes* (London, 1676), pp. 31–32.

13. Hooke, *Lectures De potentia restitutiva*, p. 1.

14. On 4, 8, 10, and 15 December 1674, 11 January and 11 February 1675: Hooke, *Diary*, pp. 133–135, 141, 146.

15. On 11 January 1675: ibid., p. 141.

16. Ibid., pp. 134–135; meeting of 10 December 1674: Birch, *History*, III, 161.

17. The case is described in a letter from James Young, a surgeon of Plymouth, to Hooke, and printed in Robert Hooke, *Lectures and collections* (London, 1678), pp. 105–112. Mayow's part in the case, which had also been treated by Lower, is given on pp. 107 and 110.

18. Bodleian Library, MS. D. D. All Souls College b. 96–99 (Stewards' Books, 1673–1680).

19. Hooke, *Diary*, pp. 219, 325.

20. Meeting of 31 January 1678: Birch, *History*, III, 381. The name of the doctor who commented on Millington's case is unrecorded, but it is almost certainly Mayow.

21. Ibid., III, 384, 442.

22. *Remarks and collections of Thomas Hearne*, ed. C. E. Doble (Oxford: Oxford Historical Society, 1885), I, 192.

23. Wood, *Athenae*, III, 1199.

24. Rate Book H460 for 1679, Parish of Saint Paul, Covent Garden, deposited in the Archives Department, Westminster Public Library, London. Mayow is not listed, most likely because he arrived after March 1679, when the parish was walked for that year, and died before March 1680.

25. Wood, *Athenae*, III, 1199; for the date ("about Michaelmas") of Mayow's death, see Wood, *Life and times*, II, 464 and 478. He was buried in the Church of Saint Paul, Covent Garden, on 10 October 1679: Douglas McKie, "The birth and descent of John Mayow," *Phil. Mag.* 33 (1942): 57.

26. Hooke, *Diary* (8 November 1679), p. 430; Hooke had not seen Boyle since 18 September 1679.

27. See, for example, Michael Foster, *Lectures on the history of physiology during the sixteenth, seventeenth and eighteenth centuries* (Cambridge: University Press, 1901), pp. 172–197, especially pp. 196–197; Francis Gotch, *Two Oxford physiologists: Richard Lower 1631 to 1691, John Mayow 1643 to 1679* (Oxford: Clarendon Press, 1901), p. 32.

28. Partington, *History*, II, 622–631.

29. Diana Long Hall, "From Mayow to Haller: a history of respiratory physiology in the early eighteenth century," unpublished dissertation, Yale University, 1966, pp. 60–246.

30. John F. Fulton, *A bibliography of two Oxford physiologists: Richard Lower, 1631–1691, John Mayow, 1643–1679* (Oxford: University Press, 1935), pp. 44–48.

31. Partington, *History*, II, 609, 613–614, 617, 619, 621; Domenico Mistichelli, *Trattato dell'appoplessia* (Rome, 1709).

32. Partington, *History*, II, 614, 616–617, 619, 628.

33. Wood, *Fasti*, II, 304.

34. Oldenburg to Wallis, 2 December 1668, and Wallis to Oldenburg, 5 December 1668: *Correspondence*, V, 217, 221. The Oxford correspondent is only referred to as Oldenburg's "countryman," whom I take to be Ettmuller.

35. Partington, *History*, II, 298–300, 617.

36. Ibid., pp. 617–619.

37. John F. Fulton, *A bibliography of the Honourable Robert Boyle, Fellow of the Royal Society*, 2d ed. (Oxford: Clarendon Press, 1961), p. 61.

38. Partington, *History*, II, 619–620.

39. Thomas Guidott, *A discourse of Bathe, and the hot waters there* (London, 1676), pp. 6–11.

40. Ibid., pp. 12–25. Other references to Mayow are on pp. 26, 33, and 35.

41. Ibid., p. 6.

42. British Library, MS. Sloane 1833, ff. 41–152, 180–183, 198–200.

43. Ibid., ff. 86r, 92v.

44. Ibid., f. 124r.

45. Ibid.

46. Robert Plot, *The natural history of Oxford-shire, being an essay toward the natural history of England* (Oxford, 1677), pp. 300–301.

47. Ibid., sig. b3v, p. 301.

48. Ibid., pp. 301–304.

49. Ibid., pp. 228–229.

50. Ibid., pp. 304–305.

51. Ibid., p. 305.

52. Partington, *History*, II, 615–616, 620.

53. Robert Hooke, "Of the True Method of Building a Solid Philosophy, or of a Philosophical Algebra," in *The posthumous works of Robert Hooke*, ed. Richard Waller (London, 1705), p. 50, cf. also pp. 30–32, 46, 53, 56, 59.

54. Ibid., p. 50.

55. Hooke, *Diary* (27 December 1677), p. 336.

56. Meetings of 6 December 1677 to 7 March 1678: Birch, *History*, III, 359–389 passim.

57. Meetings of 31 January and 7 February 1678: ibid., III, 381, 384.

58. Discussed with Wren at a coffeehouse on 9 February 1678: Hooke, *Diary*, p. 344.

59. Meetings of 25 April, 2 and 9 May 1678: Birch, *History*, III, 402–407.

60. Meeting of 9 May 1678: ibid., III, 407.

61. Meetings of 2 January to 20 March 1679: ibid., III, 453–471 passim.

62. Hooke, *Posthumous works*, pp. 71–185.

63. Ibid., pp. 92, 94, 100, 110–111, 117, 163–164, 169–170.

64. Ibid., (May, 1681), p. 111.

65. R. T. Gunther, ed., *Early science in Oxford* (Oxford: For the author, 1935), X, 185 (6 February 1690).

66. Hooke, *Diary* (9, 13, 16–20 November 1672), pp. 12–14.

67. Wood, *Life and times*, II, 253.

68. K. J. Franklin, "Some textual changes in successive editions of Richard Lower's *Tractatus de corde item de motu & colore sanguinis et chyli in eum transitu*," *Ann. Sci.* 4 (1939): 283–294.

69. Oldenburg to Dodington, 15 July 1670; Oldenburg to Malpighi, 20 December 1670; Malpighi to Oldenburg, 31 January 1670/1: *Correspondence*, VII, 69, 332–333, 429–430. *Phil. Trans.* 6 (17 July and 20 November 1671): [2211–2212], [3017]. See also Fulton, *Two Oxford Physiologists*, pp. 17–19.

70. Wood, *Athenae*, IV, 297.

71. He settled first in Hatton Garden, thence moved successively to Salisbury Court, near Fleet Street; to Bow Street (1672); and to King Street (1682), Covent Garden: ibid., and

Rate Books H 454−470 (1672−1690), Parish of Saint Paul, Covent Garden, Archives Department, Westminster Public Library, London.

72. Wood, *Athenae*, IV, 297−298.

73. Gideon Harvey, *Casus medico-chirurgicus* (London, 1678), esp. pp. 8−10, 20−22, 40, 48−57, 70−71, with reference to Lower's behavior in a case he treated in November and December 1676; Harvey took more slashes at Lower in his *Conclave of physicians* (London, 1683), pp. 17, 23, 30−33. Richard Griffith, *A-la-mode phlebotomy no good fashion* (London, 1681), esp. sigs. A7v, B5v−8r, pp. 34, 69−83, 169−209, with reference to a case Lower had treated in Griffith's hometown of Richmond, Surrey. Interestingly, both Harvey and Griffith were younger contemporaries of Lower at Oxford, and Griffith had attended one of Stahl's chemistry courses.

74. Historical Manuscripts Commission, *Calendar of the manuscripts of the Marquess of Ormonde* (London: HMSO, 1908), V, 354−362, case and postmortem of July 1680. See also the references to consultations in Edmund King's clinical notebook of 1676−1696, British Library, MS. Sloane 1589, especially ff. 40v, 76r, 80v, 117r, 144r, 182r, 188v, 196v, 199v, 202r, 207r, 220r, 231v, 242r, 248r, 249r, 257v, 258r, 260r, 276r, 286r; also MS. Sloane 1591, ff. 14v, 15r, 22r−v.

75. British Library, MS. Sloane 1731A, f. 151, contains a poem praising William of Orange, supposed to have been written by Lower.

76. J. J. Keevil, *Medicine and the navy, 1200−1900* (Edinburgh/London: Livingstone, 1958), pp. 193−194, 200−201, 248.

77. Wood, *Athenae*, IV, 298.

78. *Phil. Trans.* 7 (15 July 1672): 5007−5011; 8 (23 June 1673): 6052−6054. See the syllabi for his anatomical lectures in the British Library, MS. Sloane 631, ff. 145−154, and MS. Sloane 1761, ff. 2−14. Needham's book is MS. Sloane 656.

79. Birch contains only the following: details of eels and lobsters dissected c. 1657 (1677); relation of the foramen ovale of a dissected otter (1678); comments on microscopical animals (1680); relation of dissection of the three muscular coats of an artery (1682): *History*, III, 347, 350, 403; IV, 47, 120.

80. E.g. Bodleian Library, MS. Locke d. 9, pp. 30, 31, 36, 40−42, 52−53; MS. Locke f. 19, pp. 153, 158−159, 166, 174−175, 178.

81. John Locke, "Elements of Natural Philosophy," pp. 194−200 in *A collection of several pieces of Mr. John Locke, never before printed, or not extant in his works* (London, 1720). The *Elements* were popular, being reprinted at least seven times before 1770.

82. Hooke attended the Oration on 31 July 1679 (*Diary*, p. 419), which was described by an anonymous admirer as "incomparable . . . both for matter and language": *The character of a compleat physician* [London, 1680?], p. 2.

83. His last meeting was 26 March 1674: Birch, *History*, III, 132.

84. Once in 1680, twice in 1684, and once in 1687: ibid., IV, 60, 252, 324, 522.

85. Ibid., IV, 58.

86. Robert Boyle, *Experimentorum novorum physico-mechanicorum continuatio secunda* (London, 1680).

87. Robert Boyle, *Memoirs for the natural history of humane blood, especially the spirit of that liquor* (London, 1683/4).

88. Robert Boyle, *The general history of the air* (London, 1692). On the role of Locke in this work, see Kenneth Dewhurst, "Locke's contribution to Boyle's researches on the air and on human blood," *Notes Rec. R. Soc. Lond.* 18 (1962): 198−206.

89. Boyle, *General history of the air*, pp. 4−7.

90. Ibid., pp. 41−42.

91. Ibid., p. 42.

92. Bathurst to [Ward], 10 February 1663/4, in Thomas Warton, *The life and literary remains of Ralph Bathurst, M.D.* (London, 1761), p. 57.

93. George Hooper, educated at Westminster School and Christ Church, later Bishop of Bath and Wells.

94. See R. T. Gunther, *Early science in Oxford*, vol. IV on the Oxford Philosophical Society, which reprints the minutes for 1683–1688, and 1690.

95. Meeting of 24 June 1684: ibid., p. 76. This was prompted by a similar experiment done at the Dublin Philosophical Society.

96. Birch, *History*, IV, 173.

97. Meeting of 7 April 1686: ibid., IV, 471. Abraham Hill's extract is still extant among his papers, British Library, MS. Sloane 2903.

INDEX

All persons and major subjects are recorded here, with the exception of the material in the table on pages 64−89, in which only the name entries have been indexed. Books have been referenced under their authors' names, and given by title only when cited substantively.

Acland, Arthur, 105, 106
Acosta, Joseph d', 112
Aeson, 170, 172, 177
Air: corpuscular nature of, 116−117, 132−137, 249−250, 262−269; dissolved in liquids, 134, 147, 150, 152, 153; experiments on, 115−117, 128−137; spring of, 117, 132−137, 143, 145, 149−150, 153, 156−157, 163, 227, 257, 262−269, 277, 285; traditional ideas of, 115; variant kinds of, 161, 257; weight of, 115−117, 128. *See also* Combustion; Nitrous aerial substance; Respiration; Vacuum, pumps
Allestree, Richard, 78−79
Allestry, James, 183, 188, 209
Alvey, Thomas, 105, 106
Animal heat. *See* Heat, animal
Aristotle: on air, 115, 116, 128, 134, 238, 239; on animal heat, 12−13; on blood, 15, 19, 39, 189; natural philosophy of, 18−19, 48, 49, 90, 91, 92, 94, 96, 103, 107, 122, 127, 182; on respiration, 98, 199, 227
Aromatari, Giuseppe degli, 34
Arris, Thomas, 104, 105, 106
Arterial Blood. *See* Blood, color
Aselli, Gaspare, 140
Atomism. *See* Corpuscular philosophy
Aubrey, John: career of, 78−79; on Harvey, 19, 21, 26, 27, 32; on Oxford science, 50, 56, 106, 129, 165, 170, 179

Austen, Ralph: career of, 68−69; and Oxford science, 46, 54, 56, 60

Bacon, Francis (Baron Verulam, Viscount St. Albans), 46, 58, 127, 263; on niter, 118
Ball, John, 133
Balle, Peter, 197
Barberius, Ludovico, 279
Barchusen, Johann Conrad, 279
Barksdale, Francis, 48, 78−79
Barlow, Thomas, 47, 58, 80−81
Bartholin, Caspar, 26
Bartholin, Erasmus, 250
Bartholin, Thomas, 49, 140, 141, 155
Bate, George, 104
Bathurst, George: career of, 26, 29, 62, 80−81; and embryology research with Harvey, 26
Bathurst, Ralph: on animal heat, 109−111, 288; on blood color, 205; and Boyle, 143, 144, 145, 147, 171, 289; career of, 29, 30, 44, 47, 58, 59, 68−69, 106−107, 239, 285, 294; and Highmore, 30, 100, 156; and Oxford scientific clubs, 50, 51, 53, 54, 60, 61, 181, 289; other physiological ideas of, 41, 106, 107−108, 183; "Praelectiones tres de respiratione" [1654], 106, 108−112, 123, 156, 205, 216, 282, 285, 289; on respiration, 108−113, 164, 227; on respiratory aerial nitrous sub-

stance, 105, 109–113, 114, 117, 120–121, 127, 156, 191, 192, 216, 231, 244; and Willis, 29, 30, 106, 107, 165, 167, 250, 289

Bauhin, Caspar: physiological ideas in *Theatrum anatomicum* (1605), 3–8, 27

Beale, John, 177

Beckman, Johann Christoph, 202

Behm, Michael, 205

Bellini, Lorenzo, 246

Berkenhead (or Berkhead), Henry, 53

Berkenhead (or Birkenhead), Sir John, 29

Bernoulli, Jakob, 279

Biggs, Noah, 49

Billich, Anthor Gunther, 166, 240

Bils, Louis de, 209

Blood: air dissolved in, 150, 196, 197, 199–200, 203–208, 214–219, 267–271; color, 4–8, 15–16, 40–41, 159, 166, 168, 183–188, 189–191, 196, 205–208, 214–215, 217, 218, 236, 242, 269, 287, 289; nature of, 5–6, 9, 98–100, 105, 166–169, 183–188, 189–191, 248–250, 255, 270; origins of, 3, 9, 38–40, 104–105, 154–160; vessels, 3–5, 7–8, 155; vital spirits in, 3–9, 39–40, 98–99, 100, 144. *See also* Circulation of blood; Fermentation; Heat, animal; Nitrous aerial substance; Transfusion

Bobart, Jacob: career of, 68–69; friends of, 28, 181; as Keeper of the Oxford Physic Garden, 25, 46, 49–50, 61

Bodley, Sir Thomas, 46

Boerhaave, Hermann, 279

Bohn, Johann, 279

Bohun, Ralph: career of, 46, 68–69, 239; *Discourse concerning the origin and properties of wind* (1671), 239; on meteorological aerial niter, 239, 244

Bordeaux, Duc de, 172

Borel, Pierre, 49, 141

Botany, 25, 49–50

Boyle, Charles (Viscount Dungarvan), 128–129, 130

Boyle, Richard (3d Earl of Cork), 130

Boyle, Robert: on air, nature, and spring of, 116, 132–137, 139, 152–153, 257, 262, 263–264, 285; and Bathurst, 107, 112, 289; on blood color, 183–186, 193, 206–207, 218, 289; career of, 64–65, 92–97, 194, 195, 208, 226, 246, 284–285, 294; on combustion, pneumatic experiments about, 134, 250–255, 259, 260; on combustion and respiration, 146, 151–152, 164, 253–255, 267, 279; corpuscular philosophy of, 92–95, 122–125, 272,

273, 288, 289; and corpuscular philosophy, biological applications of, 93–97, 113; dissections by, 50, 93, 95–96, 140–142, 183, 193, 222, 289; and Highmore, 93, 95, 96, 97, 289; and Hooke, 129–137, 139, 142, 226, 284, 289; on injection and transfusion, 170–171, 172, 173, 174, 175–178, 193–194, 202, 206, 208, 225, 289; and Locke, 186–187, 188, 284–285; and Lower, 174–178, 180, 182, 183–185, 189, 194, 195, 197, 198, 201, 206, 209; and Mayow, 225, 226, 229, 230, 231, 269, 270, 273, 275, 276, 278; and Needham, 181, 191, 198, 203; on niter, composition of, 121–127, 138, 139, 259, 289; on niter, aerial, 122, 124–126, 127, 238, 239, 285; and Oxford scientific clubs, 43, 52–55, 56, 57, 58, 59, 60, 61, 114, 163, 165, 219, 224; on respiration, function of, 141, 143–148, 150, 154, 157–160, 227, 271; on respiration, pneumatic experiments about, 142–143, 147–148, 148–151, 154, 156, 162, 164, 204, 207, 235, 237, 246, 263; scientific personality of, 218, 274, 276, 291, 293; and Stubbe, 123, 226, 237, 238; vacuum pumps of, 128–134, 226, 248, 249, 250–252, 254, 269, 270, 280, 289; and Willis, 167, 169, 235, 237, 250, 255. Works (arranged chronologically): "Essay of ye Atomicall Philosophy" [1650], 94–95; "Philosophical Diaries" [1650s], 121; "History of Heat and Flame" [c. 1657–1658], 125; *New experiments physico-mechanicall* (1660), 61, 130–134, 142–148, 152, 153, 248, 260; *Certain physiological essays* (1661), 122–125, 126, 127, 130, 135, 148, 196; *The sceptical chymist* (1661), 122, 130, 135, 148; "Notes on ye essay concerning Saltpetre" [1660s], 125; *Examen* (1662), 135–137; *Defence* (1662), 135–137; *Usefulnesse of experimental naturall philosophy* (1663), 148; *New experiments and observations touching cold* (1665), 125–126; *Origins of forms and qualities* (1666), 125; *Continuation of new experiments physico-mechanical* (1669), 207, 226, 252; "New pneumatical experiments about respiration" (1670), 148–150, 204, 207; *Flame and air* (1672), 250–256, 259, 260, 266; *Continuation of new experiments physico-mechanical* (1682), 284; *Natural history of humane blood* (1684), 186, 284; *General history of the air* (1692), 285

Brady, Robert, 23

Brain and nerves, 4, 174, 182–183, 280

Brancker, Thomas, 51, 59, 68–69

Brookes, Christopher, 80–81

Brouncker, William (2d Viscount): career of, 29, 68–69; and respiration experiments 153, 161, 231, 247–248

Brown, William, 49–50

Browne, Edward: career of, 68–69, 224, 280; dissections by, 224, 280; and Mayow, 224, 227, 241

Browne, Sir Thomas, 35, 93; on nitrous aerial substance, 120, 227, 241

Busby, Richard: scientific interests of, 59, 129, 180; his students at Westminster School, 59, 129, 179, 180, 218, 293. *See also* Westminster School

Bylebyl, Jerome J., 9, 11

Cambridge Colleges: Gonville and Caius, 1, 241; Queens', 180, 181; Trinity, 59, 61, 180, 224, 243

Cartwright, William, 29

Castle, George: and anatomical research, 105, 181, 195–196; career of, 59, 68–69, 194, 218, 224, 246, 282; *Chymical Galenist* (1667), 49, 60, 195–196; on nitrous aerial substance, 196, 200, 231, 236, 259; and Oxford scientific clubs, 54, 58, 60, 61, 106; scientific personality of, 219

Cavendish, Charles, 92, 116, 128

Charleton, Walter; and anatomical research, 24, 152, 158; books by, 44, 49, 93, 128, 186; career of, 30, 64–65, 246, 278; on corpuscular philosophy, 57, 93, 117, 128; as Harvey disciple, 24, 28, 29, 41; on muscle contraction, 24, 222; on Oxford science, 57

Charterhouse Hospital, London, 282

Chemistry, 19, 50–51, 55, 102–103, 107–108, 165–169, 288. *See also* Combustion; Fermentation; Niter

Circulation of blood: discovery of, 2, 9–11; reception and elaboration of, 21, 97–98, 102, 103, 106, 154, 155, 167, 170, 195, 213–214, 287

Clarke, John, 53, 60

Clarke, Timothy: on blood, 104, 172; career of, 29, 30, 41, 44, 47, 57, 68–69, 95, 106, 141, 176, 246, 282; dissections by, 152, 208, 209; on injection and transfusion, 172–173, 175, 176, 177, 178, 205, 282, 289

Clarke, William: career of, 70–71, 238; *Natural history of nitre* (1670), 238; on niter, 238–239, 244, 259

Clayton, Thomas, Sr.: and anatomy, 26, 27, 28; career of, 25–26, 30

Clayton, Thomas, Jr., 49, 104

Clerke, Henry: and anatomy, 49, 50, 53, 107, 175, 181; career of, 47, 51, 59, 61, 70–71; and Oxford scientific clubs, 50, 53, 54

Clodius, Frederick, 122

Coffeehouses, 58, 60, 224, 276, 277, 282

Coga, Arthur, 202, 209

Cole, William, 70–71, 277; on nitrous aerial substance, 241

Colepresse, Samuel, 226

Coles, William, 47, 70–71

Colombo, Realdo, 2, 10, 13, 141

Combustion: and air, 134, 146, 151–152, 160–161, 250–258; and animal heat, 164–165, 234–235, 249–250, 289; and niter, 118, 120–121; and nitrous aerial substance, 137–139, 193, 235, 249–250, 261, 262, 263; and respiration, 146, 151–152, 160–161, 247–248, 253–255, 266–267

Connor, Bernard, 281

Conring, Hermann, 49

Conway, Lady Anne, 179

Conyers, William, 47, 50, 51, 54, 80–81

Cooper, Anthony Ashley (1st Earl of Shaftesbury), 194, 284

Copernicus, Nicolaus, 48, 190

Corpuscular philosophy, 18–19, 90–97, 122–125, 127, 132–137, 139, 165–169, 195, 231–232, 288, 291

Coventry, Henry, 224

Coward, William, 281

Cowley, Abraham, 29, 80–81; on Harvey's lost works, 33

Cox, Daniel, 161, 239; career of, 61, 70–71; on injection and transfusion, 173, 178; and nitrous aerial substance, 208

Coysh, Elisha, 105, 106

Crew, Nathaniel, 51, 59, 80–81

Croone, William, 136, 178, 181; on muscle contraction, 222, 233; on respiration, 157, 158, 163, 242, 280

Crosse, Francis, 61, 70–71

Crosse, John: Boyle's lodgings at Oxford with, 58, 60, 158, 163, 175, 178, 195, 207, 225; career of, 51, 53, 80–81

Crosse, Joshua, 48, 54, 61, 80–81, 179

Crossley, John, 259

Curteyne, Amos, 47

Danvers, Henry (Earl of Danby), 46

Davis, Richard, 46, 47, 60, 176, 227

Davisson, William, 102

Dawson, Edward, 26
Day, William, 51
Debus, Allen G., 117
Democritus, 94, 115, 116, 146, 180
Denis, Jean, 202
Descartes, René: on air, nature of, 116, 128, 133, 135, 136, 137, 138, 239; books by, 21, 33, 46, 49, 58, 90, 242; corpuscular philosophy of, 90, 91–92, 123, 179, 272, 274, 288; English followers of, 49, 58, 59, 93, 94, 102, 117, 129, 239, 250, 272, 276; on the heart, motion of, 19, 21, 141, 168, 199, 210–213, 229, 231, 233, 289; on light, 250; on niter, 242; on respiration, 143, 144, 227
Deusing, Anton, 49, 186
Deverti, 49
Dickenson, Edmund: books by, 46, 60; career of, 47, 48, 51, 70–71; and Oxford scientific clubs, 50, 54, 171, 175, 181
Diemerbroeck, Ijsbrand de, 279
Digby, Kenelm, 94; on embryology, 36, 100–101; on nitrous aerial substance, 126–127, 242
Digestion, 3, 101, 104–105, 107–108
Drebbel, Cornelius: on vital quintessence of air, 127, 146, 153
Du Chesne, Joseph, 119
Duppa, Brian, 29

Edwards, John, 104
Eliot, Peter, 179
Elsholtz, Johann Sigismund, 176
Embryology. *See* Fetus
Ent, George: and anatomical research, 24, 27, 34, 158, 160, 195; *Apologia pro circulatione sanguinis* (1641), 22, 46, 49, 108, 109, 110, 111, 120, 141, 186, 192; career of, 22–23, 24, 27; and Harvey, 22–23, 34, 41, 43; on heart, heat of, 168; on nitrous aerial substance, 110–111, 120, 192, 196, 201, 242; on respiration, 160–161, 201
Epicurus, 91, 92, 94, 96, 117, 180
Ettmuller, Michael: career of, 70–71, 279; on nitrous aerial substance, 279, 281
Euclid, 129
Evelyn, John, 34, 52, 56, 106, 126, 135, 148, 156, 201, 239

Fabricius ab Aquapendente, Hieronymus, 34, 107, 141
Fabritius, Johannes, 205
Feake, John, 226
Fell, John, 179
Fell, Phillip, 80–81

Fermentation, 165–169, 240, 280, 289, 291; in blood, 166–169, 186–188, 190–191, 229, 231, 268–271; and nitrous aerial substance, 186–188, 196, 229, 231, 268–271; as origin of animal heat, 166–169, 190–191, 210–212, 216, 229, 231, 234–237, 262, 268–272. *See also* Heat, animal; Heart; Nitrous aerial substance
Fernel, Jean, 98
Fetus: anatomy of, 8, 17, 26–27, 101, 107, 198–199; traditional physiology of, 8, 107. *See also* Respiration, fetal
Fielding, Robert, 105, 106
Finch, Sir John, 209
Fire. *See* Combustion
Fludd, Robert, 21
Foster, Samuel, 23
Fracassati, Carlo: on blood color, experiments about, 205–207, 215, 217, 269, 290
Fry, Robert, 286
Fuchs, Leonhard, 49
Fuller, Thomas, 32

Galen and Galenists, 25, 27; on animal heat, 3–4, 6–7; on blood color, 4, 6–8, 38, 39; on blood vessels, 3, 4, 7–8, 17; on brain, 4; on digestion, 3; on fetus, 8; on heart, 3–5, 10; physiological system of, 2–9, 11, 92, 181, 216; on respiration, 4, 5–6, 13, 14, 15, 17, 40, 98–99, 143, 199, 241
Galilei, Galileo, 58
Garbrand, Tobias, 104
Gassendi, Pierre: *Animadversiones in decimum librum Diogenis Laërtii* (1649), 90–91, 92, 93, 117, 128, 167; other books by, 46, 58, 91, 93, 108, 109; corpuscular philosophy of, 90–91, 96, 100, 123, 133, 138, 139, 144, 272–273; English followers of, 59, 92–95, 108, 109, 195, 223, 249, 272–273; and nitrous aerial substance, 125–126, 127, 196, 239, 242
Gerard, Peter, 104, 105, 106
Gill, John, 51
Glanvill, Joseph, 44, 59, 82–83
Glauber, Johann Rudolph, 238
Glisson, Francis: and anatomical research, 24, 30, 154, 181, 182, 195; books by, 24, 49, 106, 186; as Harvey adherent, 22–23, 24, 41, 43; and scientific clubs, 24
Goddard, Jonathan: career of, 24, 41, 70–71; on combustion, experiments about, 151, 267; and microscopes, 56; on muscle contraction, experiments about, 233, 250; and Oxford scientific clubs, 44, 54, 56, 60, 163; on respiration, experiments about,

152, 159, 160, 161, 163; and transfusion, 178

Goodall, Charles, 33

Gorges, Robert, 48, 82–83

Gornia, Giovanni Battista, 243, 250

Greatorex, Ralph, 129, 130

Greaves, Edward, 28, 30, 44, 47, 82–83

Greaves, John: career of, 22, 24, 25, 30, 47, 72–73; as Harvey adherent, 22, 26–27, 28, 40; and scientific clubs, 24, 44, 47, 116

Greene, Anne, 50, 53, 107

Gresham College: professors, activities of, 51, 61, 135, 136, 151, 158, 159, 172, 175, 178, 202, 276, 281, 282; as Royal Society meeting place, 24, 52, 126, 135, 137, 150, 153, 158, 159, 160, 173, 175, 177, 178, 185, 195, 206, 247

Grew, Nehemiah, 277

Griffith, Richard, 51, 72–73

Grosshead, "Mr.," 225

Guericke, Otto: vacuum pump of, 128–129, 136

Guerlac, Henry, 117

Guidott, Thomas: career of, 59, 61, 72–73, 194, 284; *Discourse of Bathe* (1676), 279–280; and Mayow, 225, 279–280; on niter, 238, 239, 280

Gunpowder, 118, 121–122, 252–253, 259, 260

Gwynn, Edmund, 50

Haak, Theodore, 116

Hall, Diana Long, 278

Haller, Albrecht von, 278

Hamilton, William, 82–83

Harford, Bridstock, 104

Hariot, Thomas, 59

Harrington, James, 56

Hartlib, Samuel, 58, 59, 181; and Boyle, 92, 93, 95, 112, 116, 122, 128, 130, 165; on Harvey, 30, 32, 34, 35, 141; on niter, 121–122, 126; on Oxford scientific activity, 50, 52, 53, 54, 55, 56, 108, 112, 128, 130, 165, 167, 172; on pneumatics, 92, 109, 116, 128, 130

Harvey, William: on animal heat, 12–13, 38–40, 164, 287; and Aristotle, 1, 12, 13, 17, 18; on blood color, 15–16, 40–41, 168, 184, 205, 215, 217; on blood, nature of, 12, 15, 38–39, 40, 98, 111, 189–191, 234, 287; and Boyle, 94, 95, 96, 141, 143, 146, 170; career of, 1, 25, 64–65; and circulation of blood, discovery of, 2, 9–11; and circulation of blood, reception of,

21, 102, 103, 106, 154, 155, 167, 170, 195; disciples of, 21, 23–25, 26–30, 41, 43, 102, 103, 167, 168, 169, 170, 195; on embryology, 17, 26–27, 32, 34–36, 38, 96, 101, 107, 198–199, 204; and Ent, 22, 23, 24, 25, 34, 41; and Glisson, 22–23, 24, 25, 41; and Greaves, 22, 25, 26–27, 28, 30; on heart, motion of, 2, 10–11, 97–98, 102, 141; and Highmore, 27–28, 29, 41, 97–100, 280, 289; and Lower, 184, 188–191, 209–217; natural philosophy of, 18–19, 43, 90, 287; at Oxford, 25–30, 36, 37, 51, 53, 60, 62, 170; physiological methods of, 1, 10–11, 16–18; on respiration, fetal, 14, 40, 42, 146, 149, 271; on respiration, function of, 13–15, 40, 99, 115, 143, 147, 287. Works (arranged chronologically): "Prelectiones" [1616], 1, 3, 10, 11, 12, 13, 17, 18, 19, 38; "De motu locali anumalium" [1627], 31; *De motu cordis* (1628), 1, 2, 3, 4, 10–12, 13, 14, 15, 16, 17, 18, 20, 21, 22, 30, 31, 34, 35, 36, 38, 41, 98, 170, 209; lost works, 30–33, 210; *De circulatione* (1649), 17, 30, 31, 33–34, 36, 38, 39, 40; *De generatione* (1651), 16, 18, 19, 26, 27, 29, 30, 31, 33, 34–36, 38, 39, 40, 98, 105, 107, 108, 111, 120, 143, 146, 182, 189, 190, 191, 205; unspecified, 46, 49, 58, 94, 112

Hearne, Thomas, 278

Heart: anatomy of, 210, 234; experiment of blowing into, 157, 158–160, 191, 231; ferment in, 166–167, 168–169, 190–191, 196, 199, 211–212, 216, 229, 233; motion of, 2–6, 10–11, 36, 97–98, 102, 103, 141, 150, 210–213, 229–231

Heat, animal: whether cooled by respiration, 3–7, 13–14, 39–40, 99, 100, 109, 144, 146, 151–152, 159, 162–163; and fermentation, 166–169, 190–191, 211–212, 216, 229–232, 234–237, 268–272, 281; generated in blood, 38–40, 99, 100, 109, 190–191, 212–213, 216, 234–237, 268–272, 281, 288; generated in heart, 164–169, 193, 196, 199, 211–212, 216; as innate, 3–7, 12–13, 38–40; and nitrous aerial substance, 169, 190–191, 229–232, 234–237, 244–245, 268–272, 281

Hele, Nicholas, 105, 106

Helmont, Jan Baptista van: books by, 49, 92, 94, 108, 141, 195; on combustion, 109, 263; on fermentation, 166, 187

Henshaw, Nathaniel: *Aero-Chalinos* (1664), 156–157, 158, 160, 242; on respiration, mechanical function of, 156–157, 158, 163, 185, 193, 199, 265, 282

Henshaw, Thomas, 29, 156, 282; on niter, 126, 238

Herfeld, Heinrich Gerhard, 279

Hevelius, Johannes, 202

Heyden, Joachim van der, 49

Highmore, Nathaniel: and Bathurst, 30, 106, 108, 112, 156; on blood, nature of, 97–98; on blood color, 168, 205; books by, 46, 49, 58, 186, 195; and Boyle, 59, 93, 95, 96, 97, 140, 143, 144, 147, 157, 158, 184, 289; career of, 27, 28, 29, 44, 62, 64–65, 284, 294; *Corporis humani disquisitio anatomica* (1651), 97–100, 205; on corpuscular philosophy, biological applications of, 93, 98, 99, 100–101, 113, 288, 289; on embryology, 26, 96, 100–101; as Harvey disciple, 26, 27–28, 41, 97–100, 280, 289; *History of generation* (1651), 97, 100–101; on nitrous aerial substance, 156; on respiration, 98–100, 108, 111, 112, 223, 227, 233; on respiration, mechanical function of, 99–100, 147, 156, 158, 163, 193, 199

Hill, Abraham, 287

Hippocrates, 241

Hobbes, Thomas: and Boyle's pneumatic experiments, 60, 135–136, 148, 149, 292; mechanical philosophy of, 107, 117; and Petty, 56, 92, 102

Hodges, Nathaniel: career of, 72–73, 240, 284; *Loimologia* (1671), 241; on nitrous aerial substance, 240–241, 244; and Oxford scientific community, 44, 48, 51, 54, 58, 59, 60, 61, 181; and Willis, 51, 54, 240

Holder, William, 44, 54, 82–83

Hooghelande, Cornelis van: books by, 49, 90, 196; on heart, ferment in, 168, 196, 210–213; and Petty, 92, 102

Hooke, Robert: on air, spring of, 52, 61, 135, 136–137, 277, 288; on blood color, 207–208, 215, 218; and Boyle, 51, 52, 60, 61, 129–132, 134–137, 138, 142–143, 146–147, 148, 153, 284; career of, 64–65, 129, 135, 218, 246, 294; on combustion and air, 137–139, 252, 255–258, 281–282, 289; on combustion and respiration, 247–248, 254, 267, 281–282; on injection and transfusion, 173, 175, 177, 178; and Lower, 182, 183, 195, 197–198, 200–201, 207, 209, 215; and Mayow, 227, 228, 231, 273, 276–278, 281–282; *Micrographia* (1665), 137–138, 154, 158, 160, 162, 187, 193, 282; on nitrous

aerial substance, 137–139, 154, 158, 160, 162, 186–187, 193, 197, 207, 231, 236, 237, 244, 252, 253, 259; and Oxford scientific clubs, 44, 51, 52, 54, 55, 57, 58, 60, 163, 172, 291; and respiration, fetal, 146–147, 153; and respiration, mechanical theory of, 158–160, 162, 185, 203–205; and respiration, physiological experiments on, 158–160, 162, 185, 193, 195, 197, 201, 202, 204, 207, 218, 247; and respiration, pneumatic experiments on, 142–143, 148, 150–152, 157, 161, 197, 207; and respiration, pulmonary insufflation experiments on, 197, 200–201, 203–204, 207–208, 215, 217, 218, 227, 228, 231, 242, 243, 287, 289; scientific personality of, 129, 218, 219; and vacuum pumps, 129–135, 250–252, 280–281, 289, 292; and Westminster School, 59, 129; other works by, 46, 276, 281, 282

Huet, Pierre, 209

Huygens, Christiaan, 209, 250

Injection: early history of, 17–18, 170–175, 205, 206, 225, 281, 282, 289; physiological use of, 174–175, 184, 211–212, 228, 229, 231, 286, 287, 289. *See also* Transfusion

Jeamson, Thomas, 51, 54, 59, 61, 72–73, 104, 105, 106, 181

Jeanes, Thomas, 51, 59, 82–83

Johnson, William, 49

Jones, 105, 106

Jones, Katherine (Countess Ranelagh, née Boyle), 57, 195

Jones, Richard, 57, 126, 130, 142

Joyliffe, George: and Boyle, 95, 96, 141–142; career of, 23, 30, 41, 72–73; and lymphatics, discovery of, 24, 28, 96, 280; and Oxford scientific community, 28, 44, 59; and splenectomy, 141–142

Justel, Henri, 202, 209, 227

Kepler, Johann, 49

Kerger, Martin, 187

King, Edmund: and respiration, experiments on, 197, 204, 207; and transfusion, 178, 202; and Willis, 233, 248, 283

König, Emanuel, 279

Kuffler, Abraham, 153

Ladyman, Samuel, 82–83

Lamphire, John, 51, 54, 59, 82–83, 106

Langbaine, Gerard: career of, 47, 58, 84–

85; and Oxford scientific clubs, 44, 54, 55, 58, 60

Langley, Henry, 47

Laurens, André du: *Historia anatomica* (1600), 3–8, 10; physiological ideas of, 3–8, 10

Lavoisier, Antoine Laurent, 244, 278

Lemery, Nicolas, 279

Leucippus, 94

Levinz, William, 51, 84–85

Libavius, Andreas, 170

Linus, Franciscus (Francis Hall), 135–137, 148

Littre, Alexis de, 279

Llewellyn, Martin, 29

Locke, John: on blood color, 186–188, 193, 196, 215, 289; and Boyle, 54, 186–187, 196, 284–285; career of, 72–73, 194, 218, 246, 284; and Lower, 180, 182, 186–188, 195; medical education of, 49–50, 51, 120, 155; on nitrous aerial substance, 186–188, 193, 196, 284; in Oxford scientific community, 44, 51, 54, 58, 59, 60, 181, 182, 223, 225; on respiration, 186–188, 193, 195–196; scientific personality of, 219; writings of, 49–50, 155, 182, 186–188, 195–196, 284

London "1645 Group," 23–24, 52, 53, 56

Lovell, Robert, 46, 51, 58, 60, 72–73

Lower, Richard: anatomical research of, 50, 157, 175, 180, 181, 183, 187, 193, 197, 198, 201–202, 208, 209, 213, 222; on animal heat, 190–191, 193, 199, 212–213, 216, 234; and Bathurst, 112, 191, 216; on blood, circulation of, 189–190, 213–214, 217; on blood color, 183–184, 187–188, 189–191, 193, 206, 211, 214–216, 217, 218, 235, 242, 243, 268–269, 281, 289; on blood, nature of, 189–190, 193, 199; and Boyle, 50, 174–178, 180, 182, 183–185, 189, 194, 195, 197, 198, 201, 206, 209; career of, 64–65, 176, 179–180, 194, 195, 201, 208, 218, 246, 282, 283, 294; as Harvey disciple, 188–192, 209–210; on heart, fermentation in, 106, 169, 190, 210–212, 216, 218, 229, 233, 242; on heart, structure of, 210, 280, 281; and Hooke, 182, 183, 195, 197–198, 200–201, 202–203, 207; on injection and transfusion, 174–178, 193–194, 202, 211–212, 225, 229, 281, 289, 292; and Locke, 180, 182, 186–188; and Mayow, 224, 225–226, 227, 229, 230, 231, 273, 278; on muscle contraction, 210, 232–233, 271, 289; and Needham, 181, 199–200, 283; on nitrous aerial substance, 191,

193, 215–217, 221, 236, 244; and Oxford scientific clubs, 44, 50, 51, 54, 57, 58, 59, 60; on respiration, 190–191, 193, 214–217, 286; on respiration, pulmonary insufflation experiment on, 200–201, 204, 208, 209, 214–215, 217, 218, 289; scientific personality of, 218, 219, 272, 288, 289, 293; *Tractatus de corde* (1669), 46, 208–217, 221, 227, 232, 233, 234, 241, 283; *Vindicatio* (1665), 46, 188–192, 193, 206, 212, 214, 216, 231, 289; and Willis, 50, 54, 174, 180, 182–183, 187, 188–190, 212–213, 223, 235, 236, 237, 250

Lungs: inflation and insufflation of, 15, 99, 100, 157, 158–160, 162, 193, 197, 200–201, 203–204, 207–208, 214–215, 217, 218, 227, 228, 231, 282, 287, 289; structure of, 155, 227, 246, 247, 280, 290. *See also* Respiration

Lydall, John: career of, 47, 84–85; and Oxford scientific clubs, 50, 54, 61, 106, 112, 165

Lydall, Richard: career of, 47, 84–85; and Oxford scientific community, 44, 51, 54, 60, 105, 106

Lymphatics. *See* Aselli; Digestion; Joyliffe; Pecquet

Magalotti, Lorenzo, 250

Magenus (or Magnenus), Johann Chrysostom, 94

Major, Johann Daniel, 173, 175

Malpighi, Marcello: books by, 155, 205, 206; as correspondent, 209, 243, 250; on respiration, mechanical theory of, 155, 158, 199, 227, 228, 246, 265, 289

Manget, Jean Jacques, 279

Mariotte, Edme, 279

Martyn, John, 188

Masters, John, 51, 58, 84–85, 106, 248, 283

Mayer, Julius Robert, 244

Mayow, John: on air, spring of, 227, 262–267; on animal heat, 229–232, 244–245, 267–272; on blood color, 269; on blood, fermentation in, 229, 231, 267–272; and Boyle, 225, 226, 229, 230, 231, 259, 260, 269, 273, 275, 276, 278; and Browne, Edward, 224, 227, 241, 280; career of, 64–65, 224–227, 258, 276–278, 283, 294; on combustion and air, 256, 260–263; on combustion and respiration, 263–267; corpuscular philosophy of, 231–232, 266–267, 272–273, 288; on fermentation, as source of heat, 229–232, 244–245, 262, 267–272; on the heart, fermentation in, 229; and Hooke, 227, 228, 231, 273, 276–278, 281–282; on injection and transfusion, 228, 229, 230, 231, 271; and Lower, 224, 225–226, 227, 229, 230, 231, 273; on muscle contraction, 229–231, 233, 271–272, 289; on niter, 259–261; on nitrous aerial substance, 228–232, 236, 241, 244–245, 256, 259–272, 278–282, 289; and Oxford scientific community, 44, 51, 58, 61, 224, 225, 226, 238, 258, 273, 278; on respiration, cause of, 227; on respiration, fetal, 230, 271; on respiration, function of, 228–232, 265–267, 269; on respiration, mechanical theory of, 228, 265; scientific personality of, 230–232, 272–274, 293; *Tractatus duo* (1668), 46, 227–232, 237, 241, 258, 259, 267, 270, 273, 276, 279; *Tractatus quinque* (1674), 46, 244, 258–274, 275–276; and vacuum pumps, 225, 226, 269, 273, 292; and Willis, 223, 224, 225, 227, 229, 231, 236, 237, 244, 250, 269–271, 273

Mayow, Thomas, 224, 225

Meara, Edmund, 188–190

Mechanical Phylosophy. *See* Corpuscular philosophy

Medea, 170

Merrett, Christopher: and anatomical research, 41, 151, 152, 178; career of, 24, 28, 29, 74–75; and scientific clubs, 24, 28, 41, 44, 60

Mersenne, Marin: books by, 21, 58, 90, 92, 128; and English science, 21, 53, 58, 92, 93, 116, 128, 179

Microscopy, 56, 91, 94–95, 101, 137, 155, 265

Millington, Thomas: and anatomical research, 157, 159, 169, 175, 177, 181, 182, 183, 184, 191, 198, 292; career of, 48, 59, 74–75, 284, 285, 293; on nitrous aerial substance, 231; and Oxford scientific collaboration, 44, 51, 52, 54, 58, 59, 60, 61, 107, 157, 159, 175, 177, 181, 182, 183, 184, 194, 195, 198, 209, 218, 224, 226, 231, 240, 258, 273, 278, 283, 292

Mistichelli da Fermo, Domenico, 279

Moebius, Gottfried, 49, 144, 186

Monconys, Balthasar de, 153, 179

Montgomery, Hugh (3d Viscount of the Ards), 36

Montmor, Henri Louis Habert de, 130

Moray, Robert, 152, 175, 177, 256

More, Henry, 92

Morhof, Daniel Georg, 279

Morris, Samuel, 74–75, 181

Morton, Charles: career of, 53, 61, 74–75, 239; on nitrous aerial substance, 239–240, 244

Mundy, Henry, 46, 74–75, 281

Muscle contraction, 108–109, 210, 216, 221–223, 229–234, 271–272, 289

Musgrave, William, 286

Needham, Jasper, 34

Needham, Walter: and anatomical research, 181, 198–199; and Boyle, 60, 157, 158, 159, 160, 184, 198, 203, 204; career of, 64–65, 180–181, 194, 197, 198, 242, 246, 282, 283; *Disquisitio anatomica de formato foetu* (1667), 198–200, 225, 246; on heart, experiment of blowing into, 157, 158, 159, 191, 193, 230, 231, 242; and Hooke, 200, 203–205, 207; and Lower, 181, 191, 198, 199–200, 215, 283; and Oxford scientific collaboration, 44, 51, 54, 57, 59, 157, 158, 159, 163, 169, 181, 184, 240; on respiration, mechanical theory of, 199–200, 203–205, 215, 217, 228, 247, 265, 280, 289; scientific personality of, 217, 218–219, 289

Neile, Sir Paul, 52

Neile, William, 51, 52, 53, 61, 74–75, 171, 202

Neukranz, Zacharia, 279

Newton, Isaac, 291

Niter, 118–127, 138–139, 167, 238–239, 253, 255, 257, 259–261

Nitrous aerial substance: and animal heat, 110–111, 120–121, 162–163, 169, 191, 193, 196, 208, 216, 221, 229–232, 244, 249–250, 267–271; and blood color, 169, 186–188, 215–216, 228; and cold, 118, 125–126, 244; and combustion 110–111, 118, 137–139, 160, 162–163, 193, 207–208, 235, 244–245, 249–250, 261–262, 278–282, 289; and meteorological phenomena, 118, 119–120, 238, 239–240, 244; and muscle contraction, 222–223, 229–231, 244, 271–272; and niter, 109, 117–119, 124–125, 244, 253, 259–261, 285; other properties of, 118–119, 240–241, 243–245, 249–250, 262, 279–280; and respiration, 105, 109–113, 118–121, 124–125, 127, 154, 156, 160–161, 162–163, 169, 186–188, 191–192, 193, 196, 197, 207–208, 215–216, 221, 228–232, 235–236, 238, 240–241, 242–243, 244–245, 249–250, 265–267, 278–282

Nourse, Anthony, 105, 106

Oldenburg, Henry: on blood color, 205–206, 215; book promotions by, 208–209, 222, 233, 239, 243, 248, 250, 283; book reviews by, 138, 191, 200, 203, 227, 228, 233, 239, 241, 243, 247, 248, 250, 275–

276, 283; and Boyle, 55, 57, 126, 130, 132, 157, 158, 159, 175, 176, 178, 185, 194, 201, 202, 203, 204, 206, 215, 250; career of, 74–75, 283, 293; on injection and transfusion, 173, 175, 176, 178, 194, 202, 205; and Lower, 191, 194, 201–202, 206, 208–209, 233, 283; and Mayow, 226, 227, 228, 241, 275–276; and nitrous aerial substance, 126, 275–276; and Oxford scientific clubs, 44, 55, 57, 163, 173; on respiration, 153, 159, 200, 203, 204, 246–247; scientific personality of, 219, 275–276

Ovid, 170

Owen, John, 48, 58, 60, 84–85

Oxford: booksellers, 46–47, 60, 61; medical education at, 48–51, 101–106, 107–108; scientific clubs at, 50–57, 60–61, 106–107, 112, 128, 129, 130, 132, 157–158, 163, 165, 169–188, 194–195, 286–287; scientific community at, 25–30, 45–89, 101–114, 120, 123, 126, 169–195, 217–220, 224–227, 231, 237, 244, 258, 280–287, 289–294; Visitors, 47–48, 106, 107, 170, 179

Oxford Colleges and Halls: All Souls, 30, 47, 53, 58, 61, 172, 181, 194, 195, 224, 225, 227, 231, 258, 277, 278; Balliol, 29; Brasenose, 45, 53; Christ Church, 48, 49, 50, 51, 58–59, 60, 129, 165, 172, 177, 179, 180, 181, 182, 186, 187, 195, 224, 237, 240, 248, 283; Exeter, 45; Hart Hall, 61; Jesus, 45; Lincoln, 45; Magdalen, 61; Magdalen Hall, 57; Merton, 26–27, 28, 45, 46, 48, 54, 58, 60, 224; New College, 61, 239; Oriel, 45, 238; Pembroke, 45, 47; Queen's, 58; St. John's, 45; Trinity, 26, 27, 29, 45, 48, 56, 58, 61–62, 106, 107, 170; University, 53; Wadham, 45, 47, 48, 52, 53, 54, 55, 56, 57, 58, 59, 61, 106, 129, 165, 171, 183, 194, 224, 239, 279

Oxford Institutions, 259, 285; Bodleian Library, 46, 54, 55, 58, 61, 107, 182, 283; Linacre Lecturers in Medicine, 45, 46; Physic Garden, 25, 46, 49–50, 61; Regius Professor of Medicine, 25; Savilian Professor of Astronomy, 25, 46, 47, 57, 162, 181, 285; Savilian Professor of Geometry, 25, 46, 47, 57, 285; Sedleian Professor of Natural Philosophy, 25, 46, 49, 54, 61, 179, 181, 231, 244, 248; Tomlins Reader in Anatomy, 25, 46, 49, 50, 61, 102, 181

Page, William, 105, 106

Pagel, Walter, 9, 38

Palmer, John, 47, 58, 60

Papin, Denis, 284

Paracelsus (Theophrastus Bombast von Hohenheim), 90, 92, 94, 117, 122, 127; on nitrous aerial substance, 118–119

Parigiano, Emiglio, 21, 22

Parker, Samuel, 44, 46, 84–85, 107, 183

Partington, J. R., 117, 278

Pascal, Blaise, 116, 133

Pechlin, John Nicholas, 279

Pecquet, Jean, 117, 128; on the thoracic duct, 104, 105, 117, 140, 154–155, 190, 289

Pell, John, 59

Pepys, Samuel, 161, 173, 178, 201, 208

Petit, Pierre, 116

Pett, Peter, 44, 48, 54, 58, 61, 86–87

Petty, William: and anatomy, 49, 50, 61, 95, 101–103, 106, 107, 113, 114, 140, 171, 172, 173; and Boyle, 95, 116, 128, 140, 172; career of, 53, 66–67, 92, 284; and chemistry, 102–103, 118, 289; and mechanical philosophy, 92, 93, 116; and Oxford scientific clubs, 43, 44, 50, 52–56, 60, 106, 107, 109, 114, 165, 171, 172, 175, 182, 223, 239; and pneumatics, 109, 116, 128, 182, 223

Philosophical Transactions, 47, 205, 246; articles in, 175, 176, 183, 203, 205, 206, 209, 237, 283; reviews in, 138, 191, 198, 200, 203, 213, 227, 233, 239, 241, 243, 248, 250, 259, 275–276, 283

Pierrepont, Henry (1st Marquis of Dorchester), 29, 86–87, 172

Plemp, Vopiscus Fortunatus, 181

Plot, Robert, 74–75; his chemistry course with Mayow, 51, 225, 226; *Natural history of Oxford-shire* (1677), 280–281; on Oxford anatomical discoveries, 280–281; and Oxford Philosophical Society, 285–286

Pneumatics. *See* Air

Poisons, 170–174

Pope, Walter: and anatomical research, 178, 202–203, 218; career of, 48, 76–77, 130, 135, 246; and Oxford scientific clubs, 44, 53, 54, 58, 59, 61, 106

Potter, Francis, 170

Power, Henry, 23, 117, 136, 152, 186

Priestley, Joseph, 268

Primrose, James, 21, 26, 98

Quartermaine, William, 86–87, 104, 105, 106

Rawlinson, Richard, 51, 52, 54, 58, 86–87

Renou, Jean de, 49

Respiration: and combustion, 109, 110, 146, 151–152, 160–161, 235, 247–248, 253–255, 266–267; cooling function of,

6, 13−14, 40, 99, 105, 109, 113, 141, 144, 154, 196, 228; of fish, 110, 147, 150, 152, 156, 199, 200; fetal, 8, 40, 146−147, 199, 202−203, 228, 230, 271, 276, 282; mechanical function of, 14, 40, 99−100, 154−160, 162, 193, 197, 199−200, 200−201, 203−205, 217, 228, 242, 265, 280, 282; mechanics of, 108−109, 141, 143−144, 227; and nitrous aerial substance, 105, 109−113, 118−121, 124−125, 127, 154, 156, 160−161, 162−163, 169, 186−188, 191−192, 193, 196, 197, 207−208, 215−216, 221, 228−232, 235−236, 238, 240−241, 242−243, 244−245, 249−250, 265−267, 278−282; physiological experiments on, 15, 99, 100, 141, 157, 158−160, 162, 193, 197, 200−201, 203−204, 207−208, 214−215, 217, 218, 227, 228, 231, 282, 287, 289; pneumatic experiments on, 142−143, 148−150, 152, 156, 157, 161−162, 247−248, 253−255, 289; traditional ideas of, 5−7; ventilative function of, 6, 40, 99, 105, 109, 144−145, 150, 162, 186, 191, 228, 235; and vital spirits, 5−6, 15, 40, 98−99, 111, 113, 144, 162

Rhode, Johannes, 22
Riolan, Jean, 33, 36, 107
Rivière, Lazare de la, 141
Roberval, Giles Persone de, 116, 132
Robinson, Thomas, 47, 60, 61
Rooke, Lawrence, 44, 50, 51, 53, 54, 58, 60, 61, 76−77, 107
Roy, Henrik de (Henricus Regius), 90
Royal College of Physicians, London, 179, 284; anatomical activity at, 21, 22, 24−25, 41; and Harvey, 1, 25; Lumleian Lectures at, 1, 3, 10, 11, 12, 13, 15, 16, 17, 18, 19
Royal Society of London, 182, 219, 286; activity of members of, 126, 127, 134, 135, 175, 179, 197, 198, 208, 233, 238, 243, 246, 274, 277, 283, 284; combustion experiments at, 138, 151−152, 193, 248, 253, 254, 255, 256−258, 266, 267, 282; injection, transfusion, and blood experiments at, 173, 177−178, 185, 194, 202, 206; origins of, 23, 52, 55, 134; other physiological experiments at, 187, 197, 198, 233; respiration experiments at, 148, 150−152, 153, 154, 156, 157, 158−162, 197, 200−201, 202, 203, 204, 207, 218, 242, 247, 248, 253, 254, 282, 287
Rudbeck, Ole, 49

Sala, Angelus, 49, 166
Sanctorius, Sanctorius, 95, 141

Savile, Henry, 46
Scarburgh, Charles: and anatomical research, 24, 27, 258; career of, 27, 30, 76−77; and Harvey, 27, 29, 31, 32, 41, 44; and scientific clubs, 24, 30, 47
Schlegel, Paul Marquand, 18
Schott, Gaspar, vacuum pump of, 128−129
Sendivogius, Michael, 119
Sennert, Daniel: books by, 49, 119, 120, 141; on nitrous aerial substance, 119−120, 238
Servetus, Michael, 287
Seton, Alexander, 119, 127
Sharrock, Robert: and Boyle, 53, 54, 60, 61, 126, 130; career of, 46, 48, 76−77, 194, 284; and Oxford scientific clubs, 44, 53, 54, 60, 181, 182, 183
Sheldon, Gilbert, 194
Sheldon, Sherrington, 104, 105, 106
Skinner, Robert, 107
Slare, Frederick, 284
Sluse, René François de, 202
Smart, "Mr.," 121, 122
Smith, "Dr.," 181
Smith, Francis, 51, 187
Smith, John, 47, 76−77
Smith, Thomas, 35
Sorbière, Samuel, 179
Southwell, Robert, 44, 58, 86−87, 286
Spiegel, Adriaan van den, 27, 35
Spinoza, Benedict, 126
Spleen, 105, 141−142, 183
Sprat, Thomas: career of, 86−87, 194; *History of the Royal Society* (1667), 126, 162, 238; and Oxford scientific clubs, 44, 53, 54, 58, 59, 61, 106, 107, 179
Sprigg, William, 86−87
Stahl, Peter: career of, 51, 76−77, 194, 282; chemistry courses of at Oxford and Exeter, 51, 53, 60, 118, 225, 243, 244; students of, 51, 59, 126, 181, 187, 224, 243
Starkey, George, 122
Stationers, London Company of, 46
Steno, Nicolaus, 232; on muscle contraction, 232
Stephens, Phillip, 48, 49−50, 51, 76−77; his anatomy classes at Oxford, 50, 61, 103, 106, 114
Stubbe, Edward, 105, 106
Stubbe, Henry: and Boyle, 54, 60, 123, 126, 172, 226, 237, 238; career of, 59, 61, 78−79, 237−238; on niter, 123, 126, 238, 239, 259; and Oxford scientific clubs, 44, 54, 58, 59, 60; and Willis, 51, 237
Surgeons, London Company of, 24, 30, 280, 283
Swammerdam, Jan, 49, 250

Sydenham, Thomas, 49
Sylvius, Franciscus (Frans de la Boë), 49, 174, 290

Templer, John, 246, 247
Thomas, David: career of, 54, 59, 86−87, 194, 284; and chemistry with Locke, 51, 181
Thoracic duct. *See* Pecquet
Thorndyke, 59
Thruston, Malachi: career of, 241−243, 282; *De respirationis usu* (1670), 241−243, 247; on nitrous aerial substance, 242−243; and Oxonians, 242−243; on respiration, mechanical theory of, 242−243, 265, 280
Tillyard, Arthur, 51, 58, 60, 86−87, 181, 224
Toone, Stephen, 50, 51
Torricelli, Evangelista, mercury experiment of on the weight of air, 92, 96, 115−117, 128, 133, 136, 226
Towneley, Richard, 136−137
Transfusion, 169−170, 173, 175−178, 193−194, 202, 208, 209, 225, 230, 231, 243, 271, 281, 289. *See also* Injection
Trapham, Thomas, Jr., 78−79; on nitrous aerial substance, 240
Travagino, Francesco, 202, 243
Trevor, Richard, 60, 88−89
Tulp, Nicolaas, 141
Tyson, Edward, 78−79, 286

Ussher, James, 29

Vacuum, 116−117, 128; pumps, 128−134, 139, 150, 151, 152−153, 197, 207, 226, 234, 246, 247, 250−252, 269, 280, 284, 286, 289, 292. *See also* Air; Combustion; Respiration
Valentine, Basil, 49
Velthusius, Lambertus, 49, 186
Venous Blood. *See* Blood color
Verheyen, Philip, 279
Vernon, Francis, 59, 88−89, 172
Vesalius, Andreas, 2, 99, 287
Vesling, Johann, *Syntagma anatomicum* (1651) of used at Oxford, 49, 50, 103, 107
Vigenère, Blaise de, 119
Viridet, Jean, 279
Viviani, Vincenzo, 115

Wale, Jan van der, 33−34, 102, 108
Walker, Obadiah, 88−89
Wallis, John: and anatomical research, 24, 170, 176, 177, 182, 183, 202, 292; and Boyle, 52, 132, 133, 135−136, 226, 289, 292; career of, 44, 47, 48, 61, 66−67,

285; and London "1645" Group, " 23−24, 29, 47, 116; and Oxford scientific clubs, 44, 46, 47, 48, 51, 52, 53, 54, 55, 56, 57, 58, 60, 61, 106, 175, 181, 239; and pneumatics, experiments on, 55, 93, 109, 116, 132, 133, 135−136, 226, 277, 289, 292
Ward, John, 88−89; on anatomy at Oxford, 50, 103, 140, 174, 177, 180, 181, 182, 183, 187; on Boyle, 50, 140, 174, 226; chemistry course of with Mayow, 51, 225, 226; on Lower, 177, 180, 182, 183, 187, 188; medical education of, 49−51; and Oxford scientific community, 49−51, 54, 58, 60, 61, 103, 181, 182, 240
Ward, Seth: career of, 47, 61, 66−67, 181, 243; and Oxford scientific clubs, 44, 48, 51, 52, 53, 54, 55, 56, 58, 60, 107, 112, 129, 165, 170, 175, 285; and pneumatics, 52, 112, 132, 133, 289
Warner, Walter, 59
Webster, Charles, 52, 136
Wepfer, Johann Jacob, 49, 186
Westminster School (St. Peter's College), London: former students of, 59, 135, 175, 179, 180, 181, 186, 187, 202, 210, 218, 224, 231, 240, 282, 283, 284, 293; teaching at, 59, 129, 179, 180−181
Wharton, Thomas, 24, 41, 49, 195
Whistler, Daniel: and anatomy, 29, 93; career of, 28, 29, 44, 47, 78−79, 107
Whitteridge, Gweneth, 9, 38
Wildan, William, 51, 225, 244, 280
Wilkins, John: and anatomical research, 158, 198, 202, 243; career of, 47, 61, 66−67, 224, 226, 282, 293; on combustion and air, 151, 267; and injection, 170, 171, 173, 175, 178, 243; and London "1645 Group," 23, 24; and microscopy, 56; and Oxford collaborators, 47, 48, 52, 60, 95, 112, 117, 165, 239, 289; and Oxford scientific clubs, 24, 43, 44, 52−59, 106, 112, 114, 129, 163, 165, 239; and respiration, 112, 156, 161
Williamson, Anthony, 277
Williamson, Joseph: career of, 58, 59, 88−89, 286; and Oxford scientific clubs, 44, 51, 53, 58, 59, 175, 176, 181
Willis, Thomas: and anatomical research, 50, 53, 154, 157, 181, 182, 183, 195, 240, 248, 249, 289; on animal heat, 121, 166−169, 190−191, 212−213, 234−237, 288, 289; and Bathurst, 29, 30, 50, 106, 107, 112, 156, 216, 250, 289; on blood color, 166, 168, 188, 190, 205−206, 214, 218, 235, 236; on blood, nature of, 166−169, 189, 234−236; and Boyle, 95, 123, 157,

165, 235, 250, 255, 292; on brain, 182–183, 222–223, 248, 280; career of, 28–29, 30, 47, 51, 58, 59, 66–67, 165, 179, 180, 194, 208, 246, 248, 282–283, 285, 294; and chemistry, 50, 51, 54, 55, 129, 165–169, 181, 195, 289; corpuscular philosophy of, 165–169, 222–223, 236, 248–249; on fermentation, 121, 166–169, 190, 196, 234–236, 237, 238, 240, 241, 244, 269–271, 273, 280, 289; as Harvey follower, 41, 167; on heart, ferment in, 121, 166–167, 168–169, 234; and injection, 171, 174, 178; and Lower, 50, 60, 112, 157, 174, 180, 182–183, 187, 188–190, 212–213, 216, 235, 236, 237, 250; and Mayow, 223, 225, 227, 229, 231, 250, 260, 269–271, 273; on muscle contraction, 221, 222–223, 229, 233–234, 244, 289; on nitrous aerial substance, 123, 168–169, 196, 235–236, 238, 244, 249–250; and Oxford scientific clubs, 43, 44, 50, 53, 54, 55, 57, 58, 60, 129, 165, 175, 291; on respiration, 112, 235–236, 249–250; as Sedleian Professor, 49, 179, 182, 189, 223, 248. Works (arranged chronologically): *Diatribae duae* (1659) [composed of "De fermentatione" and "De febribus"], 165–169, 179, 188, 190, 196, 205–206, 223, 236, 237, 250, 260; *Cerebri anatome* (1664), 174, 182–183, 198, 199, 203, 222; *Pathologiae cerebri* (1667), 222–223; "De motu musculari" (1670), 233–234, 246, 270; "De sanguinis accensione" (1670), 233,

234–237, 246, 270; *De anima brutorum* (1672), 248–250; nonspecific references to, 46, 49, 58, 186, 242

Willoughby, Francis, 173

Winthrop, John, 243

Wood, Anthony: on Lower, 176, 180, 195; on Mayow, 226, 258, 278; on other Oxford scientists, 27, 50, 51, 106, 181, 282

Wood, Robert, 48, 78–79, 130, 167; and Oxford scientific clubs, 44, 51, 54, 55, 56, 165, 171

Woodroffe, Benjamin, 59, 88–89, 224

Worm, Ole, 21

Worsley, Benjamin, 53, 121–122

Wren, Christopher: and anatomical research, 169, 171, 181, 182–183, 284; and Boyle, 52, 54, 128, 132, 133, 171, 226, 289, 292; career of, 58, 59, 66–67, 171, 172, 194, 224, 246, 283–284, 285, 293, 294; and injection, 170, 171–172, 173–174, 178, 225, 281, 289; and microscopy, 56, 137; on nitrous aerial substance, 154, 161, 162, 193, 231, 282, 289; and Oxford scientific clubs, 44, 51, 52, 53, 54, 56, 57, 58, 59, 61, 163, 171, 181, 182–183; and pneumatics, 52, 128, 132, 133, 226, 231, 289, 292; on respiration, 154, 161, 162, 193, 280, 281, 284

Wren, Matthew, 54, 60, 88–89, 107, 152; on microscopy, 56

Wren, Thomas, 54

Yerbury, Henry, 51, 88–89

The River Chartwell

Christ Church Coll and Walke

The Bowling Green

The Bowling Green

Magdalen

College Grove

Holywell Mill

A Bowling Green